Dynamics and Kinetics in Structural Biology

Dynamics and Kinetics in Structural Biology

Unravelling Function Through Time-Resolved Structural Analysis

Keith Moffat
Department of Biochemistry and Molecular Biology
The University of Chicago
Chicago, IL
USA

Eaton E. Lattman
Department of Materials Design and Innovation
University at Buffalo
Buffalo, NY
USA

Registered Office
John Wiley & Sons, Inc., 111 River Street, Hoboken, NJ 07030, USA
John Wiley & Sons Ltd, The Atrium, Southern Gate, Chichester, West Sussex, PO19 8SQ, UK

For details of our global editorial offices, customer services, and more information about Wiley products visit us at www.wiley.com.

Wiley also publishes its books in a variety of electronic formats and by print-on-demand. Some content that appears in standard print versions of this book may not be available in other formats.

A catalogue record for this book is available from the Library of Congress

Paperback ISBN: 9781119696285; ePub ISBN: 9781119696308; ePDF ISBN: 9781119696339; oBook ISBN: 9781119696353

Cover image: Greg Stewart, SLAC National Accelerator Laboratory
Cover design: Wiley

Set in 9.5/12.5pt STIXTwoText by Integra Software Services Pvt. Ltd, Pondicherry, India

This book is dedicated to our wives
Anne Simon Moffat and Susan Pfeiffer
and to our late colleagues
Quentin Gibson and John C. H. Spence

Contents

List of Figures

Acknowledgments

A book such as this could not have been completed without the advice and good counsel of many scientists. We're deeply indebted to colleagues – still friends – who supplied constructive comments on drafts ranging from single chapters to the (nearly) complete text: Andy Aquila, Sean Crosson, Tom Grant, Andreas Moeglich, Abbas Ourmazd, George Phillips Jr., Joseph Piccirilli, Phoebe Rice, Marius Schmidt, Peter Schwander, Tobin Sosnick, Vukica Srajer, Emina Stojkovic, and Minglei Zhao. In particular, we acknowledge Joseph Sachleben (University of Chicago), who rewrote and reillustrated Chapter 8.5 on nuclear magnetic resonance. We incorporated nearly all their suggestions. We also thank Caitlin Cash who competently prepared many of the figures, and Andreas Moeglich for Figure 3.4. All responsibility for remaining inaccuracies and clumsy wording of course rests with us. Perfection is impossible – but "damn good" is an attainable goal.

Scientists often reflect their mentors. Both authors are old enough that unfortunately, our formal mentors Max Perutz, Quentin Gibson (Moffat), and Warner Love, Carolyn Cohen, Don Caspar (Lattman) are no longer with us. Happily, Lattman's postdoctoral advisor Robert Huber is alive and well as this is being written.

Our own research has been driven by the enthusiasm they displayed which has infected us; we hope they would approve. That research in turn has been motivated and carried forward by generations of students, postdocs, and scientific colleagues, in our own labs and X-ray facilities worldwide. In addition to those listed above, Keith Moffat notes Marian Szebenyi, Don Bilderback, Michael Wulff, Dominique Bourgeois, Wilfried Schildkamp, Xiaojing Yang, Zhong Ren, Sudar Rajagopal, Ben Perman, and Spencer Anderson, together with generations of users such as Michael Rossmann, Jack Johnson, and Ada Yonath at MacCHESS/CHESS and BioCARS/APS. He's very conscious that many more names could be added to the list of users, all of whom contribute to the ongoing success of these facilities.

Eaton Lattman is also indebted to most of those named above. He has also had many invaluable collaborators and trainees whose work and creativity allowed his laboratory to flourish. Among these are Patrick Loll, Apostolos Gittis, Wayne Hendrickson, Lisa Keefe, George Rose, Mario Amzel, and Bertrand Garcia-Moreno. Many more names could readily be added.

Lattman also wants to offer special thanks to his coauthor Keith Moffat who introduced him to the delights and frustrations of SR data collection in the early days of MacCHESS.

The user consortium BioXFEL, which he directed and whose Advisory Committee Moffat once chaired, played a dominant role in informing their interest and knowledge of XFELs.

Sarah Higginbotham was the persuasive Acquisitions Editor at John Wiley and Sons who, starting with a cold call, persuaded us that we both should and could undertake this book on a novel topic, seeking to expand decades of powerful – but of necessity limited – static structure determination into dynamics and kinetics. Advice while we considered whether to proceed and if so how, was offered by two friends very experienced in publishing: John McMurry (Cornell, author of numerous textbooks in chemistry) and Garrett Kiely (Director of the University of Chicago Press). Stefanie Volk and latterly, Richa John expertly guided our efforts into print, aided behind the scenes by the copyediting and production teams at Integra. We thank them all. We particularly thank the artist Greg Stewart and the Media Manager Manuel Gnida at the Stanford Linear Accelerator Center for the spectacular cover art and permission to use it, and Tanya Domeier at Wiley who expertly coordinated the cover design.

Last but certainly not least, we owe a large debt to our wives Anne Simon Moffat and Susan Pfeiffer. They inspired our efforts, gently criticized run-on sentences and jargon, and got us back on track when writers' block struck. At a more practical but equally important level, Anne announced "lunch is ready!" when Keith became immersed - all too frequently - in the text.

Acronyms/Abbreviations

ADE	Acoustic droplet ejection
ADP, ATP	Adenosine diphosphate, adenosine triphosphate
AGIPD	Adaptive gain individual pixel detector
AI	Artificial intelligence
AMD	Accelerated molecular dynamics
APS	Advanced Photon Source at Argonne National Laboratory
ASIC	Application-specific integrated circuit
c	Velocity of light
CAT	Collaborative Access Team at the Advanced Photon Source, Argonne National Laboratory
CD	Circular dichroism
CESR	Cornell Electron Storage Ring
CEST	Chemical exchange magnetization transfer
CHESS	Cornell High Energy Synchrotron Source
C_p, C_v	Specific heat at constant pressure or constant volume
CPMG	Carr-Purcell-Meiboom-Gill
Cryo	Cryogenic
Cryo-EM	Cryoelectron microscopy
CTF	Contrast transfer function
DED	Difference electron density
D_{max}	Maximum particle dimension in solution scattering
E	Energy of an X-ray photon or an electron beam in a storage ring
ε	Emittance of an electron beam
ELA	Energy landscape analysis
ESRF	European Synchrotron Radiation Facility, Grenoble, France
EuXFEL	European X-ray Free Electron Laser, Hamburg, Germany
EXAFS	Extended X-ray absorption fine structure
F	Larmor frequency of a radiofrequency pulse in NMR
FAD	Flavin adenine nucleotide
FMN	Flavin mononucleotide
FT	Fourier Transform
FWHM	Full width half maximum

G	Gibbs free energy
GDVN	Gas dynamic virtual nozzle
GPCR	G-protein-coupled receptor
H	Enthalpy
HDX	Hydrogen-deuterium exchange
HIV	Human immunodeficiency virus
hkl	Coordinates of a point in the reciprocal lattice
HSQC	Heteronuclear single quantum coherence
IR	Infrared
ITC	Isothermal calorimetry
k	Denotes a rate coefficient
κ	Transmission coefficient across an energy barrier
KB	Kirkpatrick-Baez focusing mirror pair
k_B	Boltzmann constant
K_M	Michaelis constant
λ	X-ray wavelength
LCLS	Linac Coherent Light Source, at the Stanford Linear Accelerator Center
LCP	Lipidic cubic phase
LOV	Light-oxygen-voltage
MBA	Multiple bend achromat
MD	Molecular dynamics
MFX	Macromolecular femtosecond crystallography beamline at the Linac Coherent Light Source
MS	Mass spectrometry
MT	Magnetization transfer
NAD, NADH, $NADH_2$	Nicotinamide adenine dinucleotide
NBD	Nucleotide binding domain
NMR	Nuclear magnetic resonance
NOESY	Nuclear Overhauser Effect Spectroscopy
NSLS	National Synchrotron Light Source, Brookhaven National Laboratory
OYE	Old Yellow Enzyme
PAS	Per-ARNT-Sim domain
PD	Projection direction
PDB	Protein Data Bank
PF	Photon Factory, Tsukuba, Japan
PRC	Photosynthetic reaction center
PS-II	Photosystem II
PYP	Photoactive yellow protein
Q	Quantum yield
R	Gas constant
RDC	Residual Dipolar Coupling
RF	Radiofrequency
R_g	Radius of gyration

RLP	Reciprocal lattice point
Rms	Root mean square; rmsd Root mean square displacement
RNA	Ribonucleic acid
S	Entropy
S	Svedberg (unit of sedimentation in a centrifuge)
σ	Sigma; the rms value of the difference electron density across the asymmetric unit
SAD, MAD	Single (Multiple) wavelength anomalous dispersion
SASE	Self-amplified spontaneous emission
SAXS	Small angle X-ray scattering
SFX	Serial femtosecond crystallography
SP	Special pair
SPB	Single particles, clusters and biomolecules beamline at the EuXFEL
SPEAR	Stanford Positron Electron Accelerator Ring
SPI	Single particle imaging
SR	Storage ring
SSC	Serial sample chamber
SSX	Serial synchrotron crystallography
SV	Singular vector; rSV right singular vector; lSV left singular vector
SVD	Singular value decomposition
SX	Serial crystallography
τ	Relaxation time
T_1	Longitudinal relaxation time, in NMR
T_2	Transverse relaxation time, in NMR
TR	Time-resolved
TRX	Time-resolved crystallography
TT	Timing tool at the LCLS
U	Potential energy function or force field
UV	Ultraviolet
WAXS	Wide angle X-ray scattering
WT	Wild type
XANES	X-ray absorption near edge spectroscopy
XAS	X-ray absorption spectroscopy
XES	X-ray emission spectroscopy
XFEL	X-ray free electron laser
XSS	X-ray solution scattering
1D, 2D, 3D	One-, two-, or three-dimensional

Units

Length, Time, Concentration

m cm mm μm nm	meter, centimeter, millimeter, micrometer, nanometer
s ms μs ns ps fs as	second, millisecond, microsecond, nanosecond, picosecond, femtosecond, attosecond
M, mM, μM, nM, pM	molar, millimolar, micromolar, nanomolar, picomolar

Miscellaneous

s^{-1}	Per second (first order rate coefficient)
$M^{-1}\,s^{-1}$	Per molar per second (second order rate coefficient)
kV	Kilovolt (voltage)
KeV	Kiloelectronvolt (energy of a photon or electron)
GeV	Gigaelectronvolt (energy of an electron)
J, kJ	Joule, kilojoule (energy)
kJ/mol	kiloJoule per mole (energy in molar units)
kcal	Kilocalorie (energy)
kcal/mol	Kilocalorie per mole (energy in molar units)
Da, kDa	Dalton, kilodalton (molecular mass)
Hz, kHz	Hertz, kilohertz (frequency)
K	Kelvin (temperature)
C	Celsius (temperature)
mA	milliamp (current)
GW	GigaWatt (power)
Gy GGy	Gray, GigaGray (radiation dose)
mrad	milliradian (angle)

1

Introduction: Principles of Kinetics and Dynamics

1.1 Structure, Function, and Mechanism

Why study structure in biology? The imaging of structure has long been of interest, where structures range in length scale from an entire ecosystem to the gross anatomy of the human body, to cellular, molecular, and atomic structure. Here, we concentrate on structure at the molecular and atomic level and on the essential, dynamic variations in structure with time. Determining biological structure at this level is undeniably powerful, as evidenced by the library of more than 160,000 experimental and increasingly, computational structures in the Protein Data Bank (PDB). An arguably more interesting reason is broader: structure at the molecular and atomic level provides a powerful avenue into understanding both function and mechanism.

The three words *understanding*, *function*, and *mechanism* are critical to our arguments. *Understanding* means different things to biologists, chemists, and biophysicists. For example, a biologist seeks to understand gene transcription by identifying transcription factors, the specific DNA sequences to which they bind, the role of the RNA polymerase enzyme that catalyzes transcription, and the genes whose expression they control. A biochemist or biophysicist is certainly interested in those aspects, but also wishes, for example, to determine the atomic structures of the molecules involved, the chemical interactions that confer specificity between a transcription factor and its binding site, the intermediate structures involved in the binding of RNA polymerase and its catalytic processes, and how a particular step confers specificity or limits the overall rate of transcription. With our goal of understanding dynamic processes, we largely adopt the viewpoint of the biochemist and biophysicist. *Function* and *mechanism* are related but not identical. To most biochemists and biophysicists (and to us), *function* labels what the biological system does, and *mechanism* labels how it does it. Put less formally, mechanism denotes how the system works. Studies of structure and, as we shall see, of structural dynamics thus directly seek to identify mechanism. The scientific discipline now labeled structural biology explores the relationship between chemical, biochemical, and biophysical processes and three-dimensional (3D) structure at the atomic and molecular level, and thus underpins mechanism. The linkages between structure, mechanism, and function are strong at the atomic and molecular level: structure indeed directly determines mechanism and somewhat less directly, function.

Dynamics and Kinetics in Structural Biology: Unravelling Function Through Time-Resolved Structural Analysis, First Edition. Keith Moffat and Eaton E. Lattman.
© 2024 John Wiley & Sons Ltd. Published 2024 by John Wiley & Sons Ltd.

For example, hemoglobin molecules from species as diverse as humans, horses, and lampreys have the same overall fold. They are practically identical in their 3D structures and in fluctuations about those structures. The molecules are also closely similar in how that structure changes as hemoglobin carries out its function of transporting oxygen and carbon dioxide between the lungs and the tissues. At this level, structure, mechanism, and function are tightly conserved.

Conservation of structure holds over the length scale of a few nm, characteristic of individual protein molecules such as hemoglobin, up to a few tens of nm, characteristic of large complexes and molecular machines assembled from many proteins and nucleic acids such as a ribosome, transcriptional complexes, or an icosahedral virus. However, the linkage between structure and function weakens at longer length scales characteristic of subcellular organelles and cells. Function is largely conserved, but structure begins to exhibit a wider range. For example, organelles such as mitochondria from a given cell type all have a closely similar function, but their structures vary between individual cells of that type. Although variation is definitely present, it is not extensive enough to prevent confident identification of mitochondria in structural images obtained by electron microscopy.

We restrict ourselves for the moment to the length scale from individual molecules up to large macromolecular complexes and pose a critical question. Is structure both necessary and sufficient to fully understand mechanism and function? The fundamental thrust of our argument is that the answer to this question is: necessary, yes; but sufficient, no. The phrase "Structure determines function" is too limiting; the dynamics of structural changes must also be included. We assert that a more accurate phrase is "Structural dynamics determines function." This assertion underlies all aspects of our argument here. We recognize that the extent of structural dynamics varies from system to system. For example, in proteins whose principal function is scaffolding, dynamics generally plays a lesser role except during their assembly, disassembly, and modulation by ligand binding. The dynamics of sequence-dependent, structural changes in DNA plays a substantial role during replication and transcription.

Whatever form the experimental sample takes when exploring structural dynamics, the molecules comprising the sample should be demonstrably active. That form could be, for example, a crystal, dilute solution, within an intact cell or on a cryo-electron microscopy (cryo-EM) grid. That is, the molecules must display mechanism; they must *work* in the form used for determination of their structure. If they do not work or work only in aberrant fashion, elaborate experiments in structural dynamics will be seriously misdirected.

Consider structure determination by X-ray crystallography. A large majority of the structures in the PDB were determined by applying crystallographic techniques using synchrotron radiation emitted by storage ring (SR) X-ray sources (Chapters 5.1 and 6.2). Most crystals were frozen to around 100 K to mitigate radiation damage to the structure arising from X-ray absorption (Chapter 4.3.4). Freezing exposes three problems. First, at 100 K molecules have lost their normal function. Atomic motion is abolished, literally frozen. The molecules are inert, devoid of biological activity, unable to work (Rasmussen et al. 1992). Atomic motion and the resultant, time-dependent changes in structure – structural dynamics – are inescapably linked to mechanism and function, not merely in biology but also in chemistry and physics. "If it doesn't wriggle, it's not biology!" is a trenchant statement attributed to the British physiologist A.V. Hill in conversation with John Kendrew in the late 1940s. Kendrew was at the time a beginning research student in Max Perutz's

Medical Research Council Unit in Cambridge, England, at the dawn of protein X-ray crystallography. Hill's statement remains valid today: the absence of "wriggling" at 100 K accounts for the lack of function. Second, raw experimental data in crystallography arise from a space average over all the very large number of molecules in the crystal and a time average over the X-ray exposure time. For almost all systems, refinement against these data ultimately generates a single set of atomic coordinates to represent the structure. If the data were acquired at near-physiological temperature, refinement represents any atomic motion by the fuzziness of individual atoms, the so-called temperature factors. The atomic coordinates and temperature factors defined by refinement are both time-independent. That is, the structure is static, independent of time, and lacks any wriggling. Third, structures and more importantly, their energetics differ in detail between 100 K and more physiological temperatures (Halle 2004; Bock and Grubmüller 2022) where wriggling and function are retained (Chapter 2.6). Some of these energetic and structural differences are critical for mechanism.

Decades of crystallographers have studied structure as an essential determinant of mechanism related to function. Even when a crystal structure is determined at near-physiological temperatures where wriggling and function are retained, that structure is time-independent and does not yield the range of structures – the structural changes – inherent in mechanism and function. Although it's possible to trap the structures of normally short-lived intermediates in an overall reaction by chemical means rather than by the physical means of freezing, these static structures are limited in what they can reveal about mechanism (Chapter 2.6).

1.2 Activity in the Crystal

Since structural changes and the dynamic, time dependence of these changes are inherent in mechanism at the molecular level, a full understanding of mechanism and biological function must extend beyond inert, static structures to active, dynamic structures. This requires the ability to generate and determine short-lived, intermediate structures whose populations vary with time as the biological reaction proceeds. Going even further, it is becoming possible to determine the *functional trajectories* by which molecules pass from one intermediate structure to the next (Chapter 9.2). As we shall see, structures of intermediates can be determined experimentally by, for example, time-resolved crystallography under conditions where the molecules in the crystal can indeed wriggle. Another potential problem appears: the structural changes essential to activity may be affected by the 3D packing of molecules into a crystal, or by the solvents from which crystals were grown that occupy their intermolecular channels. Unusual, nonphysiological properties of the solvent such as extremes of pH, high ionic strength, or the absence of a key cation may substantially alter or even abolish activity and mechanism in the crystal. For time-resolved crystallography to be physiologically useful, we must confirm that mechanism in the crystal is similar to mechanism and function in the authentic, physiological environment.

This is by no means a new problem. In the earliest days of protein crystallography, a fundamental question arose: are the molecules in the crystal biologically active? The very first, near-atomic resolution structure of any protein was determined by John Kendrew and colleagues in 1960, that of myoglobin, a simpler, single subunit relative of hemoglobin (Kendrew et al. 1960). The relevance of the crystal structure of myoglobin to its physiology was at once questioned. As an oxygen storage protein, myoglobin has the relatively simple physiological function of binding and release of molecular oxygen. The crystal structure showed that the iron atom at the center of the heme group to which molecular oxygen binds was buried in the interior of myoglobin. This raised the possibility that the protein surrounding the heme might hinder access to the iron by molecules of the dimensions of oxygen.

The biophysicist Britton Chance and colleagues therefore sought to test "...the possibility that the structures of crystalline and dissolved [myoglobin] are not identical" (Chance et al. 1966). Hence, their functions might differ quantitatively or worse yet, qualitatively. Chance was the first to measure biological activity in any protein crystal and compare it directly with solution. As a pioneer in stopped-flow kinetic techniques based on rapid mixing (Chapter 3.3.2.1), he had designed the necessary mixing apparatus to fit in an elegant, fully portable leather briefcase. (Expensive apparatus and world-class competition were no barriers. Better off than most scientists, he had won a gold medal in yachting at the Olympic Games in 1952.)

Kendrew's initial crystals contained the oxidized form of myoglobin known as metmyoglobin whose ferric iron can bind and release small, oxygen-sized molecules such as cyanide and azide. Access of cyanide or azide to the ferric heme iron closely resembles that of oxygen to the ferrous heme iron in the authentic, physiological reaction. Chance applied his rapid mixing, stopped-flow kinetic techniques to initiate the reaction of metmyoglobin in a slurry of tiny crystals and in solution. He monitored progress of the reaction with ms time resolution, through the substantial change in optical absorbance that accompanies binding of azide. The relatively simple function of azide binding was indeed qualitatively retained in crystals but occurred with a reaction rate that was quantitatively different, some 20-fold lower in crystals of 5–10 μm dimensions than in solution. The reaction rate might have been lowered by hindered diffusion of azide through the solvent channels between the myoglobin molecules in the crystals, or by restriction of rate-contributing structural changes in the myoglobin molecule itself. Although it was unlikely that the structural changes in the molecule were large-scale, it was a realistic possibility that they were small-scale. For example, these changes could arise from the necessity to displace a sulfate ion derived from the crystallization solvent that was non-covalently bound to the distal histidine side chain. This histidine lies on the likely binding path of azide from the solvent to the heme. In a prescient observation, Chance concluded that "...transitional conformation changes are restricted in the crystalline state and diminish the reactivity with azide..." and went on to note that "...the problem can ultimately be explored by rapid methods for investigating protein structure in the transition [sic] state" (Chance et al. 1966). Just so! We give a modern view of the fleeting nature of the transition state that lies between two short-lived, transient intermediates in Chapter 2.5.

The determination of high-resolution crystal structures of other proteins such as hemoglobin and the enzymes lysozyme, α-chymotrypsin, and ribonuclease-S in the middle to late 1960s made discussion of possible differences in structure and reactivity between crystals and solution both more general and more urgent. In seminal research (Parkhurst and Gibson 1967) directly attacked the question of the physiological relevance of crystal structures and purified solutions, once more using heme proteins. They compared the reaction of CO with the ferrous form of horse hemoglobin under three very different conditions: in the authentic physiological environment of the erythrocyte, in the biochemical environment of a dilute solution of purified hemoglobin, and in the crystallographic environment of a polycrystalline slurry. Similar reaction properties under these three conditions would support the physiological relevance of results on purified solutions and crystals. Conversely, very different reaction properties would suggest that purification and crystallization were modifying hemoglobin, perhaps to the point of physiological irrelevance of the crystal structure. This reaction offers several experimental advantages. The CO-bound form of heme proteins such as hemoglobin and myoglobin is light-sensitive, which enables the covalent bond between CO and the ferrous heme iron to be readily broken by a light pulse in the visible region of the spectrum. The photo-dissociated CO diffuses briskly away from the heme and rebinds in the dark after the light pulse has terminated. Reaction initiation by a light pulse avoids the need for rapid mixing and the ensuing, slower diffusion processes of CO or azide into crystals (Chapter 3.3.2), both of which might influence the overall reaction course. The time resolution can be improved from the ms range characteristic of mixing reactions to the μs range by using a short light pulse, which also takes advantage of the high quantum yield of the photodissociation reaction. As with azide binding to myoglobin, the progress of the CO rebinding reaction to hemoglobin can readily be measured optically (Figure 1.1).

However, hemoglobin is considerably more complicated than myoglobin in its structure, function, and the structural changes that accompany oxygen or CO binding and release. Myoglobin is monomeric with a single polypeptide chain, but hemoglobin is tetrameric

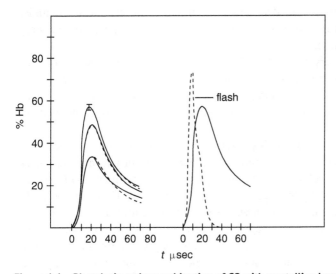

Figure 1.1 Photolysis and recombination of CO with crystalline horse hemoglobin after flash photolysis. The kinetics of recombination of CO from a thin film of polycrystalline slurry were followed optically. The ordinate shows the % of deoxyhemoglobin (Hb). Solid lines calculated, dashed lines observed; flash energies of 2025, 1500, and 900 J. Parkhurst and Gibson 1967/ ELSEVIER/Licensed under CC BY 4.0.

with four globin chains of two types denoted α and β, each of which contains a heme that binds and releases oxygen or CO. The amino acid sequences of the chains – their primary structures – are closely related to each other and to that of the single chain of myoglobin. Correspondingly, the overall folds – their tertiary structures – of the individual chains and of myoglobin are similar. The crystal structures of deoxyhemoglobin with no oxygen or CO bound, and that of methemoglobin (a proxy for the oxygen- and CO-bound forms) differ slightly in the tertiary structure of each chain but substantially in the quaternary structure of the molecule. That is, the structures of deoxy- and CO-hemoglobin differ in the spatial disposition of the four chains with respect to each other and the nature of the interfaces between the chains (Muirhead et al. 1967). Aspects of the function of hemoglobin such as the binding reactions of oxygen and CO also differ substantially from those of myoglobin.

In the erythrocyte and dilute solution, Gibson found that photodissociation of CO from the hemes was followed by two bimolecular rates of CO rebinding denoted fast and slow. In the crystal only one bimolecular fast rate was observed; the slow rate was absent. The fast rates are closely similar in magnitude in the erythrocytes, dilute solution, and crystals (see Table 1.1 of Parkhurst and Gibson 1967). Whatever structures give rise to this rate, they are for practical purposes identical under these three very different experimental conditions.

The first important conclusion is that hemoglobin in erythrocytes, solution, and crystals quantitatively retains the function represented by this fast reaction. Thus, the results confirm the physiological relevance of the crystal structure of the reactant, CO-bound form, and of the immediate photoproduct. The second conclusion follows from the absence of the slow reaction in the crystal. We now attribute the fast reaction to hemoglobin in the quaternary structure of the reactant, CO-bound form, and the slow reaction to the quite different quaternary structure of the ultimate photoproduct, the deoxy form. The transition in quaternary structure and from the fast to the slow rebinding forms in hemoglobin underlies its unusual cooperativity in oxygen and CO binding, a distinguishing, functional property not found in myoglobin. The change in quaternary structure from the fast to the slow form is unconstrained in the erythrocyte and dilute solution and occurs more rapidly than CO rebinding. However, the change is apparently constrained by the intermolecular contacts in this crystal form. That is, a functionally critical reaction of hemoglobin, namely the change in quaternary structure from the CO-bound form to the deoxy form, is qualitatively altered in this crystal.

However, many protein systems are highly polymorphic. The same protein or the homologous protein from other species can be crystallized under a range of solvent conditions in many different space groups, each with a different crystal lattice, intermolecular interfaces, spatial constraints, and sensitivity to large structural changes. It is often possible to discover a crystal space group compatible with whatever large structural changes the system exhibits. This is indeed the case with hemoglobin. One of the best pieces of evidence that crystallization does not interfere with tertiary and quaternary structure comes from the many cases in which a given molecule – such as hemoglobin – shows the same overall structure, though crystallized under very different conditions.

All is not lost but the cautionary tale remains. Activity and mechanism in the crystal, or indeed in solution, on a cryo-EM grid or in other forms, may be altered from the physiological state.

1.3 Other Structure-informing Techniques

These classical experiments emphasize the strong desirability of preceding or at least, paralleling dynamic structural studies by optical, spectroscopic studies in solution and in polycrystalline slurries of tiny single crystals, in relevant space groups. Exploring a variety of sample conditions and techniques enables crystals to be selected in which the molecules clearly retain the desired reactivities. Selection may also identify a time range when spectroscopically interesting changes do NOT occur. In such a time range only a single intermediate structure may be present, which simplifies determination of the crystal structure of that intermediate (Chapter 5.1.4).

A more general conclusion is that for dynamic studies to be most relevant, the sample and experimental conditions must permit the desired reaction to proceed with minimum constraints. This conclusion applies to all techniques and does not depend on the structural technique actually used. In principle, any static structural technique can be extended to the time domain, thus adding a dynamic aspect. This emphasizes the desirability of dynamic studies that offer the near-atomic structural resolution of crystallography but do not require such non-biological aspects as unusual solvents, high vacuum, dehydration, or crystallization. In addition to time-resolved X-ray crystallography, we therefore consider in more detail a number other structure-based, experimental techniques and their application to dynamic studies: time-resolved solution X-ray scattering (Chapter 5.2), nuclear magnetic resonance (Chapter 8.5), single particle imaging by cryo-EM (Chapter 8.2), X-ray spectroscopies (Chapter 8.4), and hydrogen-deuterium exchange (Chapter 8.6). The last can identify flexible, solvent-accessible regions. In a powerful combination of experiment and computation, interpretation of today's successful experimental approaches to dynamics by structure and spectroscopy can also be aided by recently-developed theoretical and computational approaches such as energy landscape analysis (Chapter 8.3). Extension of powerful, artificial intelligence-based predictions of static protein structures (Jumper et al. 2021) to dynamic structures offers a new challenge (Chapter 9).

Quasi-structural techniques such as spectroscopies complement these directly structure-based techniques. Optical spectroscopic studies are based on electronic or vibrational transitions in a naturally occurring chromophore such as flavin mononucleotide (FMN) or heme that absorbs or fluoresces in the visible region of the spectrum; or of an aromatic side chain such as the indole of tryptophan that absorbs in the near-ultraviolet (UV) region. They may probe a general spectral property such as absorbance in the infrared (IR) region (Chapter 3.3.2.3). Since heme and flavin are near-planar and optically anisotropic, their orientation in the crystal may be determined by illumination with plane-polarized light. Circular dichroism (CD) probes the overall secondary structure of a protein; fluorescence spectroscopies probe the immediate environment of the fluorophore, which may for example be an indole side chain or an added dye; fluorescence resonance energy transfer reveals the distance between donor and acceptor dyes; and vibrational spectroscopies in the IR offer wider scales of structure and span many atoms whose chemical nature and even their identity in the primary structure may be established. Electron paramagnetic resonance spectroscopy of an added paramagnetic species – a spin label – can directly reveal its extent of dynamic motion and relate that motion to structure (Moffat 1971).

Almost all static, quasi-structural spectroscopic techniques are speedy to apply, at least by the standards of crystallography. They don't require crystals, and directly detect the rates of change in the structural property and sometimes their extent. The limitations of quasi-structural techniques usually do not lie in experimental detection of the change, but in their structural interpretation. If the technique can also be applied to a polycrystalline slurry or even to a single larger crystal or a single cell, so much the better. Recapitulation of the Parkhurst and Gibson experiment remains powerful. For example, single crystal optical studies can be conducted by absorption spectroscopy in the near-IR (say, 700 – 1000 nm) where the extinction coefficient and hence the crystal absorbance is low despite the protein concentrations, often as high as 1–50 mM, found in crystals.

Structural and quasi-structural techniques may also be classed into those which reveal global information about an entire protein such as X-ray scattering, CD and vibrational IR spectroscopies; and those which provide local information about the environment of a specific component, such as optical absorption spectroscopy of a flavin, tryptophan fluorescence, or electron paramagnetic resonance spectroscopy of a spin label. This introduces a note of caution. The dynamics of the observables in a local technique may not exactly parallel those in a global, directly structure-based technique. The techniques probe quite different scales of structure. For example, the different protein structures of the reaction intermediates of a flavin-containing enzyme may all exhibit closely identical optical absorption spectra of the flavin, a local property that depends very largely on its redox state.

1.4 Dynamics, Kinetics, Movies, Pathways, and Functional Trajectories

Much depends on the consistent use of terminology in structural dynamics, as in many scientific areas. We expand here on our usage of key words and phrases, their meaning and applicability.

The most general scientific usage of dynamic means varying with time or time-dependent. The exact meaning depends on the discipline. For example, in mathematics a system in which a function describes the motion of a point in space is denoted a dynamical system. In the fundamentals of X-ray crystallography (Rupp 2010), dynamical diffraction characterizes in detail the repeated, multiple scattering of an electromagnetic X-ray wave by the electrons associated with spatially separated atoms. This approach contrasts with kinematic diffraction, a treatment of simpler, single scattering by crystals which is nevertheless adequate to describe the position of diffraction spots in reciprocal space and the intensity of each spot. Single scattering in kinematic diffraction underlies most crystallography as we know it and all of static and dynamic structure determination in chemistry and biology. Despite the adjectives dynamical or kinematic, atomic motion is not explicitly considered in either treatment of diffraction (!). The X-ray wave is moving, not the atoms and their electrons in the lattice. Time-resolved X-ray scattering or more specifically, time-resolved crystallography is based on a straightforward extension of kinematic diffraction to incorporate atomic motion (Chapter 5.1).

The distinction between dynamics and kinetics is critical and extends beyond terminology to the fundamentals of experimental approaches. An experiment in dynamics

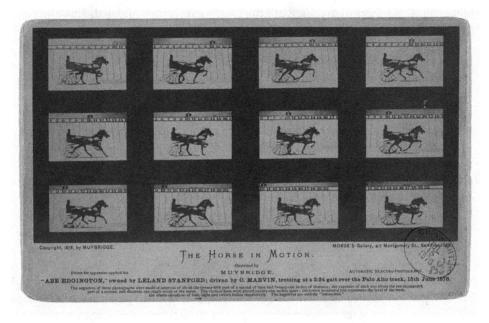

Figure 1.2   The gait of a horse. Photograph by Edward Muybridge; reproduced with permission from the US library of Congress.

probes the variation with time of a property of a single molecule or entity such as the famous Muybridge horse (Figure 1.2). In contrast, an experiment in kinetics probes the variation with time of a property measured as an average over a statistically large population of molecules. *A single molecule or entity exhibits dynamic behavior; a statistically large population of molecules exhibits kinetic behavior.* Kinetic behavior is of course based on the underlying – but not directly observable – dynamic behavior of the individual molecules that make up the population. For example, molecular dynamics refers to the time-dependent variations in structure within a single molecule, which may be followed experimentally or more usually, by computation (Chapter 8.2). Chemical kinetics refers to the time-dependent changes in the populations of each of the states (for example, reactants, intermediates, and products) that comprise a chemical or biochemical mechanism (Chapter 2.3). (We consider an explicit thermodynamic definition of *state* in Chapter 2.2.) However, only the average value can be observed experimentally of a property that depends on a statistically large number of molecules in the sample. Both 3D structure and optical absorbance are examples of such a property. Each state within the sample also contains a statistically large number of molecules. Although the average structure of each state is time-independent, that average differs detectably from state to state. Hence as the reaction proceeds from the reactant via intermediate states to the product, the experimentally measurable quantity of the average structure over all molecules in the sample is time-dependent.

In more specific terms, consider a reaction which proceeds via a set of states S_1, S_2, $S_3...S_j...$. Let P_j be the time-independent structure of molecules in state S_j and $f_j(t)$ the population of molecules in that state, at a time t after a reaction is initiated. Then the average structure $P_{AV}(t)$ of all molecules in the sample is

$$P_{AV}(t) = \frac{\sum_j f_j(t) P_j}{\sum_j f_j(t)}. \tag{1.1}$$

where the summations are taken over all states S_J. The structure P_j in each state is likely to be chemically sensible (though it may be strained; Chapter 7.6). However, the average structure $P_{AV}(t)$ over all molecules and states need not be chemically sensible. As Equation 1.1 shows, $P_{AV}(t)$ is a time-dependent mixture of the time-independent structures in the individual states. This time-dependent mixture affects the total X-ray scattering by the sample (Chapter 5.1).

The raw data of experiments in crystallography, solution scattering or ultrafast spectroscopies thus arise from the simultaneous observation of the properties of a statistically large number of molecules in the sample, averaged over all molecules. A quite large protein complex may have a volume of 500 nm³. Even a tiny nanocrystal of $50 \times 100 \times 200$ nm will thus contain around 2000 molecules of this complex, a statistically large number. It follows that all such time-dependent experiments explore chemical kinetics, not dynamics.

To reemphasize: in chemical kinetics the overall time dependence arises from the variation with time of the populations of individual states, in which each individual state has its own time-independent, average structure. An experiment in chemical kinetics thus observes the formation and decay of states. A chemical kinetic mechanism consists of a definite number of states (e.g. reactant(s), a set of intermediates, and product(s)) connected by reactions describing the formation and decay of each state. Each reaction has an associated forward (formation) and reverse (decay) rate coefficient (Chapter 2). That is, a chemical kinetic mechanism describes states and how they are connected. Figure 1.3 illustrates some simple mechanisms with four states A, B, C, and D.

The overall goal of structural experiments in chemical kinetics is to decide which mechanism(s) are compatible with the data, to characterize the structure of each of the states in the mechanism(s), and to determine how those states are connected and the rates with which they interconvert.

An experiment in dynamics is distinctly different. Here, the overall time dependence arises from the variation with time of the structure or spectroscopic properties of an individual molecule (more precisely, of a statistically small number of molecules).

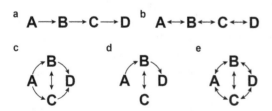

Figure 1.3 Some simple chemical kinetic mechanisms. Each mechanism has four states A, B, C, and D. Conversion between states is shown by arrows.

Dynamics indeed deals with states and more broadly, with pathways between states that can be directly probed by structural or spectroscopic experiments. Kinetic experiments can establish that two states are connected and the magnitudes of the forward and reverse rate coefficients. Dynamic experiments further establish the structural (or spectroscopic) pathways by which this connection occurs.

Use of the terminology *movie* to describe the results from chemical kinetics is tempting but scientifically misleading. (We acknowledge being tempted; (Moffat 1989).) Movies conjure up visions of motion, of time-dependence. However, the essence of chemical kinetics lies in the time-independent structures of the states that constitute a chemical kinetic mechanism. Authentic movies can however be appropriately presented. For example, the time-dependent X-ray intensity over a range of scattering angles or a time-dependent electron density map both arise from an average over all molecules present in the sample, in all states. Both vary continuously with time as the populations of the constituent states and the average X-ray scattering or electron density associated with each state rises and falls. Movies can also be made by computer-based morphing between the structures of each state. Such movies may guide the eye and help in appreciating the structural differences between states. Unfortunately, they do not contain any structural information beyond that in the time-independent structures themselves. In contrast, molecular dynamics simulations do provide new information and generate genuine movies that illustrate the continuous structural variation of a single molecule.

If it were possible to examine experimentally the structure of an individual molecule as it executed its biochemical reaction e.g. of an enzyme as it bound a substrate, catalyzed its conversion to product, and released the product, the time-varying, dynamic structure of the molecule could be continuously detected by structural or spectroscopic approaches. The more formal terminology *functional trajectory* denotes a pathway by which a single molecule passes from one state to the next. The functional trajectory (or more likely, the set of functional trajectories followed by many molecules) could in principle be visualized and depicted for each individual molecule as an authentic movie of structural dynamics. We could title this movie the functional trajectory of the biochemical reaction, or the pathway. This highly desirable result has not yet been achieved for an enzyme (but see a functional trajectory for the ribosome; Chapter 8.3.2). In marked contrast to experiments in chemical kinetics, authentic movies directly result from a molecular dynamics calculation. When this calculation is repeated over many individual molecules, it explicitly represents the candidate functional trajectories – with a deliberate use of the plural – between states.

However, it is becoming feasible to obtain raw experimental data on the structure of an individual molecule (or at least at present, of a larger complex) at a single point on the pathway. By repeating the experiment on each of many individual molecules and identifying the point on the reaction coordinate (Chapter 2.5) or more generally, on the high-dimensional energy landscape (Chapter 8.3) associated with each structure, the structures of the various states and the functional trajectories between them could in principle be identified (Ourmazd 2019). We consider experimental approaches to the structures of single molecules and complexes by cryo-EM (Wang et al. 2016) and X-ray scattering in Chapters 8.2, 9.2, and 9.4.

We end in Chapter 9.5 with the interesting and still open question: what is the evolutionary relevance of functional trajectories, mechanisms, and states?

1.5 The Time-resolved Experiment: An Overview

We introduce here several of the general concepts of the time-resolved experiment in structural dynamics, expand on their experimental details in Chapters 3, 4, and 5, and on their analysis and interpretation in Chapter 7. Chapter 6 considers synchrotron radiation and its production in storage ring and free electron laser X-ray sources. Chapter 8 considers non-X-ray-based structural approaches, and Chapter 9 suggests promising undersolved (or unsolved) problems in dynamics and kinetics.

1.5.1 Proof and Disproof of Mechanisms in Structurally-based Chemical Kinetics

A comprehensive structural experiment in chemical kinetics will span the complete time range of the reaction. When the data are of high quality, careful data analysis by, for example, singular value decomposition (SVD) can identify the number of distinct, orthonormal structures and the relaxation times between them (Chapter 7.4). However, each orthonormal structure is a linear combination of the structures of the authentic reactants, intermediates, and products. The coefficients that describe this linear combination are not identified by SVD itself but by the specific chemical kinetic mechanism. The mechanism depends on the linkage between the orthonormal structures, the number of states, relaxation times, and the rate coefficients for the formation and decay of each state, where each state represents a reactant, intermediate, or products. That is, the mechanism depends on how the states are connected (Figure 1.3).

All experiments in chemical kinetics have a fundamental limitation which makes identification of the specific mechanism nontrivial. Although the experiments can **disprove** candidate mechanisms by demonstrating that they are incompatible with the time-resolved data, they cannot **prove** that a particular chemical kinetic mechanism holds. Proof of that mechanism demands that all other candidate mechanisms with the observed number of orthonormal structures can be disproved. Since there are many distinct mechanisms with 4, 5, or more states, disproving all but one is a challenging task which has rarely, if ever, been achieved.

Nevertheless, two powerful routes to disproving candidate mechanisms are available. First, a candidate mechanism compatible with the orthonormal structures and associated rate coefficients (Chapter 7) will yield a candidate structure for each of the states. Crystallographic refinement can demonstrate whether the structures in that mechanism are structurally homogeneous. Structural heterogeneity – that is, the presence of a mixture of structures – in one or more states in a candidate mechanism offers convincing disproof of that mechanism. Second, all structures must also be chemically and structurally plausible: they must be satisfactorily refinable using established bond lengths and angles. This route to disproof is weaker than the first since much hinges on the interpretation of "plausible." All such established constraints on refinement of static structures are derived from unstrained small molecules, proteins, and nucleic acids at thermodynamic equilibrium. The difficulty in establishing implausibility is that intermediate structures in chemical kinetics are by definition short-lived and not at equilibrium. The shorter their lifetime, the less stable they are, the further from equilibrium and the more likely to be strained.

Even if the candidate mechanism is correct, strained structures are unlikely to be satisfactorily refinable with established, plausible bond lengths and angles.

These two routes to disproof are specific to structurally-based chemical kinetics. They offer potentially powerful means of excluding those candidate mechanisms that may be compatible with the chemical kinetic data but whose intermediates are structurally heterogeneous and/or chemically implausible. However, these routes are untested and remain as an experimental challenge. No equivalent disproof routes are available to mechanisms derived from, for example, chemical kinetics based on optical spectroscopies.

A third route to at least partial proof/disproof is available if structural data over the complete time range of the reaction can be obtained. Time ranges may be identified during which no significant changes can be detected. That is, the raw structural data are time-invariant in these time ranges. During each such range, it is very likely that only a single, time-independent structural state is present, which should be satisfactorily refinable. Candidate mechanisms that fail to account for this time invariance are disproved.

A fallback but deservedly popular strategy is to adopt Occam's Razor: in the absence of other, nonstructural information, adopt the simplest mechanism compatible with the time-resolved data. In the words of Alfred North Whitehead: "Seek simplicity but distrust it" (Whitehead 1920). Simplicity is in the eye of the beholder, who typically sees a linear mechanism for an irreversible reaction (illustrated in panels a and b of Figure 1.3). In practice, this generally means that a linear mechanism is assumed first and if largely satisfactory, subsequent analysis seeks to identify any minor modifications to that mechanism such as a weakly-populated side-path (panel d of Figure 1.3), or structurally-related, parallel paths (panel c of Figure 1.3).

Finally, we note that detection of a particular intermediate in a candidate mechanism ultimately depends on its peak population. When ultrafast or fast reactions precede slower reactions in a mechanism, intermediates form faster than they break down and can attain a high peak population. These intermediates can readily be generated and characterized if their lifetimes exceed the temporal resolution (Chapter 1.5.3). Indeed, a characteristic of many biological systems, including those initiated by absorption of light, is that fast reactions precede slower reactions. However, if fast reactions follow slower reactions in a mechanism, intermediates break down more rapidly than they form. The peak populations attained by these later, fast-decaying intermediate states may be so low that they are undetectable by any structural or spectroscopic technique. These intermediates may be plausible and present in the sense that they are believed on chemical grounds to lie on the reaction coordinate (Chapter 2.3), and all molecules must pass through them. An example is that of an enzyme-catalyzed reaction in which slow substrate binding is followed by a series of more rapid, chemical steps that separate short-lived intermediates, and ended by slow product release. However, such intermediates could only be characterized by studying the dynamics of individual molecules. They remain invisible to experiments in chemical kinetics. Thus, a chemical kinetic mechanism considers only detectable states.

1.5.2 Spatial Resolution

Designing a time-resolved experiment in chemical kinetics begins by establishing the spatial and temporal resolutions demanded by the scientific problem and matching these to the sample and structural technique.

If the goal of a dynamic experiment is to explore mechanism at the atomic and molecular level, a crystallographic resolution of 0.2 nm or better is generally required for an initial, static structure of, for example, an enzyme or a photoreceptor. In favorable cases, these crystals lead to refined structures at 0.2 nm resolution with a precision in atomic coordinates of around 0.015 nm. However, time-resolved crystallography is more concerned with the precision and accuracy of the structural changes – the structural variation with time – rather than the precision of the underlying static structure. The precision in changes in atomic positions obtained from refinement of difference electron density (DED) maps is often less than 0.01 nm, superior to the precision of the underlying structures themselves (Chapter 7.6; Appendix 1).

Although raw data to high spatial resolution can be obtained by wide angle X-ray scattering (WAXS) in solution (Chapter 5.2) and by nuclear magnetic resonance (NMR) in solution and the solid state (Chapter 8.5), *de novo* structure determination at high resolution by these techniques is less developed than in crystallography. However, two newer routes to *de novo* static structure determination at high resolution are advancing very rapidly and offer much promise: cryo-EM (Chapter 8.2) and sequence-based structure prediction by sophisticated artificial intelligence (Chapters 9.3.1, 9.3.2, and 9.3.3). At present, cryo-EM is largely static or restricted to cryo-trapped structures of relatively long-lived (ms) intermediates. Extension of these static approaches to dynamic structure by electron microscopy at near-room temperature where activity is retained, poses an attractive – but large – challenge (Chapter 9.4.1). WAXS has been effectively applied to time-resolved difference measurements (Chapter 3.3.2). Various forms of NMR spectroscopy inherently offer dynamic information but again, structural data on short-lived intermediate states are difficult to access (Chapter 8.5).

X-ray crystallography continues to offer the strongest approach to structure determination at high spatial resolution of the intermediates that populate a chemical kinetic mechanism.

1.5.3 Temporal Resolution

Many dynamic measurements require an experimental approach to initiating a reaction in the molecules in the sample, an approach conventionally known as the pump. The course of the reaction is followed by another experimental approach known as the probe. In structural dynamics, the probe is a structure-determining approach such as time-dependent X-ray crystallography or time-dependent X-ray solution scattering. That is, these are pump–probe experiments. A single pump pulse may be followed, after a user-determined time delay, by a single probe pulse (pump once, probe once), or by a train of probe pulses over an extended time (pump once, probe many).

The temporal resolution required is established by the shortest time scale over which the interesting reactions are to be followed; that is, by the fastest reaction. For example, key chemical processes such as isomerization and electron transfer by the chromophore in light-sensitive systems such as a photosynthetic reaction center or bacteriorhodopsin occur on the ultrafast time scale of fs to a few ps (Martin and Vos 1992). Their observation requires a temporal resolution less than a few tens of fs. These ultrafast processes are followed by much slower changes in tertiary structure associated largely with the protein or RNA that may extend, remarkably, over 12 decades in time from a few ps to s. Observation of these slower changes requires less stringent temporal resolution, say 100 ps or 10 ns. On even longer time scales, the turnover number (Chapter 2.4) of most enzymes is low, certainly less than 1000 s^{-1} and often as low as 1 s^{-1} or even 0.1 s^{-1}. In these cases, an even more modest temporal resolution of 20 ms will be adequate. Time scales are illustrated in Figure 1.4.

The duration of the pump process must be shorter than the required temporal resolution (Chapter 3.3.1). This ensures that the true time course closely matches the observed time course and is not masked by convolution with the time course of the pump process itself. For example, if turnover by an enzyme occurs in ms and longer, then reaction initiation should occur in 500 μs or shorter. This is achievable by rapid mixing of a solution of an enzyme with its substrate (Chapter 3.3.2.1). If the interesting tertiary structural changes in a light-sensitive photoreceptor occur on the time scale of 1 ns and longer, then a laser pulse of 100 ps or less should be used. If isomerization of the light-absorbing chromophore in a photoreceptor is to be examined on the time scale less than 5 ps, then a laser pulse in the fs range is required.

The temporal resolution also imposes strong constraints on time-dependent structural measurements. However, see (Fung et al. 2016). The minimum exposure time of the probe

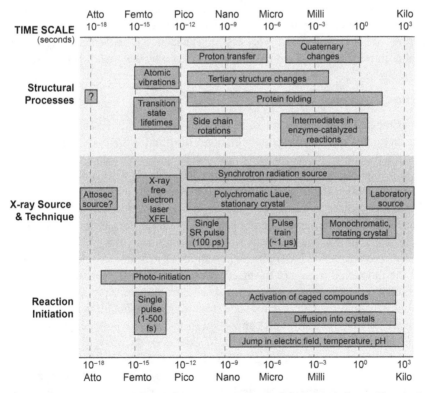

Figure 1.4 Time scales of structural processes and experimental techniques. Time is shown on a logarithmic scale at the top. The left and right edges of each box illustrate the time range roughly appropriate to the process or technique noted in the box.

required to give a good signal in X-ray, optical or indeed in any measurements must be less then (or at least, comparable with) the required temporal resolution. For example, hard X-ray free electron lasers (XFELs; Chapter 6.3) are both exceptionally intense (small, weakly scattering samples can generate a good signal) and deliver X-ray pulses of short duration in the few fs to ps time scale (rapid reactions can be followed). XFELs are therefore essential for all experiments that require ultrashort temporal resolution of fs to a few ps. In contrast, SR X-ray sources are substantially less intense than XFELs (somewhat larger, more strongly scattering samples are required to generate a good signal) and their X-ray pulses are of longer duration, around 100 ps (Chapter 6.2). All experiments that use the full X-ray pulses at SR sources therefore have a minimum temporal resolution of 100 ps. The time scale of ultrafast chemical processes lies in the ps to fs range (Figure 1.4), too short to be accessible at SR sources. However, SR sources are very well suited to all time-resolved experiments where the temporal resolution required is >100 ps. SR sources also support time-resolved experiments which utilize many X-ray pulses in a pulse train, where the temporal resolution is set by the duration of the pulse train. Other factors being equal, the longer the pulse train, the longer the temporal resolution, but the weaker the scattering of the sample required to obtain good diffraction. We contrast XFEL and SR sources in Chapter 6.

The experimental temporal resolution is thus influenced by three times: the time course of the pump, the time course of the probe, and the jitter in the time delay between the pump and probe (i.e. the variation over many pump–probe pairs of the true time delay between them). The time courses of the three processes are independent and unless they are of roughly equal duration, the overall temporal resolution is set by the longest of the three.

1.5.4 Reaction Initiation

Dynamic experiments fall into two classes. In the first class considered above known as pump–probe, an explicit means of reaction initiation – the pump – drives the system along a nonequilibrium functional trajectory by the addition of energy. Initiation is achieved by processes such as binding of a reactant, temperature jump, or absorption of a photon. In the second class, no explicit means of reaction initiation is required and no pump is needed. The system remains at thermodynamic equilibrium throughout and undergoes spontaneous, dynamical fluctuations around the equilibrium position that are detected by a structural or spectroscopic probe. The two classes of experiments are connected. A system at equilibrium will visit, albeit with lower probability, all conformations occupied by the molecules along the functional trajectory traversed in nonequilibrium, pump–probe experiments. Put another way, the set of reactant, ground state structures at equilibrium contains at low probability the intermediate, excited state structures, even though these may be far from the ground state equilibrium. This connection forms the basis of energy landscape analysis (ELA; Chapter 8.3).

For dynamical experiments in crystallography and solution scattering at both XFEL and SR sources, the initial perturbation or pump that drives the system along the nonequilibrium functional trajectory may be any physical or chemical variable on which structure or chemical equilibria depend (Chapter 3.3). Perhaps the simplest approach to reaction initiation is by a jump in concentration of a chemical species. The jump is achieved by mixing of reactants such as an enzyme with its substrate, or in a pH-dependent reaction, by a variation in pH. The rates of mixing are usually limited to the ms time range by slower diffusion processes. Better time resolution in the µs range or even less can be achieved by relaxation approaches (Chapter 3.3.2.2), a further example of the first class of reaction initiation. Here, perturbation of the system by, for example, a rapid temperature jump or electric field pulse perturbs the equilibrium to produce structural changes across almost all of the structure of the molecules. After the pulse, the molecules relax as equilibrium is reestablished at the new temperature or electric field, in a potentially complex set of structural reactions (Hekstra et al. 2016). In light-sensitive systems, the best time resolution is readily achieved by light from a laser pulse. Most proteins and nucleic acids do not absorb light in the visible region of the spectrum. Naturally light-sensitive proteins therefore bind a small organic molecule known as a chromophore that can absorb a photon in the visible (Chapter 3.3.2.3). The resultant electronic and structural changes in the chromophore propagate across the molecule, which slowly reverts to its dark, ground state via a series of electronically and structurally distinct intermediates. The time range accessible is set by the duration of the initiating light pulse which typically ranges from ns to fs. Although initiation of a reaction by light is experimentally convenient, the number of naturally light-sensitive biological systems is small. This has led to successful efforts to confer sensitivity to light on otherwise light-inert systems and extend the range of suitable, light-sensitive targets by optogenetic engineering and photopharmacology (Chapter 4.4).

1.5.5 Conclusion

In an ideal world of structural dynamics, we seek to determine 3D structure at the atomic level, under sample conditions where activity is retained, where the structural changes associated with mechanism can be determined over an appropriately wide time range, where quantitative measurements of mechanism can be made on the same sample in parallel with structure determination, and where mechanism can be related to function. Structural change – dynamics at the molecular level – is fundamental to this relationship. Static structure alone is powerful but insufficient. Understanding results from the experimental world requires parallel studies in theoretical and computational worlds, subject to fewer constraints but with very different challenges. There is a powerful interplay among the three worlds. The experimental world informs the theoretical and computational worlds; the theoretical and computational worlds both seek to understand the current experiments and to suggest new experiments that test their theories.

Experiments in structural dynamics also make demands on those carrying them out. Relatively standard, static experiments at SR sources are now conducted remotely. However, dynamic experiments are still sufficiently complex that an expedition by the scientific team to the SR or XFEL beamline itself is essential. They require the presence of the team during data collection on a 24/7 schedule under intense time pressure. The team includes other collaborators who contribute to areas such as the design of experiments and data analysis in real time, to assess the presence and magnitude of the signal and adjust experimental parameters as needed. Yet other collaborators conduct offline data analysis and interpretation, and parallel spectroscopic experiments and theoretical calculations. Finally, diplomatic skills help to deal with the inevitable crises, at the beamline and later!

References

Bock, L.V. and Grubmüller, H. (2022). Effects of cryo-EM cooling on structural ensembles. *Nature Communication* 13 (1): 1–13.

Chance, B., Ravilly, A. and Rumen, N. (1966). Reaction kinetics of a crystalline hemoprotein: an effect of crystal structure on reactivity of ferrimyoglobin. *Journal of Molecular Biology* 17 (2): 525–534.

Fung, R., Hanna, A.M., Vendrell, O. et al. (2016). Dynamics from noisy data with extreme timing uncertainty. *Nature* 532 (7600): 471–475.

Halle, B. (2004). Biomolecular cryocrystallography: structural changes during flash-cooling. *Proceedings of the National Academy of Sciences* 101 (14): 4793–4798.

Hekstra, D.R., White, K.I., Socolich, M.A. et al. (2016). Electric-field-stimulated protein mechanics. *Nature* 540 (7633): 400–405.

Jumper, J., Evans, R., Pritzel, A. et al. (2021). Highly accurate protein structure prediction with Alphafold. *Nature* 596 (7873): 583–589.

Kendrew, J., Dickerson, R., Strandberg, B. et al. (1960). The three dimensional structure of myoglobin. *Nature* 185: 422–427.

Martin, J.-L. and Vos, M.H. (1992). Femtosecond biology. *Annual Review of Biophysics and Biomolecular Structure* 21 (1): 199–222.

Moffat, J.K. (1971). Spin-labelled haemoglobins: a structural interpretation of electron paramagnetic resonance spectra based on X-ray analysis. *Journal of Molecular Biology* 55 (2): 135–146.

Moffat, K. (1989). Time-resolved macromolecular crystallography. *Annual Review of Biophysics and Biophysical Chemistry* 18 (1): 309–332.

Muirhead, H., Cox, J.M., Mazzarella, L. et al. (1967). Structure and function of haemoglobin: III. A three-dimensional fourier synthesis of human deoxyhaemoglbin at 5·5 Å resolution. *Journal of Molecular Biology* 28 (1): 117–150.

Ourmazd, A. (2019). Cryo-EM, XFELs and the structure conundrum in structural biology. *Nature Methods* 16 (10): 941–944.

Parkhurst, L.J. and Gibson, Q.H. (1967). The reaction of carbon monoxide with horse hemoglobin in solution, in erythrocytes, and in crystals. *Journal of Biological Chemistry* 242 (24): 5762–5770.

Rasmussen, B.F., Stock, A.M., Ringe, D. et al. (1992). Crystalline Ribonuclease A loses function below the dynamical transition at 220 K. *Nature* 357 (6377): 423–424.

Rupp, B. (2010). *Biomolecular Crystallography: Principles, Practice, and Application to Structural Biology*. Garland Science.

Wang, R.Y.R., Song, Y., Barad, B.A. et al. (2016). Automated structure refinement of macromolecular assemblies from cryo-EM maps using Rosetta. *Elife* 5: e17219.

Whitehead, A.N. (1920). The concept of nature. *The Tarner Lectures*, Dover.

2

Physical Chemistry of Reactions

2.1 Introduction

The fundamentals of thermodynamics, equilibria, and kinetics are essential to structural dynamics and indeed, to dynamics in all disciplines. These topics are presented in depth in textbooks of physical chemistry, many of which are specifically aimed at biologists, biochemists, and biophysicists. For the general details, we refer the reader to such texts (Eisenberg and Crothers 1979; McQuarrie and Simon 1997; Dill et al. 2010). Readers already well-grounded in physical chemistry may prefer to skim or omit this chapter.

We define here important terms and concepts in thermodynamics and kinetics and offer some examples specific to structural dynamics. Thermodynamics enriches understanding of dynamics and mechanisms in many ways. For example, applications to protein folding have been numerous, including the use of funnel-shaped energy landscapes to illuminate folding pathways. The energetic costs of processive motion rather than random walks – say, in DNA repair – are only apparent when the thermodynamic concept of entropic cost is included. The large energy fluctuations seen in macromolecules (Chapter 8.3.2) were revealed by statistical thermodynamics.

2.2 Thermodynamics: States and Equilibria

Thermodynamics is an abstract and completely general approach to the properties and transformations of matter, which deals with concepts such as *energy, heat, temperature, and work*. Concepts are derived by considering the physical reactions of an ideal gas such as its expansion and contraction, chemical reactions in an isolated system, or phase transitions such as liquid to solid. These concepts do not depend on the detailed structure of matter. As a result, thermodynamics does not deal directly with the 3D structure of molecules and the reactions they engage in. Although thermodynamics can identify whether a particular reaction will occur spontaneously and the direction it will take, it provides no information on the rate of that reaction. The study of rates constitutes *kinetics*, considered below (Chapters 2.3 and 2.4). Thermodynamics applies to a *system*, defined as a small part of the world separated by a boundary from its surroundings. Systems may be open or closed

Dynamics and Kinetics in Structural Biology: Unravelling Function Through Time-Resolved Structural Analysis, First Edition. Keith Moffat and Eaton E. Lattman.
© 2024 John Wiley & Sons Ltd. Published 2024 by John Wiley & Sons Ltd.

(put more formally, *isolated*). A closed system such as a sealed capillary containing a protein crystal does not allow the passage of matter across its boundary, though it does allow the passage of heat or work. An isolated system is closed and in addition, does not exchange energy. The capillary and contents can be heated or cooled; it is not isolated. In contrast, a small droplet of aqueous protein solution sitting on a tiny, exposed conveyor belt is an open system. A second droplet can be added to mix with the first, or gas exchange can occur between the droplet and its surroundings.

A system is associated with a small number of *properties*, also known as *variables of state* or *functions of state*. A system is in a defined state when the magnitude of all its properties can be specified. Importantly for structural dynamics, *the magnitude of a property does not depend on the pathway by which the state was attained*. The magnitudes are therefore independent of the history of the system. In particular, they do not depend on the pathway (more formally known as the functional trajectory introduced in Chapter 1.4) which the system followed in moving toward that state. If the magnitudes of the properties of a system are independent of time, the system is said to be at thermodynamic *equilibrium*.

Properties can be *intensive or extensive*. An intensive property is unchanged if the system is divided into parts; examples are temperature or the concentration of a homogeneous solution. In contrast, the magnitude of an extensive property such as volume is changed if the system is divided. In biochemistry and biophysics, concentration is expressed in molar units: such-and-such per gram or per mole. All such molar properties are intensive. An important example is molar energy, expressed as kJ mol^{-1}.

Two central properties of a system are its *entropy* (denoted by the symbol S) and its *enthalpy* (denoted H). For a reaction of an isolated system in which a reactant spontaneously converts to a product, the value of the entropy change ΔS between products and reactants must be positive. A consequence is that the entropy of an isolated system is a maximum when the system reaches equilibrium, after which no further reaction occurs. In biology, reactions generally occur at constant temperature and pressure. Truly isolated systems are almost unknown.

More useful for our purposes is to take advantage of the physical meaning of *entropy as a measure of disorder* in a system. For example, the flexible side chain of an amino acid could adopt many orientations, each of low occupancy, and would have a higher entropy than if it had a single orientation of high occupancy. The same is true for the polypeptide chain as a whole. The number of ways a state of fixed total energy can be achieved is known as the *degeneracy* of that state (denoted Ω). The larger the degeneracy, the more disordered is that state and the higher its entropy. Degeneracy Ω and entropy S are directly linked by Boltzmann's equation:

$$S = k_B \ln \Omega. \tag{2.1}$$

where k_B is known as Boltzmann's constant, with units of kJ/degrees Kelvin (usually abbreviated as deg K or simply K, if no confusion of K with an equilibrium constant K will arise). The change in entropy ΔS associated with a reaction is then related to the difference in degeneracy between products and reactants:

$$\Delta S = k_B \ln \left(\frac{\Omega_{products}}{\Omega_{reactants}} \right). \tag{2.2}$$

The change in enthalpy ΔH associated with a reaction largely arises from the changes in covalent and non-covalent chemical bonding between the products and the reactants. For measurements at constant pressure, $\Delta H = C_p \Delta T$ where C_p denotes the heat capacity at constant pressure. For example, thermally driven protein unfolding – denaturation by heat – occurs with large changes in non-covalent bonding. A well-ordered native structure with extensive non-covalent bonding (the reactant) is converted to a substantially disorganized, unfolded form lacking this bonding (the product). The large enthalpy decrease on unfolding is mostly counterbalanced by an increase in entropy. Protein unfolding occurs over a relatively narrow temperature range around a melting temperature. Unfolding is associated with a substantial enthalpy change ΔH of several hundred kJ mol^{-1}, which largely arises from a sharp increase in C_p near the melting temperature.

Since both entropy and enthalpy are properties of a system, the changes in entropy and enthalpy associated with a reaction are independent of the pathway taken between reactants and products. A third important property of a state of a system, also independent of the pathway, is the *Gibbs free energy* (denoted G) that combines enthalpy and entropy by the definition $G = H - TS$ where T is the temperature. For a reaction occurring with enthalpy change ΔH and entropy change ΔS at constant T, we have:

$$\Delta G = \Delta H - T\Delta S. \tag{2.3}$$

The terminology free energy reflects the fact that ΔG is the maximum amount of work that can be driven by the reaction. Thus, "free" describes the sense of fully available to do work. For a reaction at constant temperature and pressure to be spontaneous, ΔG must be less than zero.

The properties of a state such as S, H, and G each have a specified magnitude at equilibrium, in which there is no net flow of matter or energy. Importantly, no net flow does not imply that the equilibrium state is static. Consider a system in thermodynamic equilibrium at constant temperature and pressure containing two states X and Y that interconvert: $X <-> Y$, for which the free energy difference between the X and Y states denoted $\Delta G_{X,Y}$ is very small (Figure 2.1). The rates at which molecules in X convert to Y and from Y to X must be equal since by the definition of equilibrium, there can be no net flux of matter – here, of molecules – between the states. However, molecules in state X are constantly and spontaneously converting to state Y and vice versa. The system at equilibrium is definitely not static but rather, truly dynamic. Since both the forward reaction $X -> Y$ and the reverse reaction $Y -> X$ are spontaneous, $\Delta G_{X,Y}$ must be less than or equal to zero for both the forward and reverse reactions. This can only be true if $\Delta G_{X,Y}$ is zero. If $\Delta G_{X,Y}$ is zero at thermodynamic equilibrium, the value of G must be a minimum. That is, a system at equilibrium is generally in its state of lowest free energy. However, that state may only be local. There may be other, distant states of even lower energy that are not accessible at a realistic rate in a realistic time. This emphasizes the fact that the rates at which transitions between states can occur is not the province of thermodynamics but rather of kinetics, as we shall see below.

The free energy difference between reactants and products is directly related to the equilibrium constant K, where square brackets [] denote concentrations and K = [products] / [reactants]:

$$\Delta G^0 = -RT \ln K = -2.303 RT \log K. \tag{2.4}$$

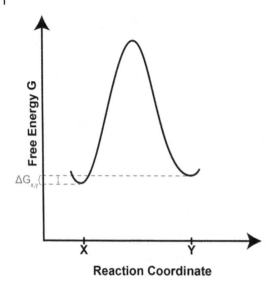

Reaction Coordinate

Figure 2.1 Interconversion of X and Y state. Two states X and Y are separated by an energy barrier, crossed when moving from X to Y or from Y to X. State Y is of slightly more positive free energy than state X; the free energy difference $\Delta G_{x,y}$ between the states, shown in red, is small.

Here, the superscript in ΔG^0 denotes the standard free energy change in which the reactants and products are in what is known as their standard conditions (see e.g. Chapter 4 in (Eisenberg and Crothers 1979)), R is the gas constant with the value of 8.314 J K^{-1} mol^{-1}, and T is the temperature in deg K. At 300 deg K, 2.303 RT has the value 5.74 kJ mol^{-1} = 1.37 kcal mol^{-1}. If the superscript is omitted, standard conditions are understood and ΔG^0 may be simply written as ΔG. Since K is the ratio of the concentrations of products and reactants, both of which are positive, K must also be positive. A value of $K > 1$ means that products are favored and $\Delta G^0 < 0$; a value of $K = 1$ when reactants and products are at equilibrium means that ΔG^0 is zero, as is required; and a value of $K < 1$ means that reactants are favored and $\Delta G^0 > 0$.

As Equation 2.3 shows, ΔG^0 can be positive or negative depending on the signs and values of ΔH^0 and $T\Delta S^0$. A large value of K is associated with a negative value of ΔG^0 and can arise from a large negative value of ΔH^0, a large positive value of $T\Delta S^0$, or a combination of a negative value of ΔH^0 with a positive value of $T\Delta S^0$. Likewise, a small value of K can arise when ΔH^0 is positive, $T\Delta S^0$ is negative and the values of ΔH^0 and $T\Delta S^0$ are nearly equal. In cases which are common in protein folding and in many biochemical reactions, K is nearly 1, ΔG^0 is nearly zero and the reaction is said to exhibit enthalpy–entropy compensation. When $|\Delta H^0| \gg |T\Delta S^0|$, the reaction is said to be enthalpy-driven; when $|T\Delta S^0| \gg |\Delta H^0|$, the reaction is said to be entropy-driven.

The exponential relation between K and ΔG^0 in Equation 2.4 means that a small numerical change in ΔG^0 is associated with a large change in K. The value of 2.303 RT at 300 deg K, 5.74 kJ mol^{-1}, corresponds to log $K = 1$, $K = 10$.

In structural dynamics, time-dependent structural changes ultimately arise from a time-dependent difference in free energy between an intermediate (or product) at time t and reactant at time 0. That is, $\Delta G(t) = G(t)_{intermediate} - G(0)_{reactant}$. However, the ability to detect those structural changes requires a significant difference in structure between the intermediate and the reactant. Significance depends on the magnitude of the structural change and the noise level σ in that change. For example, in an X-ray DED map between an intermediate and reactant (Chapter 7.3), difference densities less than 3σ are typically taken to be noise and not significant.

In addition to the significant difference in structure between intermediate and reactant, there must also be a significant difference in their occupancies. A large change in the equilibrium constant K may not produce a significant difference in occupancies. Consider for example a protein side chain. In the reactant, assume that two spatially distinct structures A and B of the side chain are in equilibrium with equilibrium constant $K = [B]/[A]$. The occupancy O of state A is $1/(1 + K)$ and of state B is $K/(1 + K)$. When a reaction is initiated, energy is put into the system, the intermediate is formed, the value of K changes, and the occupancies of states A and B change. The dependence of occupancy O on K (i.e. the value of dO/dK) is largest at $K = 1$, $\Delta G^0 = 0$ when the equilibrium is poised. However, the dependence on occupancy approaches zero when $K \ll 1$ or $K \gg 1$; changes in occupancy are not detectable. Changes are only readily detectable when $K \sim 1$ for the equilibrium between the reactant (here, state A) and the intermediate (here, state B). The ability to detect a structural change between two states thus requires that both the magnitude of structural changes between the states (such as displacement of the backbone or of a side chain) and the changes in their occupancies be sufficiently large.

Conversely, for systems with K ~ 1, little free energy is required to produce a substantial change in occupancy. An example arose in the early crystallographic studies of deoxyhemoglobin and parallel studies of the kinetics of ligand binding to its α and β chains. Kinetics identified the β chain as having a slightly faster, bimolecular rate of ligand binding than the α chain. "The kinetic studies have miss-identified the chains," said the crystallographer. The crystal structure of deoxyhemoglobin showed that a valine side chain partly obstructed the ligand binding site of the β chain, but the ligand binding site of the α chain was open and unobstructed. The valine in the β chain of deoxyhemoglobin would have to be displaced before ligand binding could occur. The crystallographer asserted this should result in a lower rate of ligand binding to the β chain than to the α chain. However, these were early days of protein crystallography in which structures were not refined. The occupancy of the obstructing valine side chain in the β chain was at best a guess and may well have been substantially less than 100%. If so, the free energy necessary to displace this side chain, open the ligand binding site in the β chain and allow ready access to the incoming ligand would be very low and insignificant kinetically. The unrefined crystallographic data were not sufficiently robust to contradict the kinetic studies, and crystallographers did not appreciate how small the energy difference might be between side chain rotamers.

Measuring the temperature dependence of K and hence of ΔG^0 enables the entropy change ΔS^0 and enthalpy change ΔH^0 associated with a reaction to be separately obtained. By Equation 2.4, changes in free energy ΔG^0 depend on temperature T. At constant pressure, this dependence can be expressed in two forms of the Gibbs-Helmholtz equation:

$$\frac{d(\Delta G^0)}{dT} = -\Delta S^0 \tag{2.5}$$

and

$$\frac{\mathrm{d}\left(\Delta G^0 \Big/ \mathrm{d}T\right)}{\mathrm{d}T} = -\frac{\Delta H^0}{T^2}$$

which is equivalent to

$$\frac{\mathrm{d}(lnK)}{\mathrm{d}T} = \frac{\Delta H^0}{RT^2} \tag{2.6}$$

Equation 2.6 is known as the van't Hoff equation. Thus, measuring the temperature dependence of K and hence of ΔG^0 enables the entropy change ΔS^0 and enthalpy change ΔH^0 associated with a reaction to be separately obtained.

More recently, the method of isothermal titration calorimetry has become a principal tool in measuring experimentally the thermodynamics of ligand binding to proteins (Atri et al. 2015). The method utilizes paired cells, one containing buffer and the other buffer plus protein, both within an adiabatic jacket. A small aliquot of ligand is added to the protein cell. In the case of an exothermic reaction a small amount of heat is generated upon ligand binding, and the instrument adds a measured amount of heat to the water cell to keep the temperatures of the two cells equal. This process is repeated until the injection of ligand evokes no additional heat release. The data from this experiment are a series of narrow peaks showing the heat evolved for each injection. The envelope of these peaks looks like a conventional binding isotherm from which the equilibrium constant K for binding can be extracted. The integral under the set of peaks is the total heat evolved, which provides ΔH^0 for the reaction. Using this information in Equations 2.3 and 2.4, ΔS^0 and ΔG^0 can be determined for the reaction, and its stoichiometry.

Reactions are often initiated by mixing reactants in the liquid state, for example of an enzyme and a ligand (Chapter 3.3.2.1). Other systems are inherently light-sensitive or can be made so by optogenetic approaches, thus allowing the reaction to be initiated by absorbing a photon (Chapter 3.3.2.3). However, the thermodynamics of reaction initiation by ligand binding or by light are very different. Even for ligands that bind with relatively high affinity, the net free energy associated with binding and available to drive subsequent structural changes is likely to be small. Although all binding reactions release energy by binding a specific ligand, they also absorb energy to displace solvent from the site at which the ligand is to bind. The net free energy available is reduced. In contrast, if a reaction is initiated by absorption of a photon (Chapter 3.3.2.3), the net free energy available to drive subsequent structural changes is large. A large energy is associated with a photon of visible light, and no solvent must be directly displaced. For example. the absorption of 500 nm photons injects 239 kJ mol^{-1} energy into the system. Some of this energy is very rapidly dissipated as vibrational energy (heat) and in other nonproductive reactions, but much remains to drive productive structural and chemical changes. Binding of a high affinity ligand with nanomolar dissociation constant is concentration-dependent but typically injects much less energy, only about 50 kJ mol^{-1}.

The structural changes that follow absorption of a photon are very substantially downhill in energy, largely irreversible and far from equilibrium, at least during the initial stages of the reaction. In marked contrast, the structural changes that follow ligand binding for example by rapid mixing, occur with relatively small energy changes, remain near equilibrium and are potentially reversible. In thermodynamic terms, reaction initiation by light quantitatively differs from initiation by ligand binding. After initiation, the subsequent reactions may proceed in a qualitatively different manner.

2.3 Kinetics, Rates, and Rate Coefficients

In marked contrast to thermodynamics, *the concepts of time and reaction rate are central to kinetics.* The goal of kinetics in structural dynamics is to establish a plausible, structure-based reaction mechanism by which reactants are transformed into products. Complex biochemical reactions proceed from reactants to products via several transient intermediate states. These states are short-lived and not readily isolated by chemical approaches. The progress of these complex reactions may be regulated by the binding of other small molecules or solution components that activate, modulate, or inhibit the system. That is, these molecules or components adjust the rates of one or more individual steps in the mechanism.

Again, details of kinetics are found in most textbooks of physical chemistry and biochemistry. See, for example, chapter 8 in Tymoczko et al. (2011) and Chapter 6 in Eisenberg and Crothers (1979).

Kinetics deals with systems containing a statistically large number of molecules and the variation with time of the concentrations of these molecules as they undergo chemical, biochemical, and/or structural reactions. The rate of a reaction is expressed as the rate of change with time of the concentration of a component of a system. Each type of molecule constitutes a component, which may be a reactant such as an enzyme substrate denoted S, a product P, a transient intermediate I, the free enzyme E, or a component of the solute such as hydrogen or magnesium ions. Thus, for the appearance of product molecules at concentration $[P]$,

$$\text{Reaction rate } R = \frac{d[P]}{dt}$$

where t denotes time. The rate is positive since the concentration of product $[P]$ increases with time. For each product molecule generated one substrate molecule is consumed, and the concentration of substrate $[S]$ decreases with time. Hence the identical reaction rate R is

$$R = -\frac{d[S]}{dt}.$$

The value of $d[S]/dt$ is negative since the concentration of substrate $[S]$ decreases with time.

Suppose the reaction is more complicated and involves the enzyme-catalyzed combination of two substrates A and B to form a single product P: $A + B <=> P$. By examining experimentally how the observed rate depends on the concentrations of the reactants, a rate law may be established:

$$R = \frac{d[P]}{dt} = k[A][B],$$

where k is defined as the rate coefficient for the reaction. The coefficient k is often known somewhat misleadingly as the rate constant although unlike the Boltzmann constant, k is not a constant of nature. The value of rate coefficients varies with other experimental conditions such as temperature and pH. We therefore prefer the terminology *rate coefficient* to denote k.

More generally,

$$R = \frac{d[P]}{dt} = k[A]^v[B]^w$$

where the exponents v and w denote the order of the reaction with respect to A and B. If v or $w = 1$, the reaction is said to be *first order* in A or B, respectively. If v or $w = 2$, the reaction is second order in A or B; and so forth. The total reaction order is the sum of the exponents, here $(v + w)$. Thus, if both $v = 1$ and $w = 1$, the total reaction order is 2, a *second-order* reaction. Structural dynamics and indeed, biochemical kinetics in general are almost entirely concerned with the kinetics of first- and second-order reactions, since collisions of three or more molecules A, B, and C to form a reactive, ternary complex are exceedingly rare.

Biochemical and chemical reactions consist of a *series of elementary reaction steps*, in each of which the reactant involves either only a single molecule – a *unimolecular step* – or the collision of one molecule with another – a *bimolecular step*. No intermediate can be detected between the reactants and products of an elementary step. The overall reaction combines all the elementary steps.

Consider an elementary, irreversible, unimolecular step $A \rightarrow B$. If the molecules of A behave independently of each other, the fraction $d[A]/[A]$ that transforms to B in a time dt is proportional to dt where the proportionality constant is the rate coefficient k_1:

$$\frac{d[A]}{[A]} = -k_1 dt, \tag{2.7a}$$

and

$$\frac{d[A]}{dt} = -k_1[A] = -\frac{d[B]}{dt}. \tag{2.7b}$$

Equations 2.7a and 2.7b demonstrate that unimolecular reactions have a first-order rate law (here, $v = 1$). Separating the variables and integrating both sides of Equation 2.7a,

$$\ln[A] = -k_1 t + \ln[A]_0, \tag{2.8a}$$

where $[A]_0 = [A]$ at $t = 0$. That is,

$$[A] = [A]_0 \exp(-k_1 t), \tag{2.8b}$$

in which $[A]$ decays exponentially with rate coefficient k_1.

Now consider a bimolecular step $A + B \rightarrow C$. Here, the rate $d[C]/dt$ depends on the frequency of collisions of one molecule of A with one of B, and hence is proportional to the concentrations of both A and B. The proportionality constant is the rate coefficient k_2. This gives:

$$\frac{d[C]}{dt} = -k_2[A][B], \tag{2.9}$$

which demonstrates that bimolecular reactions have a second-order rate law (here, $v = 1$ and $w = 1; v + w = 2$).

The units must be identical on both sides of Equations 2.7a and 2.9. Hence, rate coefficients for a first-order, unimolecular reaction have units of $(\text{time})^{-1}$, such as s^{-1}. Rate coefficients for a second-order, bimolecular reaction have units of $(\text{concentration})^{-1}(\text{time})^{-1}$, such as $M^{-1} s^{-1}$.

The rate law for a simple second-order reaction such as Equation 2.9 can be integrated analytically to give $[C]$ as a function of time t, $[C(t)]$, but the resulting equation is lengthy and not very informative. Rather than seeking the analytical expression for potentially more complicated reactions, rate laws today are typically integrated numerically.

The rate coefficients for an elementary, reversible step are directly related to the equilibrium constant K for that step, which is given by the ratio of the rate coefficients for the forward (k_1) and reverse (k_{-1}) reactions:

$$K_1 = \frac{k_1}{k_{-1}}. \tag{2.10}$$

If the total reaction is a series of elementary reversible steps, the equilibrium constant K_{total} for the total reaction is the product of the equilibrium constants for each of the elementary steps:

$$K_{total} = K_1 K_2 K_3 \ldots, K_n. \tag{2.11}$$

Since K is a pure number and dimensionless, the dimensions of k_1 and k_{-1} must be identical in Equations 2.9 and 2.10, namely s^{-1}.

We saw in Equation 2.6 that the temperature dependence of an equilibrium constant K is given by the van't Hoff equation:

$$\frac{d(\ln K)}{dT} = \frac{\Delta H^0}{RT^2}.$$

Arrhenius noted that an equilibrium constant is the ratio of the rate coefficients for the forward and reverse reactions, as in Equation 2.10. He proposed that the temperature dependence of a rate coefficient such as k_1 or k_{-1} should be describable by an equation of the same form as that for an equilibrium constant:

$$\frac{d(\ln k)}{dT} = \frac{E_a}{RT^2} \tag{2.12}$$

in which the equilibrium constant K in Equation 2.6 is replaced by the rate coefficient k and ΔH^0 by an energy term E_a known as the energy of activation. Equation 2.12 is integrated to give:

$$\ln k = -\frac{E_a}{RT} + \ln A \tag{2.13}$$

where A at this stage is simply an integration constant, developed further in Chapter 2.5.1. Equation 2.13 is known as the Arrhenius equation after its originator. By experimental measurement of the rate coefficient k as a function of temperature T, the Arrhenius equation enables the values of E_a and A to be determined.

2.4 Enzyme Kinetics

2.4.1 Steady-State and Pre-Steady-State Kinetics

In the simplest enzyme-catalyzed reaction, a single substrate S is converted to a single product P. Substrate S binds to free enzyme E, forms a complex ES which is chemically transformed and dissociates to liberate product P and regenerate free enzyme E. Enzymes are very efficient catalysts, but a rate for this reaction must be established low enough to be readily measured, over a few seconds. The total enzyme concentration $[E]_{tot}$ must be very low, such that $[E]_{tot} \ll [S]$. A common way to monitor the course of an enzymatic reaction is the progress curve, which plots the concentration of product $[P]$ against time. Although enzymes catalyze both the forward reaction $S \rightarrow P$ and the reverse reaction $P \rightarrow S$, at the early stages of the forward reaction $[P] \sim 0$. The reverse reaction $P \rightarrow ES$ (with rate coefficient k_{-2} in Equation 2.14) can thus be neglected and the initial rate of the reaction at its earliest stages, known as its *initial velocity*, can be measured from the roughly linear segment of the progress curve that appears at short times. The computer program ENZO analyzes the whole progress curve to derive the kinetic constants of the Michaelis-Menten mechanism discussed below, or of more elaborate mechanisms (Bevc et al. 2011).

The simplest mechanism is then:

$$E + S \underset{k_{-1}}{\overset{k_1}{\rightleftarrows}} ES \underset{k_{-2}}{\overset{k_2}{\rightleftarrows}} E + P. \tag{2.14}$$

Assume now that the rate of the elementary step for release of product, $ES \rightarrow E + P$, is much lower than the rate of formation of ES from E and S. That is, ES breaks down slowly; the rate of its breakdown controls the overall rate of the reaction and is said to be *rate-limiting*. Under these assumptions, the initial velocity v is:

$$v = \frac{d[P]}{dt} = k_2[ES] = -\frac{d[S]}{dt}. \tag{2.15}$$

The total enzyme concentration $[E]_{tot}$ is conserved:

$$[E]_{tot} = [E] + [ES] \tag{2.16a}$$

$$[E] = [E]_{tot} - [ES] \tag{2.16b}$$

and

$$\frac{d[ES]}{dt} = k_1[E][S] - k_{-1}[ES] - k_2[ES]. \tag{2.17}$$

All of $[E_{tot}]$, $[ES]$, and $[E]$ are much less than $[S]$, which supports the final assumption that a steady state exists in which $[ES]$ does not vary significantly with time. That is, $d[ES]/dt \sim 0$. With this assumption, rearranging Equations 2.15, 2.16a, and 2.17 gives:

$$[ES] = \frac{k_1[E_{tot}][S]}{(k_{-1} + k_2 + k_1[S])}, \tag{2.18}$$

which with Equation 2.14 gives the initial velocity v:

$$v = k_2[ES] = \frac{k_2[E_{tot}][S]}{\left[\dfrac{(k_{-1}+k_2)}{k_1} + [S]\right]}.$$ (2.19)

Define the maximum initial velocity

$$V_{max} = k_2[E_{tot}]$$

and

$$K_M = \frac{(k_{-1}+k_2)}{k_1}.$$

Then substitute V_{max} and K_M into Equation 2.19 to give the classic Michaelis-Menten equation:

$$v = \frac{V_{max}[S]}{(K_M+[S])},$$ (2.20)

where K_M is known as the Michaelis constant. A graph of v against $[S]$ as in Equations 2.19 and 2.20 is a rectangular hyperbola. The maximum velocity V_{max} occurs when all molecules are in the form of the ES complex. That is, the enzyme is saturated with substrate. When $[S] = K_M$, $v = \frac{1}{2} V_{max}$. At low $[S]$, the initial velocity v is directly proportional to $[S]$, resembling a first-order reaction; at high $[S]$, v is nearly constant at V_{max}, resembling a zero-order reaction.

Under the steady-state assumption, Equation 2.17 becomes

$$k_1[E][S] = (k_2 + k_{-1})[ES]$$

and

$$K_M = \frac{(k_2 + k_{-1})}{k_1} = \frac{[E][S]}{[ES]} \equiv K_S.$$

Thus, under the assumed conditions of initial velocity with $[P] \sim 0$ and steady state, the Michaelis constant K_M resembles the dissociation equilibrium constant K_S for the binding of substrate to enzyme.

The quantity $V_{max} / [E_{tot}]$, sometimes denoted k_{cat}, has units of s^{-1} and is known as the *turnover number*. For typical enzymes the turnover number is quite low, lying in the range of $1-10^4 \, s^{-1}$ with an observed upper magnitude (for carbonic anhydrase) of $10^6 \, s^{-1}$. That is, the overall rate of a typical enzyme-catalyzed reaction is low although still very much greater than the rate of the corresponding reaction in the absence of any enzyme – the uncatalyzed reaction.

Since the rate of the overall reaction cannot be faster than that of the rate of any elementary step in the reaction, the turnover number establishes a lower bound for the magnitude of the first-order rate coefficients of each elementary step in the reaction.

The ratio k_{cat}/K_M combines the two parameters k_{cat} and K_M of the Michaelis-Menten equation and is known as the *catalytic efficiency*, with units of $M^{-1}\,s^{-1}$. Catalytic efficiency is the biologically relevant variable when considering how "good" an enzyme is, and not merely a shorthand combination of simplified enzymatic concepts. Catalytic efficiency can be directly altered by mutation. For example, the small GTPase switch protein Gsp1 is highly regulated and interacts with numerous biological processes by interfacing with different downstream effectors. Plausibly, different effectors were believed to interact with Gsp1 by binding to distinct interfaces. However, a series of mutations at candidate interfaces showed this is not the case. Rather, individual mutations at distinct interfaces exerted different biophysical effects on the catalytic efficiency of Gsp1. The mutations thus allosterically modify the kinetics of the GTPase (Perica et al. 2021).

The classical treatment for enzyme-catalyzed reactions is unfortunately more descriptive than informative about actual mechanisms. Equation 2.14 is clearly a vast oversimplification but even when a more realistic mechanism with a series of enzyme–substrate intermediate complexes between substrate and product is introduced, an equation for the initial velocity of exactly the same form as the Michaelis-Menten equation can be derived. In a realistic mechanism, the definitions of V_{max} and K_M are much more complicated since they contain numerous rate coefficients associated with all the elementary steps in the mechanism. However, the dependence of the initial velocity v on $[S]$ remains a rectangular hyperbola as in Equation 2.20.

How then can enzyme kinetics give information about the number and nature of intermediates that lie between substrate and product, and the detailed mechanism which they populate? This general question is independent of whether the kinetics of the reaction are to be followed chemically, spectroscopically, or structurally. The answer lies in observing the earliest stages of the enzyme-catalyzed reaction, prior to the attainment of the steady state. Suppose a reaction that follows the scheme of Equation 2.14 is initiated by rapidly mixing free enzyme E at an initial concentration $[E_0] = [E_{tot}]$ with substrate S, whose initial concentration $[S_0] \gg [E_0]$. All molecules of E will bind S, pass through ES, then break down to release product P and free enzyme E. The buildup of ES is not instantaneous. Prior to establishing the steady state, there is a *pre-steady-state phase* whose time course depends on $[E_0]$, $[S_0]$ and the four rate coefficients in Equation 2.14. Figure 2.2 illustrates schematically the buildup of the steady state when $[S_0] \gg [E_0]$. Essentially all molecules of E bind a molecule of S and traverse all intermediates before liberating product. (For simplicity, Equation 2.14 and Figure 2.2 contain only the single intermediate ES.) In this example, each enzyme molecule passes through all intermediates in the mechanism in about the first 100 ms, when product P begins to appear. A prolonged steady state typically requires $[S_0] \gg [E_0]$, or that buildup of product is sufficient to produce a corresponding back reaction.

As a practical matter, following the kinetics of an enzyme-catalyzed reaction by X-ray-based structural techniques requires high values of $[E_0]$ in the few mM (solution) or even tens of mM (crystal) range, and correspondingly higher values of $[S_0]$. If spectroscopic techniques are used, this requirement is absent since such techniques are sensitive to lower values of $[E_0]$, in the μM range or even lower.

When considering many enzymes, the second-order rate coefficient k_1 associated with the initial, bimolecular step of substrate or ligand binding is found to vary experimentally

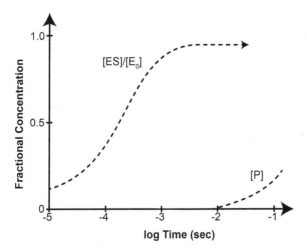

Figure 2.2 **Formation of the pre-steady state and buildup of the steady state.** Here $[S_0] \gg [E_0]$. In this schematic example, all enzyme molecules bind substrate to form ES in about the first 50 ms, the pre-steady state. Thereafter, product P begins to appear as the system moves toward steady state in which [ES] is effectively constant and product appears at a constant rate.

by a factor of roughly 10^2, from a few times 10^5 $M^{-1}s^{-1}$ to a few times 10^7 $M^{-1}s^{-1}$. In contrast, the equilibrium binding constant varies over a much larger factor of roughly 10^8, from 50 pM to 5 mM. Most of the experimental variation in the equilibrium binding constant K thus arises from variation in the first-order rate coefficient for dissociation, the off rate k_{-1}.

Most studies of the chemical kinetic mechanism of enzymes by rapid mixing generate and probe the pre-steady state and the earliest stages of product release. These are known as *single turnover conditions*: each enzyme molecule binds a single molecule of substrate and converts it to product. However, mixing of reactants is limited by the rates of turbulent diffusion and requires a nonzero physical mixing time before true chemical reaction initiation can occur. The mixing time can be as short as 100 μs when very small solution volumes are mixed in stopped flow experiments but can easily extend to a few ms. Pre-steady-state kinetics are thus limited by the time scales of diffusion to reaction times greater than 100 μs. This reaction time is much longer than the lifetime of many chemically distinct enzyme intermediates. Thus, the early chemical steps in enzyme-catalyzed reactions are often inaccessible to rapid mixing approaches.

One way of circumventing the limitation imposed by diffusion is to photoactivate a reactant such as a substrate or an enzyme, by uncaging an unreactive, inert precursor (Chapter 4.4). In principle, the time taken to achieve photoactivation can be as short as the duration of a visible or near-UV laser pulse, which can easily be ns or even ps. In practice, the immediate photochemistry is often followed by slower rearrangements before the substrate or enzyme is liberated, free of the cage and active. Completion of these rearrangements determines the time zero of photoactivation.

A further limitation on both rapid mixing approaches and photoactivation approaches is present because fast elementary steps are only detectable if they precede slower elementary steps (Chapter 1.5). When they do so, the product of a fast elementary step accumulates to a detectable value before that product decays in the next, slower elementary step. If in contrast the fast elementary step follows a slower elementary step, the product of the slow step does not accumulate. This is the reason why in all models for the mechanism of photocycles (such as those of bacteriorhodopsin and photoactive yellow protein), the rate coefficients between intermediates progressively decrease as the photocycle proceeds and correspondingly, the lifetimes of intermediates progressively increase. This does not mean that fast steps never follow slow steps in these photocycles. Rather, if that circumstance should occur, these fast steps are not readily detectable by kinetic methods and do not appear in the chemical kinetic mechanism.

2.4.2 Relaxation Kinetics

A completely different form of chemical kinetics known as relaxation kinetics can readily access even more rapid reactions. *Relaxation kinetics is based on perturbation of a reaction initially at equilibrium.* The perturbation can be achieved, for example, by a rapid change in temperature, pressure, or electric field. All are physical parameters which if changed, affect the position of the equilibrium.

Consider for example applying a temperature jump experiment to the simplest bimolecular reaction:

$$E + S \underset{k_{-1}}{\overset{k_1}{\rightleftharpoons}} ES, \tag{2.21}$$

with equilibrium constant K:

$$K = \frac{[E_{eq}][S_{eq}]}{[ES]_{eq}}$$

and

$$\ln K = \ln[E_{eq}] + \ln[S_{eq}] - \ln[ES_{eq}]$$

where the subscripts *eq* denote the concentration of that species at equilibrium.
By making use of the van't Hoff relation (Equation 2.6) it may be shown (see Chapter 16 of (Cantor and Schimmel 1980)) that:

$$\Delta[E_{eq}] = \Gamma\left(\frac{\Delta H^0}{RT^2}\right)\Delta T \tag{2.22}$$

where $\Delta[E_{eq}]$ is the change in the equilibrium concentration of enzyme $[E_{eq}]$ from its value at the initial, lower temperature to its value at the final, higher temperature, and

$$\Gamma = \frac{1}{\left\{[E_{eq}]^{-1} + [S_{eq}]^{-1} + [ES_{eq}]^{-1}\right\}}.$$

Equation 2.22 expresses the relation between a nonzero increase ΔT from the initial temperature T and the associated change $\Delta[E_{eq}]$ in the concentration of the reactant E. $\Delta[E_{eq}]$ is directly proportional to both the standard enthalpy change ΔH^0 of the reaction and the value of Γ. The latter is maximized when the reaction is poised, i.e. $[E_{eq}] \sim [ES_{eq}]$.

The rate at which the new equilibrium is established at the final temperature depends on the magnitude of the rate coefficients associated with the equilibria. In the simplest mechanism of Equation 2.21 it may be shown that

$$\Delta[E] = \Delta[E_{eq}]\exp\left(-\frac{t}{\tau}\right) \tag{2.23}$$

where τ is known as the *relaxation time*; $1/\tau$ is given by

$$\frac{1}{\tau} = k_1([E_h] + [S_h]) + k_{-1} \tag{2.24}$$

where the subscript h denotes the concentration of that species when equilibrium is reestablished at the final, higher temperature. Equation 2.24 shows that a plot of $1/\tau$ versus $([E_h] + [S_h])$ is linear with slope of k_1 and intercept of k_{-1}. This enables both the rate coefficients k_1 and k_{-1} to be directly determined by experimental measurement of τ.

Equations 2.23 and 2.24 hold only when the perturbations in concentration arising from the temperature jump are "small." Small here means that any terms in $(\Delta[E])^2$ may be ignored by comparison with terms in $\Delta[E]$. That is, a small perturbation simplifies interpretation of the kinetics. However, by Equation 2.22 the total magnitude of the signal $\Delta[E_{eq}]$ is proportional to ΔT. This raises the temptation to enlarge the perturbation and enhance the signal during the experiment by increasing ΔT, at the expense of introducing complexities in the quantitative interpretation.

Equation 2.21 presents the simplest single step enzyme reaction. For mechanisms with (n + 1) steps which are both more realistic and much more complicated, Equation 2.23 expands from a single exponential to the more general form of a sum of exponentials:

$$\Delta[E] = \sum_{j=1}^{n+1} A_j \exp\left(-\frac{t}{\tau_j}\right) \tag{2.25}$$

where the coefficients A_j are constants, τ_j are relaxation times, and the summation is taken over values of j from $j = 1$ to $j = (n + 1)$. Equation 2.25 describes a relaxation, now typically handled by matrix methods (see Chapter 16.2 in Cantor and Schimmel (1980)).

If the relaxation experiment is carried out over a wide time range and the relaxation times happen to be well separated, the relaxations may be distinguished experimentally. By repeating the experiment over a range of concentrations, it may be possible to identify a candidate mechanism and extract individual rate coefficients for steps in that mechanism from the relaxation times.

An illustrative study that knits together many of these ideas examines the lipid chain-melting transition using pressure jump calorimetry in lieu of the more common temperature-jump (Grabitz et al. 2002). Membrane lipids undergo phase transitions that echo protein folding and are characterized by large changes in enthalpy, a sharp maximum in specific heat C_p and substantial isothermal volume fluctuations. This transition is driven by the disordering of the chains of individual lipid molecules with increasing temperature. These phase changes produce physiologically important changes in macroscopic membrane characteristics such as fluidity, and the separation of lipids into separate phases within the membrane, sometimes manifested in the formation of so-called lipid rafts (Leventhal et al. 2020).

The authors introduced pressure perturbation because of the experimental difficulties encountered by others in determining the precise magnitude of temperature jumps. Their study applied pressure jumps of 40 bar while controlling temperature to within 10^{-3} K. Such a pressure jump is quite modest; pressure jumps greater than 1 kbar are typically used to unfold proteins. They derived equations analogous to those for temperature jump for pressure jump and extended these to steady-state conditions. Using membrane vesicles, they monitored heat evolved as a function of time in response to pressure jumps at various temperatures near the phase transitions. As predicted by thermodynamic fluctuation analysis, they found that the temperature variation of the characteristic vesicle relaxation time τ closely followed the trace of C_p with temperature. A heuristic insight to this behavior comes from the fact that the mean square fluctuation of enthalpy in a system is proportional to C_p. At the maximum of C_p, large fluctuations in the enthalpy H dampen the ability of the system to reestablish equilibrium. A thoughtful review (Cooper 1984) discusses the role of thermodynamic fluctuations in biological systems, which we present in more detail in Chapter 8.3.2.

2.4.3 Conclusions

A chemical kinetic mechanism is characterized by the interconversion of states. In structural dynamics, each state lying between reactant and product states is associated with an intermediate structure. Adjacent states are connected by elementary steps, where each step is associated with a forward and a reverse rate coefficient. The overall goal of kinetic studies in structural dynamics is to derive the chemical kinetic mechanism. Specifically, the goal is to identify the number of states, the structure of each state, the elementary steps by which adjacent states interconvert, and the magnitude of the rate coefficients associated with each elementary step. Attaining this goal is a tall order. Studies of kinetics, whether steady state, pre-steady state, or relaxation, directly yield experimental quantities such as the Michaelis constant K_M or relaxation times τ_j. These quantities are lumped, which is to say that each combines the forward and reverse rate coefficients associated with each elementary step. The exact nature of these combinations depends on the number of states and how they are connected in the kinetic mechanism.

2.5 Transition State Theory and Energy Landscapes

2.5.1 Transition State Theory

What determines the magnitude of the rate coefficients associated with each elementary step in a chemical kinetic mechanism? The underlying framework to understand reaction rates is known as transition state theory. This theory, originally proposed by Eyring (Wynne-Jones and Eyring 1935), remains conceptually useful.

As an example, we describe in words the reaction in solution between a substrate and an enzyme. If any reaction is to occur, the substrate must encounter and collide with the enzyme to form a potentially reactive complex. The encounter may be intermolecular as in the initial stages of binding, or intramolecular, as the substrate and enzyme subsequently reconfigure themselves in the active site. Many encounters are unproductive, no reaction occurs, and the system reverts to free substrate and enzyme, spatially distant from each other. Since few encounters are productive, most reactions are slow.

Transition state theory is based on a hypothesis: the probability that an encounter is productive depends on the free energy changes – that is, on the enthalpy and entropy changes – associated with the elaborate changes in position, orientation, and conformation of the substrate and enzyme. Many variables are required to fully describe the structures and free energies of the substrate and enzyme. Each independent variable corresponds to a degree of freedom. In principle for N atoms, there are a maximum of 3N–6 variables, where N for a small molecule substrate can be 50 and for an enzyme, easily several 1000s. Although the number of independent variables is substantially reduced from this maximum by chemical and structural constraints such as the covalent nature of the polypeptide backbone and its fold into secondary structures as a self-avoiding chain, the number of independent variables is increased by including those associated with those solvent molecules that also contribute to the free energy changes.

By any reckoning, the number of independent variables is so large that it has long been unfeasible to consider them all together in a computation. (However, see Chapter 9.3.1, 9.3.2, and 9.3.3 on the computational approach to protein folding based on artificial intelligence, exemplified by the programs AlphaFold2 and RosettaFold.) A great conceptual simplification is achieved by collapsing this very large number of variables to a single variable known as the reaction coordinate. Although described as a coordinate, this variable is descriptive and chemical rather than quantitative, spatial, or explicitly structural. As an example in words, the first step along the reaction coordinate might be a diffusion-limited collision encounter between the substrate and the surface of the enzyme; second, movement of the loosely-bound substrate across the enzyme surface toward the entry to the active site; third, snuggling of the substrate into tighter, more intimate contact with the enzyme and associated motion of residues in the active site; fourth, detailed positioning of key residues relative to the substrate, which allow a chemical step such as bond rupture to occur; fifth, relaxation of the enzyme–nascent product complex; and sixth and finally, release of the products and return of the enzyme to its initial, substrate-free conformation. Rearrangement of solvent molecules may be associated with all steps. A 2D diagram (Figure 2.3) presents *the elementary steps associated with the total reaction* along the x-axis, known as the *reaction coordinate*; and along the y-axis, the free energy associated with the initial and final states of each elementary step and of the total reaction.

A major insight of transition state theory was introduced by Eyring. Each state along the reaction coordinate is characterized by a local minimum in free energy, and the two adjacent states constituting an elementary step are separated by a peak in free energy. The peak of this energy barrier is known as the *transition state* (Figure 2.3). Since each elementary step has a transition state, the total reaction consisting of several elementary steps has a series of transition states along the reaction coordinate.

(Beware! The transition "state" is not a state in the strict thermodynamic sense of the term; see Chapter 19 in Dill et al. (2010).) The two adjacent (and authentic, thermodynamic) states flanking each transition state are unstable and often short-lived. They may plausibly be described as transient states, intermediate states, or simply intermediates. Confusion arises if the scientifically distinct adjectives *transition* and *transient* are used as though they were interchangeable. To preserve the distinction, we do not use *transient state*

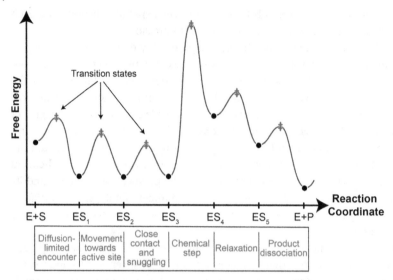

Figure 2.3 Dependence of free energy on the reaction coordinate in a schematic reaction. The chemical states through which all molecules pass as they traverse the reaction coordinate (shown along the x-axis) differ in their free energy (the y-axis). Reaction intermediates correspond to the free energy minima, each denoted by a filled circle; barriers between adjacent intermediate states correspond to free energy maxima at the transition states, each denoted ‡. Free energies can be measured experimentally for the minima at reactant, intermediate, and product states, and for all transition states. The rate-limiting step crosses the barrier of highest free energy, here the chemical step between ES_3 and ES_4. The solid line is not experimental and as a guide to the eye, joins the experimental data points.

and restrict use of *transition state* to denote the peak of the energy barrier separating the intermediates that constitute the reactant and product states of each elementary step.

The values of the free energies of reactants, intermediates, and products required for Figure 2.3 can in principle be obtained from experimental measurement of the thermodynamics of the overall reaction and of each of its elementary steps. In practice not all intermediates and elementary steps are likely to be experimentally accessible. Approximations must be made, for example by substituting the thermodynamic parameters for a relatively stable enzyme-inhibitor complex for those of a short-lived enzyme-product complex.

To complete Figure 2.3, candidate values of thermodynamic parameters for the transition states must also be obtained. Equation 2.13 showed that the temperature dependence of a rate coefficient k is expressed in the Arrhenius equation:

$$\ln k = -\frac{E_a}{RT} + \ln A, \tag{2.26}$$

where E_a is the activation energy and the constant A is known as the frequency factor. By taking the exponential of both sides of this equation,

$$k = A\exp\left(-\frac{E_a}{RT}\right). \tag{2.27}$$

In Equation 2.27, the frequency factor A may be thought of as the frequency of reactive encounters that proceed to the transition state. The value of A is typically as large as 10^{13} s^{-1}, not too remote from the frequency of diffusion-induced collisions in liquids. The exponential factor expresses the fraction of molecules that have the appropriate energy for the forward reaction to occur. In accord with experiment, higher temperatures T are associated with larger rate coefficients k. The activation energy E_a for the forward reaction is the difference between the energies of the reactant and of the transition state; and for the reverse reaction, the corresponding value $E_{a'}$ is the difference between the energies of the product and the same transition state. Thus, the overall thermodynamic energy change for the reaction is $E_a - E_{a'}$. From Equation 2.27, a plot of ln k against $1/T$ is linear with slope $-E_a/R$. Experimental measurement of the temperature dependence of the rate coefficients for the forward and reverse reactions thus enables E_a and $E_{a'}$ to be obtained, and completes the experimental data needed to construct Figure 2.3.

The data points are almost always presented as in Figure 2.3 with a smooth line joining them. The experiments themselves provide only a set of points, namely the thermodynamic parameters at the set of minima associated with each reactant, intermediate and product state (denoted by the filled circles in Figure 2.3), and the set of maxima associated with each transition state between them (denoted by $\ddagger$). No data exist on the line itself which represents the pathway between the states. The line is simply a guide to the eye. The concept of a reaction coordinate and its relation to thermodynamic and structural parameters has expanded beyond Eyring to encompass the pathways between states in energy landscape analysis (ELA), which we consider in more detail in Chapter 8.3.

Eyring also proposed that the transition state lying between the two states in each elementary step may be treated as a *quasi-thermodynamic state*. He envisaged this state to be an *activated complex* with which the (thermodynamically authentic) reactant and product states are separately in equilibrium. The thermodynamic and other parameters of the transition state are denoted by the superscript $\ddagger$. In this formulation, the reaction rate from reactant to product is given by the product of the concentration of the activated complex, the rate at which the complex moves across the energy barrier and the fraction κ of productive complexes that traverse the barrier and do not return. κ is known as the transmission coefficient. Eyring showed that in Equation 2.27, the value of the activation energy E_a is $(\Delta H^{\ddagger} + RT)$ and that of the frequency factor A is $(\kappa k_B T / h) \exp(\Delta S^{\ddagger} / R)$, where k_B is the Boltzmann constant and h is Planck's constant. The prefactor $(\kappa k_B T / h)$ expresses the frequency of vibration along the reaction coordinate and the exponential term $\exp(\Delta S^{\ddagger} / R)$ expresses an entropic factor, plausibly related to orientation of the reactants and their fluctuation. Formally,

$$k = \frac{\kappa k_B T}{h} \exp\left(\frac{\Delta S^{\ddagger}}{R}\right) \exp\left(-\frac{\Delta H^{\ddagger}}{RT}\right). \tag{2.28}$$

Transition state theory provides an insight into how enzymes are such efficient catalysts. Since a rate coefficient depends exponentially on the free energy of activation, the reaction rate can be greatly increased if the free energy of activation is diminished. This can be achieved by reducing the free energy of the transition state, for example by binding reactants in proximity to each other at the active site of the enzyme, ensuring the appropriate distance geometry and orientation of catalytic groups such as acidic or basic side chains, cofactors, and water molecules with respect to the reactants, and providing a desirable electrostatic environment for formation of the activated complex, the transition state for the reaction. Strong discussions at varying levels of detail are provided in Chapter 19 of Dill et al. (2010) and in Jencks (1987); Schramm (2011).

In the absence of an enzyme, the uncatalyzed reaction typically occurs via a single transition state of high free energy with few or no transient, short-lived intermediates. In contrast, an enzyme-catalyzed reaction breaks up the overall reaction into a series of sequential intermediates and elementary steps along the reaction coordinate. These intermediates are separated by transition states, each of whose free energy is substantially reduced from that of the sole transition state in the uncatalyzed reaction. That is, the single transition state at very high free energy characteristic of the uncatalyzed reaction is replaced in the enzyme-catalyzed reaction by several transition states of comparable, greatly reduced free energy. Since the rate-determining step in the overall reaction is given by the elementary step whose transition state has the highest free energy, the enzyme-catalyzed reaction replaces the single rate-determining step in an uncatalyzed reaction by a series of elementary steps, several of which are rate-contributing.

2.5.2 Energy Landscapes

We emphasized above that an extremely large number of variables are required to fully describe a protein structure. At thermal equilibrium, fluctuations are occurring in each of these variables. It is therefore not strictly accurate to describe a crystallographic structure as *the* structure. Rather, *the* structure is *a time- and space-average structure*. The time average is taken over the duration of the observation (e.g. an X-ray exposure time, or the time step in a computation). The space average is taken over all molecules observed (e.g. all molecules in a crystal fully illuminated by the X-ray beam). If a reaction is occurring, both the structure and its energy are varying with time, represented in the 2D representation of Figure 2.3 by progress along the reaction coordinate.

Another means of illustrating the structure-energy relationship is to replace the 2D representation by a 3D energy landscape in which all conformational variables are compressed into two dimensions (conventionally depicted horizontally) and the third dimension (conventionally depicted vertically) represents the energy at that position in conformational space. An energy landscape is often presented as a contour map in which height above sea level (energy) is indicated by 2D contours at fixed heights/energies, where each height/energy is a specific value of the third dimension. The map contains mountain passes that separate even higher mountain peaks, and valleys at lower heights. The energy landscape concept remains an instructive way to look at structure, energy, and reaction. We illustrate this concept in Figures 8.1, 8.2, and 8.3 of Chapter 8.3 and discuss there in more detail the role of functional trajectories in such a landscape,

2.6 Trapping of Intermediates

Any approach to reaction initiation can be followed after a controlled time delay by trapping in which further reaction progress is halted, usually by rapid freezing in which the sample, surrounding liquid, and its mount are plunged into liquid propane or other cryogen (Moffat and Henderson 1995; Neutze and Moffat 2012). Trapping allows the structure and properties of the frozen sample to be probed at leisure, employing techniques which routinely require cryo temperatures such as cryo-EM (Chapter 8.2) or electron paramagnetic resonance. The time course of the reaction can be followed by varying the time between reaction initiation and the rapid freezing.

This simple experimental approach has been widely applied but does have substantial complications. The time resolution is set by the rate of freezing of the sample, which in turn is controlled by its dimensions and the heat transfer characteristics to the cryogen from the sample, any surrounding liquid and its mount. Vitrification of the liquid surrounding or within the sample is essential during freezing to avoid the damaging changes in volume associated with formation of ice crystals. Very rapid freezing would ensure vitrification and lock in the structural distribution present at the initial temperature. In contrast, very slow freezing would freeze in the structural distribution present at temperatures around 200 K (if vitrification could indeed be assured). At these temperatures structural changes in proteins are arrested, and all biological activity is literally frozen (Rasmussen et al. 1992). Experimental values of freezing rates of macroscopic samples such as microcrystals lie between the two extremes of very rapid and very slow freezing, in the range of 50–300 K s^{-1} (Teng and Moffat 1998). Nevertheless, vitrification occurs rather than ice formation. Quantitative analysis of the temperature dependence of structural changes and the biophysics of freezing (Halle 2004) concluded that freezing from 300 to 200 K typically requires times in the ms range. In preparation of cryo-EM samples where vitrification is also essential, freezing of individual protein molecules and complexes apparently occurs much more rapidly, up to 10^4 K s^{-1}. However, much longer experimental time resolutions in the 100 ms range are obtained (Dandey et al. 2020). These longer times are set by mixing of reactants and by the time taken to plunge the cryo-EM sample into the ethane cryogen. For both cryo-EM and single crystal samples, experimental freeze-trapping is restricted at present to the later stages of reactions: intermediates whose lifetimes are tens of ms and longer can be successfully trapped.

A thorough description of the thermodynamics, kinetics, and complex structural consequences of freezing of tiny cryo-EM samples (Bock and Grubmüller 2022) remains to be extended to the freezing of single crystals of much larger dimensions, up to tens of μm at SR sources and around 1 μm at XFEL sources.

Freezing occurs by heat transfer from the sample to the cryogen across the interface between them. Thus the exterior of the sample, liquid and mount freezes first and fastest and the interior, last and slowest. The exterior and the interior have different thermal histories. Although their structural distributions equilibrate to different extents (Bock and Grubmüller 2022), both contribute to scattering by the structural probe. The structural distribution finally observed is thus a complex mixture of the structures encountered between the initial temperature and around 200 K, and varies with position in the sample. Details of the structural distribution with position in any protein crystal or between individual molecules at different positions on a cryo-EM grid do not appear to have been determined.

The Mpemba effect describes families of experiments in which two equal volumes of water, one at a higher temperature than the other, are placed in identical freezing conditions. In some cases, certainly counterintuitive and for some impossible, the initially hotter volume freezes before the colder one. These matters have been the subject of a recent, thoughtful review (Burridge and Hallstadius 2020), which notes that the field shows real confusion. The relevant lesson here is that very subtle alterations in conditions, such as the potential for nucleation, may have significant effects on the processes of heat transfer central to all freeze-trapping approaches in structural dynamics.

Comparison in 15 protein crystals of the structural distribution of backbone non-hydrogen atoms at cryo temperatures with that at room temperature (Halle 2004) showed that the structural differences are 0.2 – 0.8 A. This is generally taken as evidence that the cryo structures which form the large majority (around 90%) of those in the PDB are relevant to the physiological structures at near-room temperature, even though all cryo structures are profoundly inactive (Chapter 1). A specific counterexample where cryo structures do not account for room temperature activity experiments is provided by the mitochondrial enzyme glutaminase C at cryo and physiological temperatures. Various allosteric inhibitors of glutaminase C differ markedly in their potency when evaluated at physiological temperature, but cryo structures of these inhibitor-enzyme complexes show identical binding modes (McDermott et al. 2016; Huang et al. 2018). In contrast, structures of the same complexes at physiological temperature show different binding modes and suggest chemical factors that influence the binding modes and inhibitor potency (Milano et al. 2021). A review provides further examples and earlier studies such as (Keedy 2019) in the analysis of ligand binding (Bradford et al. 2021). In the same vein, very recent results compare the binding of a set of ligands – candidate drug precursors – to the therapeutic target enzyme, protein tyrosine phosphatase 1B, in X-ray structures carefully trapped and determined at cryo- and near-physiological temperatures (Mehlman et al. 2022). They identify significant, temperature-dependent, quantitative differences such as unique binding modes at common sites, changes in solvation, new binding sites and distinct allosteric conformational responses in the enzyme, and reasonably conclude that "…existing cryogenic-temperature protein-ligand structures may provide an incomplete picture."

Although the structural differences between cryo- and room temperature structures may be small, the energetic differences between them can be very large. For example, if freezing reduces the rate of decay of a short-lived intermediate and increases its lifetime from 10 μs at 300 K to 100 s, a factor of 10^7, the increase in the free energy of activation for its decay is 2.303 RT log (10^7) = 40.2 kJ mol^{-1}. The energetic difference is of course even larger if the lifetime at 300 K is shorter: 57.4 kJ mol^{-1} if the lifetime is 10 ns, and 74.7 kJ mol^{-1} if the lifetime is 10 ps. These significant values introduce a serious note of caution when using cryo-trapped structures to explore structural dynamics on time scales less than a few tens of ms, and indeed when using static, cryo-trapped, inactive structures to explore and predict ligand binding to and mechanism of an active structure at physiological temperature. To put it dramatically, one could say that when describing a structural technique, the prefix "cryo" is, if not the kiss of death, at least an indication of a potentially fatal illness. Put more diplomatically, interpretation of dynamics and the details of ligand binding from structural experiments on frozen, inactive samples should proceed cautiously.

References

Atri, M.S., Saboury, A.A., and Ahmad, F. (2015). Biological applications of isothermal titration calorimetry. *Physical Chemistry Research* 3 (4): 319–330.

Bevc, S., Konc, J., Stojan, J. et al. (2011). ENZO: a web tool for derivation and evaluation of kinetic models of enzyme catalyzed reactions. *Plos One* 6 (7): e22265.

Bock, L.V. and Grubmüller, H. (2022). Effects of cryo-EM cooling on structural ensembles. *Nature Communication* 13 (1): 1–13.

Bradford, S.Y., El Khoury, L., Ge, Y. et al. (2021). Temperature artifacts in protein structures bias ligand-binding predictions. *Chemical Science* 12 (34): 11275–11293.

Burridge, H.C. and Hallstadius, O. (2020). Observing the Mpemba effect with minimal bias and the value of the Mpemba effect to scientific outreach and engagement. *Proceedings of the Royal Society A* 476 (2241): 20190829.

Cantor, C.R. and Schimmel, P.R. (1980). *Biophysical Chemistry: Part III: The Behavior of Biological Macromolecules*. Macmillan.

Cooper, A. (1984). Protein fluctuations and the thermodynamic uncertainty principle. *Progress in Biophysics and Molecular Biology* 44 (3): 181–214.

Dandey, V.P., Budell, W.C., Wei, H. et al. (2020). Time-resolved cryo-EM using Spotiton. *Nature Methods* 17 (9): 897–900.

Dill, K.A., Bromberg, S., and Stigter, D. (2010). *Molecular Driving Forces: Statistical Thermodynamics in Biology, Chemistry, Physics, and Nanoscience*. Garland Science.

Eisenberg, D. and Crothers, D. (1979). *Physical Chemistry with Applications to the Life Sciences*. Menlo Park, CA: Benjamin/Cummings.

Grabitz, P., Ivanova, V.P., and Heimburg, T. (2002). Relaxation kinetics of lipid membranes and its relation to the heat capacity. *Biophysical Journal* 82 (1): 299–309.

Halle, B. (2004). Biomolecular cryocrystallography: structural changes during flash-cooling. *Proceedings of the National Academy of Sciences* 101 (14): 4793–4798.

Huang, Q., Stalnecker, C., Zhang, C. et al. (2018). Characterization of the interactions of potent allosteric inhibitors with glutaminase C, a key enzyme in cancer cell glutamine metabolism. *Journal of Biological Chemistry* 293 (10): 3535–3545.

Jencks, W.P. (1987). *Catalysis in Chemistry and Enzymology*. Courier Corporation.

Keedy, D.A. (2019). Journey to the center of the protein: allostery from multitemperature multiconformer X-ray crystallography. *Acta Crystallographica Section D: Structural Biology* 75 (2): 123–137.

Levental, I., Levental, K.R., and Heberle, F.A. (2020). Lipid rafts: controversies resolved, mysteries remain. *Trends in Cell Biology*.

McDermott, L.A., Iyer, P., Vernetti, L. et al. (2016). Design and evaluation of novel glutaminase inhibitors. *Bioorganic & Medicinal Chemistry* 24 (8): 1819–1839.

McQuarrie, D.A. and Simon, J.D. (1997). *Physical Chemistry: A Molecular Approach*. Sausalito CA: University Science Books.

Mehlman, T.S., Biel, J.T., Azeem, S.M. et al. (2022). *Room-temperature crystallography reveals altered binding of small-molecule fragments to PTP1B*. bioRxiv 2022.11.02.514751.

Milano, S.K., Huang, Q., Nguyen, T.-T.T. et al. (2021). *New insights into the mechanisms used by inhibitors targeting glutamine metabolism in cancer cells.* bioRxiv: 2021.2009.2020.461106.

Moffat, K. and Henderson, R. (1995). Freeze trapping of reaction intermediates. *Current Opinion in Structural Biology* 5 (5): 656–663.

Neutze, R. and Moffat, K. (2012). Time-resolved structural studies at synchrotrons and X-ray free electron lasers: opportunities and challenges. *Current Opinion in Structural Biology* 22 (5): 651–659.

Perica, T., Mathy, C.J., Xu, J. et al. (2021). Systems-level effects of allosteric perturbations to a model molecular switch. *Nature* 599 (7883): 152–157.

Rasmussen, B.F., Stock, A.M., Ringe, D. et al. (1992). Crystalline Ribonuclease A loses function below the dynamical transition at 220 K. *Nature* 357 (6377): 423–424.

Schramm, V.L. (2011). Enzymatic transition states, transition-state analogs, dynamics, thermodynamics, and lifetimes. *Annual Review of Biochemistry* 80: 703–732.

Teng, T.-Y. and Moffat, K. (1998). Cooling rates during flash cooling. *Journal of Applied Crystallography* 31 (2): 252–257.

Tymoczko, J.L., Berg, J.M., and Stryer, L. (2011). *Biochemistry: A Short Course.* Macmillan.

Wynne-Jones, W.F. and Eyring, H. (1935). The absolute rate of reactions in condensed phases. *The Journal of Chemical Physics* 3 (8): 492–502.

3

The Experiment

3.1 Introduction

Experimental, X-ray-based approaches to structural dynamics have been transformed by special purpose beamlines at SR sources and now by the advent of XFEL sources (Chapter 6). The properties of XFEL X-ray beams are sufficiently distinct from those of SRs to class XFELs as a truly disruptive innovation in technology (Bower and Christensen 1995). This opens up new possibilities for dynamic experiments (Moffat 2017; Spence 2017). SRs enable dynamics to be probed with a time resolution from seconds down to the duration of a single X-ray pulse, around 100 ps. With their generation of exceptionally intense, fs X-ray pulses, XFEL sources extend the time resolution down to the important chemical timescale of fs while retaining the biological timescale of ns to s. Structural biology at the molecular level is of course based on and directly follows from ultrafast chemical processes such as electron transfer, isomerization, and the making and breaking of covalent bonds (Kupitz and Schmidt 2018). "Femtosecond Biology" is not merely the catchy title of an early review of ultrafast spectroscopic data (Martin and Vos 1992). However, XFEL sources are sufficiently intense that a single tiny sample (a crystal or solution) yields only a single diffraction or scattering pattern before it is destroyed. Diffraction before destruction (Chapter 4.3.3) dominates all experiments (Chapman et al. 2011). All aspects of experimental design at SR and particularly XFEL sources remain under rapid development by scientists in many disciplines, not least in structural biology and dynamics.

Dynamic experiments have unusual features that distinguish them from a relatively straightforward assembly of static experiments. This chapter presents several general principles. Chapter 4 presents samples, their introduction to the X-ray beam, radiation damage, and conferral of sensitivity to light; Chapter 5, X-ray-based crystallographic and solution experiments; Chapter 6, SR and XFEL sources, detectors and beamlines; and Chapter 7, data analysis and interpretation.

3.2 Signal and Noise

3.2.1 Signal, Accuracy, and Systematic Errors

The importance of accurate, quantitative measurement of X-ray scattering cannot be overemphasized. The true scattering signal is inherently small in all time-resolved

Dynamics and Kinetics in Structural Biology: Unravelling Function Through Time-Resolved Structural Analysis, First Edition. Keith Moffat and Eaton E. Lattman.
© 2024 John Wiley & Sons Ltd. Published 2024 by John Wiley & Sons Ltd.

measurements. It can be easily masked by experimental systematic errors, considered below. Their magnitude can be comparable with or even larger than the true signal.

Scientific experiments aim at both *accuracy*, i.e. the truthful measurement of the quantity of interest known as the signal, and *precision*, i.e. the certainty with which that quantity is known, for a defined number of individual measurements or amount of precious sample. More precise measurements are more certain and display less spread among the individual measurements. Total measurement of the quantity of interest is accompanied by background which must be identified and subtracted from the total measurement to obtain an initial estimate of the signal:

$$\text{Signal} = \text{Total measurement} - \text{Background measurement.} \tag{3.1}$$

Total and background measurements in X-ray experiments are determined by the number of photons per measurement. In single crystal diffraction, the raw measurement is the integrated intensity $I(hkl)$ of Bragg spots (hkl) and the background near them; and in solution scattering, the intensity $I(q)$ as a function of scattering angle q, cylindrically averaged about the X-ray beam. If photons during total and background measurements arrive randomly in time, the random error that determines the precision varies as the square root of the total number of photons in each measurement. Precision is readily controlled by adjusting the intensity of scattering and the exposure time.

The background measurement has two components: a true background measurement, e.g. the X-ray counts surrounding every spot that arise from non-Bragg, continuous scattering from the sample and its environment, and a component known as the *systematic error* which is nonrandom. Thus:

$$\text{True background} = \text{Background measurement} +/- \text{Systematic errors.}$$

The presence of systematic errors thus affects the accuracy of the signal by introducing an offset in the measurement. Equation 3.1 expands to:

$$\text{True signal} = \text{Total measurement} - (\text{Background measurement}$$
$$+/- \text{Systematic errors).} \tag{3.2}$$

The systematic errors (deliberately plural) in Equation 3.2 originate in specific details of exactly how the experiment is conducted. Examples include crystal-to-crystal or solution sample to sample variation; pulse-to-pulse jitter in the intensity of the X-rays; radiation damage to the sample; inadequately corrected variation in properties of individual pixels in the detector; or in a solution scattering experiment, parasitic scattering from slits that define the X-ray beam (Svergun et al. 2013). These are examples; a complete list for a real beamline and experiment certainly contains other sources of error. To improve the accuracy of the signal and establish the true signal, the origins of all systematic errors must be tracked down. If the errors can be eliminated (for example by improved slits in the beamline or careful pixel by pixel calibration of the detector), the true signal is established. If more realistically elimination is not possible, the magnitude of systematic errors must be minimized.

3.2.2 Difference Experiments

Almost all crystallographic and solution scattering experiments, both static and dynamic, depend on the accurate measurement of small differences in an X-ray scattering signal rather than on a single measurement. For example, *de novo* phasing by isomorphous replacement is based on small differences in structure amplitudes between a native crystal and its heavy atom derivatives. Substantial systematic error is introduced by lack of isomorphism among the native and derivative crystals. Phasing by anomalous scattering is based on even smaller differences between the structure amplitudes of Friedel opposites |$F(hkl)$| and |$F(-h\ -k\ -l)$| or their symmetry mates on one crystal, at X-ray energies spanning the absorption edge of a heavy atom. The ability to make anomalous scattering measurements on one crystal eliminates two common, often substantial forms of systematic error, non-isomorphism between native and heavy atom crystals and crystal-to-crystal variation within a heavy atom derivative. However, other forms of systematic error may be introduced such as the dependence of the source intensity and detector sensitivity on the X-ray energy.

In dynamic experiments, few atoms move as a reaction proceeds and those that do move, do not move far. True signals arise from changes in X-ray scattering in time by the atoms in the sample as they move, and the populations of structurally different intermediates change. True signals are therefore also small in magnitude and prone to masking by new, time-dependent systematic errors such as radiation damage (Chapter 4.3) or an unanticipated change in temperature. Examples of true signals in a dynamic experiment are the change with time t in crystal structure amplitudes for the reflection hkl: $\Delta F(t) = \{|F(hkl,t)| - |F(hkl,0)|\}$, or in solution scattering for the intensity at q: $\Delta I(t) = \{I(q,t) - I(q,0)\}$.

Careful experimental design can ensure that the systematic error in the time-dependent differences in dynamic experiments, $\Delta F(t)$ or $\Delta I(t)$, can be lower than the systematic error in a single static experiment that measures |$F(hkl)$| or $I(q)$. This desirable and at first glance paradoxical situation arises when the systematic errors in measurement at a single time point t are identical to (or at least, strongly correlated with) those in all other time points, notably including time 0 immediately prior to initiation of the reaction. The systematic error term (Equation 3.2) is then identical across all time points, constant in time. Such systematic errors drop out when the time-dependent difference scattering is calculated:

$$\text{True difference signal}(t) = \Delta F(t) = \text{True}\{|F(hkl,t)| - |F(hkl,0)|\} =$$
$$\text{True}\{\text{Total measurement}(t) - \text{Total measurement}(0)\} -$$
$$\{\text{Background measurement}(t) - \text{Background measurement}(0)\}. \quad (3.3)$$

Replacing F by the intensity I throughout Equation 3.3 gives an identical expression for $\Delta I(t)$.

3.2.3 Experimental Design

A complete data set in dynamic experiments containing N time points is N times larger than a single static experiment. This introduces a fundamental problem in experimental design. How shall a large, 4D dynamic data set |$F(hkl,t)$| or a 2D dynamic data set $I(q,t)$ be collected on many samples, to minimize the impact of all systematic errors that vary with time?

An approach to this question may be illustrated by a real example from one of the authors (K.M.). Early time-resolved crystallographic experiments explored the photocycle of the blue-light bacterial photoreceptor known as photoactive yellow protein, PYP (Srajer 1996; Ren et al. 2001); see also (Perman et al. 1998; Ren et al. 1999). The beamtime schedule of an experimental run at the European Synchrotron Radiation Facility (ESRF) was generous but did not allow enough time to collect all reflections *hkl* over the unique volume in reciprocal space over the entire duration of the photocycle, which spans 10 decades in time from 100 ps to 1 s. Data were usually limited by radiation damage (Chapter 4.3) to a single time point *t* per crystal. If all went well, data sets could be collected at several time points per experimental run. That is, time was the *slow variable*. The entire photocycle of the E46Q mutant of PYP was finally assembled from 54 data sets at 30 time points obtained from several experimental runs over about two years (Anderson et al. 2004). The *data movie* spanning the photocycle had to be built up frame by frame, data set by data set. Data at each time point were of necessity obtained on different crystals, grown at different times from different protein samples, on experimental runs under slightly different conditions of the beamline, apparatus, and data processing software. These differences introduce systematic errors which are uncorrelated from time point to time point. For example, the extent of initiation into the photocycle (and hence the signal) varied between data sets over a wide range from ~6% to 34%. This necessitated extensive averaging over the 30 time points into 8 separate time bins, with some loss of time resolution. In those early days, we didn't sufficiently appreciate the effects of systematic errors in the time variable. True signal varies smoothly with time (Chapter 7.2), but systematic errors uncorrelated in time vary very rapidly (that is, they jitter) from time point to time point.

The final content of each movie frame is a difference electron density (DED) map $\Delta\rho(t)$ at each time point *t* (Chapter 7.3; also known as a difference Fourier map), which displays the difference in electron density $\Delta\rho$ between the structure at time *t* after initiation of the photocycle and time 0, immediately prior to initiation (Figure 3.1). The DED maps at adjacent time bins are broadly similar but despite best efforts at scaling the maps together, the magnitude of the signal and the distribution of the noise did not present a smooth movie. Although our intended observation was the variation with time of the structural signal over the reversible photocycle of PYP, this signal is masked by the systematic errors. Good data at each individual time point and time bin was the best that could be achieved but the movie of the photocycle, displayed as time-dependent changes in DED maps, was compromised.

If handed a lemon, make lemonade. This result drove reconsideration of systematic errors. The procedure for collection of dynamic data across the entire, 4D data space $|F(hkl, t)|$ was redesigned to minimize the systematic errors. In all dynamic experiments, the critical chemical and biological signal lies in the time variable. Systematic errors in the time variable *t* must be minimized, even if systematic errors are increased in other variables such as structure amplitudes in crystallography and scattering intensity in solution scattering.

Achieving that minimization requires that time be the variable scanned most rapidly (the *fast variable*) in a single experimental run. This ensures closely similar conditions for data collected at each time point within that run. Variables in reciprocal space (the *slow variables*) can be scanned more slowly, even if necessary in different experimental runs. As a practical matter at an SR source, Laue diffraction patterns (Chapter 5.1.2) are collected at a given crystal orientation. At that crystal orientation, the time after reaction initiation is varied across all N desired time points throughout the complete course of the reaction,

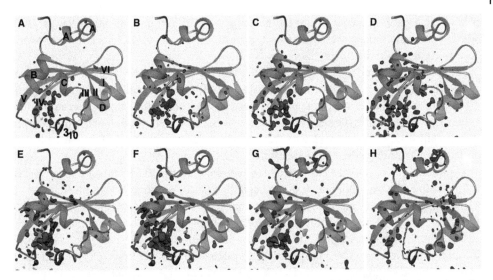

Figure 3.1 DED maps of the entire PYP molecule. The DED map at 1.6 A resolution is shown in red (−) and blue (+) contours at 2.5σ and 3.5σ (where σ is the RMS value of the DED across the asymmetric unit), superimposed on a ribbon tracing of the structure of E46Q PYP prior to entry into the photocycle. Panels **A−H** show the average DED over different time bins: **A**, 10−40 ns; **B**, 100−150 ns; **C**, 300−500 ns; **D**, 700 ns−5 μs; **E**, 10−30 μs; **F**, 75−300 μs; **G**, 900 μs−3 ms; **H**, 7−30 ms. Anderson et al. 2004/Reproduced with permission from ELSEVIER.

including time 0. This approach with time as the fast variable ensures that data at nearby time points are collected on the same crystal at very closely spaced wall clock times and very similar dose. The last minimizes the impact on the true signal of the time variation of radiation damage (Chapter 4.3). If radiation damage is sufficiently small to allow further collection of good data on that crystal, its orientation (rotation angle) is advanced, and collection of the N patterns in the time variable is repeated. Advancing the orientation continues until that crystal cannot provide further good data. The crystal is then replaced and the entire experiment is repeated from a new initial orientation. When the unique volume in reciprocal space has been measured at all time points with sufficient redundancy (which may require more than one experimental run), the raw, 4D data set is complete. At each of the N time points, data from all crystals must be scaled together in reciprocal space. This scaling is a well-understood crystallographic process presented in Chapter 8.3 of (Rupp 2009). Systematic errors in scaling are present but confined to the scattering variables in reciprocal space. Of greatest importance, scaling introduces no errors into the time variable.

This experimental design works well. Systematic errors in time are greatly reduced, the DED maps in the individual movie frames vary smoothly with time, and the structural changes in the protein during its photocycle can be readily visualized (Ren et al. 2001; Ihee et al. 2005). The design was used by us subsequently in all dynamic experiments at SR sources.

As ultrafast X-ray detectors are introduced at SR sources that can acquire scattering data continuously with the X-ray shutter open (Chapter 6.4), the same goal remains: arrange data collection to ensure that systematic errors in the time variable are minimized. In practice at SR sources, this means collecting as many diffraction patterns as possible on one crystal. Each pattern is collected at a different time point, on one (or only a few) crystal orientations, where the patterns are closely spaced in laboratory, wall clock time, in X-ray dose and extent of radiation damage.

The source characteristics are entirely different at XFEL sources where each diffraction pattern is obtained on a different microcrystal at a different crystal orientation (Chapter 4.3.3). The fast variables are automatically those in reciprocal, diffraction space; a nominal value of the time point is the slow variable. However, the actual time point both jitters from pulse to pulse and drifts more slowly. Ways in which this apparent defect can be exploited on the ultrafast, sub-ps time scale are discussed in Chapter 7.3.4.

A cautionary note must be considered for experiments at both SR and XFEL sources. If at all possible, the experimental design must allow the collection of at least one data set at a single time point and complete in reciprocal space, even if total beam time is cut short by, for example, unexpected equipment failure. Data at only a single time point may be a success. At the least, it can support a future beam time request and may be individually publishable. This suggests a hybrid experimental design at SR sources in which complete data in reciprocal space are first collected at a single time point, then a switch is made to time as the fast variable but with a carefully chosen value of N (the number of time points to be scanned fast). Too large a value of N runs the risk that partial data sets will be obtained that span all N, but each set is incomplete in reciprocal space. No DED maps or publishable results can be obtained.

3.3 Reaction Initiation

3.3.1 Principles

Almost all time-resolved experiments require both a means of initiation, the pump, which starts the reaction and defines time zero, and a means of following the reaction course, the probe. (The exceptions are experiments in which the sample remains at equilibrium and fluctuations about the equilibrium position are observed (Frank and Ourmazd 2016)). Initiation requires that the thermodynamic equilibrium of the sample be perturbed in some way, either physically or chemically, which then results in structural changes as equilibrium is reestablished. Examples of a pump perturbation are a visible laser pulse, temperature jump, electric field jump, or rapid mixing of an enzyme with its substrate. Examples of a probe are illumination of a crystal or a solution by an X-ray pulse train or a single pulse, or by a light beam in a spectroscopic experiment. The equilibrium to be perturbed may be unimolecular e.g. the conformational equilibrium of an individual protein molecule or complex; or bimolecular e.g. the binding equilibrium of an essential ion, substrate or cofactor to a protein. The pump–probe approach to dynamics and kinetics was pioneered by the physical chemist George Porter in the 1940s, continues to be applied by ultrafast spectroscopists, and has been adapted very effectively to structural dynamics of biological molecules.

We emphasize that the pump process synchronizes only the start of the reaction in the molecules in the sample. Thereafter, probes by spectroscopy, X-ray solution scattering, or single crystal diffraction show that molecules in solution and in a crystal proceed along their reaction coordinate independently of each other. There is no synchrony in their structural changes. Simply put, if one molecule changes from structure A to structure B, its neighbors do not change to structure B (or to related structures A' or B') in lockstep. In reactions where the pump is an fs laser pulse, synchrony between molecules is initially present (Chapter 9.4.4) but decays within 1 ps (Hosseinizadeh et al. 2021). The absence of synchrony was fully expected in dilute solution where intermolecular interactions are both infrequent and very weak but had to be confirmed experimentally in crystals. As in dilute solution, time-resolved crystallographic data can be fitted by chemical kinetic models based on exponential growth and decay of concentrations that assume independence of molecules (Chapter 2). Though the intermolecular interactions that stabilize crystal lattices in macromolecules are stronger than in dilute solution, they are still not strong enough to substantially influence, let alone control, the structure and reactivity of neighboring molecules. In this respect the interactions in crystals of macromolecules differ from the much stronger interactions in crystals of inorganic species or small molecules. Their crystals are hard, crack when prodded, and contain no liquid water. In contrast, crystals of macromolecules are soft, squishy when prodded and contain substantial liquid water.

However, we do not assert that the reactivity of the molecules in a crystal is identical to that in dilute solution. Intermolecular interactions in the crystal lattice may be strong enough to restrict larger-scale structural changes important to mechanism. Examples are the quaternary structural change in hemoglobin (Chapter 1.2), or the light-dependent displacement of the C-terminal Jα-helix in a small signaling photoreceptor of the light-oxygen-voltage (LOV) domain family (Halavaty and Moffat 2007). Although the rate coefficients in the chemical kinetic model for reactivity in the crystal may be similar to those in dilute solution, in some systems such as PYP the differences may be significant (Yeremenko et al. 2006; Konold et al. 2020). If significant, care must be taken to ensure that the differences do not simply arise from different conditions of the solution and the crystallization buffer, or from the selection of different mechanisms in solution and in the crystal from several possibilities, all of which are consistent with the kinetic data. Detailed quantitative studies of activity in solution and in crystals are limited in number and should become more routine.

Selecting the best approach to reaction initiation for a new biological system is an important aspect of design of dynamical experiments and perhaps the most challenging. The possible approaches are many and varied. The pump may be a change in any physical variable on which structures or their equilibria depend, such as temperature, pressure, electric field, or, in specific cases, light. It may be a change in a general solution variable such as pH or ionic strength. It may be a change in a chemical variable such as the concentration of a component specific to the system such as a substrate, inhibitor, cofactor, metal, or anion. The approach selected for initiation depends critically on the nature of the biological system, the scientific question to be addressed, and the temporal and spatial resolution needed to do so. If the detailed chemical mechanism of the system is sought, the concentration of a specific reactant or of a solution variable such as pH on which the activity of the system depends can be perturbed. If more general questions of folding or stability are posed, then perturbation of a physical variable such as temperature or pressure is appropriate and also applicable to exploration of mechanism. To give a more specific example, if the transport properties of a voltage-dependent ion channel are sought, then perturbation by applying an electric field pulse – a voltage jump – will be very informative (Hekstra et al. 2016).

As we shall see, the pump–probe approach to structural dynamics requires measurement of a single reaction time point per initiation: a sequence of pump once, delay to the required time point, probe once. In effect the movie is being acquired one frame at a time which is tedious, less accurate, and wasteful of precious sample. This contrasts with the sequence used by ultrafast spectroscopists who use optical absorbance or fluorescence to probe the time course, where radiation damage by the probe is not a significant issue. They apply the movie camera approach of pump once, probe many time points continuously, thus acquiring a movie frame at all time points. This approach is more accurate since data at all time points are acquired on the same sample. It makes more efficient use of a rare sample and offers more rapid overall data acquisition (Chapter 9.4.3).

Restriction of the probe in structural dynamics to a single, controllable time point is therefore a major limitation, largely required by two factors. First, radiation damage limits the number of time points at which good data can be collected on a single sample. Diffraction before destruction at XFELs (Chapter 4.3.3) ensures that all experiments are limited to one scattering or diffraction pattern per sample, at a single time point. Second, experiments that pump-once, probe-many time points require a detector with special characteristics such as a fast frame rate, large area, and numerous small pixels. The detector must be backed by fast data analysis and large storage capability. Until very recently such capabilities did not exist.

The situation is being transformed by advances in detectors (Chapter 6.4). Many dynamic experiments into the μs time range at SRs are less constrained by radiation damage than at XFELs and can adopt the pump-once, probe-many approach with detectors offering these characteristics.

The hardware and control software necessary to conduct initiation with a mixing device, pulsed visible laser and sample injector are complex to set up and operate successfully. Fortunately, they are almost always provided by the staff at the SR or XFEL beamline where the experiments will be conducted (Chapter 6.5). Hardware and software will be optimized by the experienced staff in close consultation with the user prior to the experiment. Beamlines specialize in different approaches to initiation. These are often approaches that their staff have developed and brought into common use, in which they are naturally expert. However, not all approaches are available at every beamline, which restricts the user's choice. Hardware and software tailored to reaction initiation of the specific biological system may of course be provided by the user.

Several features are common to all effective approaches to reaction initiation. The observable time course arises from the convolution of the time course of the pump with that of the reaction process, which complicates interpretation of the latter. The effects of the convolution can be eliminated by making the pump much faster than the fastest desired reaction. (Sprints at the Olympic Games are initiated by a starting gun, not by a roll of drums!) For example, if the desired structural changes extend from 10 ms to 10 s, the time course of the pump should require 1 ms or less.

The pump should achieve maximum spatial uniformity across the sample, while recognizing that complete uniformity cannot be achieved. Nonuniformity excites distinct volumes within the sample to differing extents, may vary from pulse to pulse, may cause molecules in distinct volumes to follow different time courses, and tends to disorder a crystal.

To maximize the structural signal, the reaction should be initiated in most molecules illuminated by the probe X-ray beam. The temptation is to increase the scattering power of the sample e.g. crystal size or solution concentration and thus the signal. There can be a downside. If a laser light pulse is the pump, this requires higher light intensity, increases spatial nonuniformity, and may introduce artifacts from the absorption of more than one photon per molecule (Chapter 3.3.2.3.1).

Repetition of the pump should match the properties of the reaction. For example, if the reaction is fully reversible in 1 s, all experimental processes including the pump and detector readout should also be repeatable at 1 Hz or higher frequencies. At SR sources, this enables more data to be collected efficiently on a single sample, thus minimizing systematic errors arising from sample-to-sample variation (Chapter 3.2.1). At XFEL sources where all experiments are irreversible, the overall repetition rate depends on the slowest of the rate of arrival of the X-ray pulses, the rate at which fresh samples can be introduced and the rate of detector readout.

Fast structural changes on the 100 ps to ns time scale are often probed by a single X-ray pulse isolated by a fast shutter system from the train of pulses in an SR. To minimize systematic errors from timing jitter, the pump and shutter systems should be synchronized with the pulse clock of the source, which monitors the location of each bunch of electrons and hence of the X-ray pulse they generate. At XFEL sources where structural changes on the ultrafast fs time scale are probed, timing jitter can be experimentally significant and harder to minimize. However, a specialized detector (the timing tool; Chapter 7.3.4) sensitive to both a visible laser pump pulse and an X-ray probe pulse can be inserted into the beamline. By measuring the time delay between these pulses, a time stamp is obtained for each diffraction pattern or scattering curve, which can subsequently be sorted into a more accurate time series. Perhaps counterintuitively, timing jitter and the spiky temporal structure of XFEL pulses can themselves be exploited to substantially improve the time resolution at XFEL sources (Fung et al. 2016; Hosseinizadeh et al. 2021).

For dynamic studies of crystals, pump approaches depend on whether subsequent structural changes vary with the orientation of the pump relative to the crystal axes. Approaches such as a temperature jump or change in a solution property such as pH or reactant concentration do not depend on orientation of the crystal, do not require orientation space to be explored, retain the symmetry of the crystal, and are simpler to apply. They are scalar; only the amplitude of the perturbation is important. The structural changes subsequent to other approaches such as illumination by plane-polarized light or application of an electric field are polar. Their amplitude is important, but the structural changes also depend on orientation, require orientation space to be explored, lower the symmetry of the crystal in very informative ways but are much more difficult to apply experimentally. A detailed example of the application and impact of a polar, electric field pulse is given by (Hekstra et al. 2016), and illustrates how the lowering of crystal symmetry can be exploited to minimize systematic errors. In their example, these errors arise from the (scalar) temperature jump that inescapably accompanies the (polar) electric field pulse.

A further powerful example is offered by crystals of the voltage-gated ion channel NaK2K. Two ion channels, each a tetramer, are packed head-to-head in a single asymmetric unit in the crystal. A single electric field pulse along the fourfold crystallographic axis is thus oriented *up* with respect to one channel and *down* with respect to the other. This fortuitous – and powerful – circumstance greatly minimizes systematic errors in the structural differences between the up and down orientations induced by the electric field pulse.

3.3.2 Experimental Approaches

Three approaches to dynamics have been widely used in biophysics with optical probes: concentration jump by mixing, relaxation after temperature jump, and light (Martin and Vos 1992; Roder et al. 2006; Toh et al. 2010). All are now being adapted to X-ray-based probes.

3.3.2.1 Mixing: Chemical Initiation

The simplest approach to reaction initiation is to produce a concentration jump by rapidly mixing two reactants such as an enzyme with its substrate, or a weakly-buffered protein with an acid or base. Both reactants may be in solution or the protein may be crystalline, suspended in a small droplet of liquid. In the first case, initiation is controlled by diffusion processes in the turbulent or laminar hydrodynamics of mixing two liquids; in the second, by diffusion of the substrate into the interior of the crystal through the aqueous channels between molecules. Diffusion in aqueous solution and within a crystal depends on concentration gradients, the square of the linear dimension over which diffusion occurs, and viscosity. With typical dimensions and viscosity, diffusion equilibrates over the ms time range in most aqueous solutions, which limits the applicability of mixing to reactions that proceed over longer time scales. Fortunately for structural enzymologists, many enzymes are slow catalysts, illustrated by low turnover numbers in the range of $0.1–100$ s^{-1} with a median of $10 \, s^{-1}$ (Bar-Even et al. 2011). Although the chemical steps of enzymatic reactions and their associated structural changes often occur on time scales as short as ns to hundreds of µs and are therefore inaccessible to mixing experiments, long-lived tertiary and quaternary structural changes and steps associated with product release are readily accessible. These reactions in slow enzymes offer very suitable targets for initiation by mixing. Since many biological reactions are pH-dependent, pH jumps are both readily achieved by mixing and a very general way to initiate reactions. An example followed by solution X-ray scattering is (Rimmerman et al. 2018) .

Mixing can be speeded up by reducing the dimensions over which diffusion occurs, such as mixing in microfluidic channels (Park et al. 2006) and restricting crystal sizes to a few microns (Schmidt 2013). Mixing can be followed by continuous flow from the point of mixing to the probe beam, which is easily achieved experimentally by varying the flow velocity or the distance between the mixing point and the probe. This approach couples the output of a simple T-mixer to the gas dynamic virtual nozzle (GDVN) sample injector widely adopted at XFEL sources (Chapter 3.3.2.1; (DePonte et al. 2008)). However, continuous flow has two serious limitations. A large amount of sample is required, most of which is never illuminated by the probe beam and effectively wasted. The hit rate (Chapter 4.2) of good diffraction patterns or scattering curves is unacceptably low, the total number of good patterns is reduced and systematic error is increased. In contrast, stopped-flow offers a "mix once, probe many" approach and has been widely used in biophysics to study reactions in solution over ms time scales. Stopped-flow is much more economical than continuous flow in sample amount and significantly, could offer a high hit rate with lower systematic error since data is acquired on the same volume element of the sample at different time points. The downside is that the data are prone to radiation damage. This rules out the use of stopped-flow at XFELs where experimental design is dominated by radiation damage, but could be effectively adapted to SR sources (Chen et al. 2014).

Mixing to initiate the reaction is being successfully extended to dynamic, serial synchrotron crystallography (SSX) at SR sources (Pearson and Mehrabi 2020). Sufficient diffraction signal is built up by repetition over many small crystals or solution volumes. Today's SR sources are sufficiently brilliant (Chapter 6.2) to enable ms exposures at microfocus beamlines particularly if polychromatic, pink X-ray beams are used (Chapter 5.1). The higher-brilliance magnetic lattices installed at SR sources such as the European Synchrotron Radiation Facility (ESRF) and the Advanced Photon Source (APS) further enhance this capability. Many more scientific problems would become accessible to mixing approaches at SR sources if the exposure could be reduced even further, perhaps to the µs time range. However, the background resulting from slits and other beamline components would have to be carefully minimized (Meents et al. 2017). The SR source would have to be very intense and in addition, the sample would have to scatter strongly. Reaching the desirable, single X-ray pulse range of 100 ps in data at SR sources would require the ability to analyze diffraction patterns with extremely low signal to noise. Although larger crystals obviously scatter more strongly, their use might be self-defeating. Diffusion would be slowed and the µs range would remain inaccessible. Radiation damage remains a significant issue which may limit each sample to a single scattering pattern. Enhancing the time resolution below the ms range with mixing techniques thus poses a substantial challenge.

Mix and inject approaches (Schmidt 2013) are also applicable at XFEL sources. Examples at the Linac Coherent Light Source (LCLS) are (Kupitz et al. 2017; Stagno et al. 2017). Since slow diffusion during mixing is likely to remain limiting, the ultrashort fs X-ray pulses of XFEL sources offer no advantage in time resolution. However, their exceptionally high peak brilliance enables tiny crystals or a very small volume of dilute solution to be used (Chapter 6.3), and their small linear dimensions speed up diffusion and thus enhance the time resolution. Most systematic errors are reduced by averaging a very large number of microcrystal diffraction patterns or solution scattering curves. The introduction of XFELs with very high pulse repetition rates such as the European XFEL (EuXFEL) and the upgraded LCLS-II (Chapter 6.3) will allow more effective use of the sample and the beamtime (Pandey et al. 2021). For example, high throughput screening of drug binding on the ms time scale may require only a few seconds per dataset and enable hundreds of datasets of high quality to be acquired per shift.

3.3.2.2 Relaxation: Physical Initiation
Application of a physical perturbation such as a fast temperature jump displaces the system from conformational and/or binding equilibria and initiates structural changes as the molecules relax to the new equilibrium position (Chapter 2.4.2). Relaxation approaches have the big advantage of being completely general. No special knowledge of details of the system such as its reactivity is required, nor must it possess an unusual property such as sensitivity to visible light or response to a specific small molecule. A disadvantage is that those spatially

localized structural changes directly related to biological activity must be teased out from spatially distributed structural changes across the entire molecule by the temperature jump. Since structure and reactivity depend on both temperature and pressure, jumps in either are applicable. Temperature jumps have been the most widely applied and we emphasize them here. Electric field jump is also generally applicable and promising (Hekstra et al. 2016), and experimental challenges are being overcome. The temperature jump may be obtained by Joule heating from a current pulse, or by a single light pulse from an IR laser. Absorption of IR photons excites the O–H stretching modes of water molecules, whose increased vibrational energy effectively heats the solvent and the solute macromolecules.

Two approaches to relaxation differ largely in the amplitude of the energetic perturbation. In the first, the pump delivers only a small energetic perturbation to the system, which remains close to thermodynamic equilibrium. The system is tickled by a small temperature jump of, for example, 5 K from which it relaxes to the new equilibrium position in a linear fashion. In the second, the pump delivers a larger energetic perturbation on the system, sufficient to drive it far from the initial equilibrium. The system is kicked by a large temperature jump or the absorption of a highly energetic photon in the visible region of the spectrum, and may relax to equilibrium in a nonlinear fashion. The second approach generates larger structural changes which are easier to detect and may be more interesting – but nonlinear relaxations are more difficult to interpret.

A recent example illustrates the second approach (Thompson et al. 2019). This group (Box 3A) used time-resolved X-ray small angle solution scattering at a SR source to follow the structural relaxation of the enzyme cyclophilin A, a proline isomerase, after a temperature jump. They detail the experimental aspects and a wealth of data. Although they encounter limitations on structural interpretation at the atomistic level, these are not fundamental. Later experiments should improve the signal-to-noise in the wide-angle X-ray scattering region and enable more direct information on structural changes.

Box 3A

Structural changes ((Thompson et al. 2019) and its Supplementary Methods) were pumped in a solution of cyclophilin A (CypA) by a 7 ns mid-IR laser pulse at 1443 nm, and probed by simultaneous small-angle (SAXS) and wide- angle (WAXS) X-ray solution scattering. WAXS experiments yield signal arising from protein + solvent and provide information at low structural resolution. In WAXS experiments the total signal has a very small protein component, arising mostly from the solvent.

The X-ray probe at the 14ID-B double undulator BioCARS beamline at the APS was polychromatic (a "pink" beam; Chapter 5.1.2) with a wide bandwidth of ~3%. Monochromatic sources have been more commonly used. However, a monochromator rejects most of the incident X-ray intensity which nevertheless could have contributed effectively to the SAXS/WAXS scattering curves. A polychromatic source offers much higher intensity and shorter exposures with better time resolution than a monochromatic source. A bandwidth of ~3% is still sufficiently narrow that the scattering curves are not degraded (by smearing in scattering angle) in almost all protein studies. The much higher X-ray intensity allowed a total exposure duration of ~500 ns to be achieved from only three X-ray pulses.

Box 3A (Continued)

The strong temperature dependence of the solvent WAXS signal provides an accurate internal thermometer. The temperature jump induced by absorption of the IR laser pulse was around 10 K, a relatively large perturbation. Although overall scattering was dominated by the much larger solvent and instrumental scattering components, careful control experiments and temperature-dependent measurements enabled the very small protein SAXS signal to be cleanly isolated. Interleaved laser on (temperature jump) and laser off (no temperature jump) measurements minimized systematic errors in the difference scattering curves obtained by subtracting the laser off and laser on curves. Since these measurements are very closely spaced in time on the same sample, they are also closely spaced in dose and in radiation damage. The influence of sample-to-sample variation and of radiation damage on the difference scattering curves is very nearly eliminated (Chapters 3.2.2 and 3.2.3). The structural relaxation evident in the scattering curves after the temperature jump was sampled at 27 time points, distributed nearly uniformly in the logarithm of time across four decades from 562 ns to 1 ms, with three or more points per decade in time to ensure efficient sampling. Sampling uniformly in the logarithm of time (as distinct from linearly in time) is statistically appropriate when the time range is large and there is no prior information about the number and time scale of likely kinetic processes, as was the case in this example. Several background corrections were made to the raw scattering data, described in the Supplementary Methods. Time-dependent changes in scattering were confined to the SAXS region of low q (Chapter 5.2). Attempts to fit these corrected, time-dependent data on wild-type CypA with a single first-order relaxation process were unsatisfactory. Extending the kinetic model to two independent, first-order relaxation processes was far superior and identified two processes in the μs time regime. The first was associated with fast expansion of the CypA particle (protein plus tightly bound solvent) within a few μs of the temperature jump, followed by slower contraction. The temperature (after the 10 K jump) was varied from 279 to 303 K to explore the temperature dependence of the rate coefficients associated with each process. Linear Arrhenius behavior was observed from which the thermodynamic properties of the two transition states could be obtained (Chapter 2.5). Both processes showed a substantial positive value of the activation enthalpy; the activation entropy of the faster process was positive and that of the slower process was negative. Comparison of time-resolved scattering of wild type CypA with two mutants (a single point mutant S99T in the core region of the enzyme and a double mutant D66N/R69H in the loops region) showed that the slower phase was absent in the enzymatically inactive S99T core mutant, and the faster phase was absent in the D66N/R69H double mutant. Data on each mutant could be fitted by a single first-order process.

By comparing these dynamic results with earlier NMR and static crystallographic results, the faster process was tentatively attributed to loop motions associated with the expansion process and the slower process to internal structural rearrangements. The inability to observe time-dependent WAXS scattering was attributed to the limited signal-to-noise at higher q. Atomistic interpretation from the dynamic results alone was therefore not accessible. Nevertheless, the authors are optimistic about the prospects of applying relaxation approaches to functionally relevant structural dynamics, exploring conformational equilibria of evolutionary significance.

In a third, quite different approach to relaxation, no pump is needed. The system – usually a dilute solution – remains at thermodynamic equilibrium throughout, and the dynamics of fluctuations around its equilibrium position are observed directly. In a sense, the system is being continuously pumped by thermal energy and continuously relaxing. Since no explicit initiation process is required, this experimental approach is much simpler than the first two. Some continuous relaxation approaches require that only a small number of molecules are probed, in which the fluctuations arise from changes in the number of molecules as they diffuse in and out of the volume probed. The molecules may for example be in a monomer – dimer equilibrium where the species differ in their diffusion coefficients. The structural signal is very small and its fluctuations even smaller. Fluctuational X-ray scattering has apparently not been applied to structural dynamics, although using fluorescence as an optical probe with good signal-to-noise is successful.

Other approaches (Kam 1977; Donatelli et al. 2015) examine a statistically large number of molecules in solution scattering and are better suited to X-ray probes. Solution scattering arises from spherical averaging over all molecular orientations in a long X-ray exposure and as a result has low information content (Chapter 5.2). However, an X-ray probe pulse of duration less than the rotational diffusion time of the illuminated molecules effectively eliminates their rotational motion. This introduces intensity angular correlations within each ring at a fixed scattering angle – correlated X-ray scattering. The correlations substantially increase the overall information content and can be related to the 3D diffraction volume of a single molecule. Additional interest comes from the fact that the correlations change as the structure changes during a reaction such as ligand binding or absorption of a photon (Pande et al. 2015). Thermal fluctuations in structure in the ground state may be sufficiently large to sample structures along the reaction coordinate, though with much lower probability. The more interesting case of fluctuations in the scattering of the individual molecules or complexes such as the ribosome as their structures change in a tiny sample is considered later (Chapter 8.3).

3.3.2.3 Light

Only a small number of light-sensitive biological systems are known to exist in nature, but all are central to important processes. Many use light for imaging (e.g. rhodopsin in the eye), or to control transport of ions across a membrane (e.g. bacteriorhodopsin). Others use light as an energy source (e.g. light-harvesting complexes, photosystem II and reaction centers in photosynthesis), or are photoenzymes such as the Old Yellow Enzyme (Chapter 4.4). In yet others, light is an information source (e.g. signaling photoreceptors such as phytochromes that control plant development, or the related bacteriophytochromes that control swimming in certain bacteria and development in fungi). Some organisms have the ability to sense IR or UV radiation directly (Sakai et al. 2017; Meier et al. 2018). A few inherently light-sensitive, chemical complexes are also known such as the CO complexes of iron porphyrins or carotenoid-containing enzymes, but the biological relevance of their sensitivity to light is unclear.

If restricted to naturally occurring systems, light clearly lacks generality as a means of reaction initiation. Four reasons make it worthy of extensive consideration in structural dynamics. First and most importantly, several approaches have been developed that confer sensitivity to light on the reactions of otherwise light-inert systems. For example, artificial signaling photoreceptors can be designed, genetically expressed and shown to control a normally light-inert biological function such as catalysis or binding of a transcription factor (Möglich and Moffat 2010; Moffat 2014). This suggests that reaction initiation and the activity of your favorite biological system can be made sensitive to light by biological or chemical means. Your system can be engineered with an eye on subsequent structural dynamics by approaches such as optogenetics, caging, and photopharmacology (Chapter 4.4). Second, many of the experimental techniques for structural dynamics have been developed on light-sensitive systems. The techniques are both experimentally convenient and applicable to a wider range of systems including those which are not sensitive to light. Third, the mechanisms of light-sensitive signaling photoreceptors such as PYP and Light-Oxygen-Voltage (LOV) domains are representative of wider, light-inert Per-Arnt-Sim (PAS) domain proteins. More broadly, the signaling mechanisms of cytoplasmic photoreceptors may resemble those of light-inert, integral membrane chemoreceptors. For example, α-helices and coiled coils act in both photoreceptors and chemoreceptors as structural elements capable of transmitting signals over a long range. Fourth, reaction initiation by light is inherently very rapid and extends into the chemical time scale of fs to ps. In principle and often in practice, this can make the earliest, chemical steps in a reaction accessible to structural dynamics. These steps are usually much faster than the ms accessible to initiation by mixing, or even the μs to a few hundred ns accessible to relaxation. At present, no other approach to initiation makes ultrafast chemical steps accessible to structural dynamics. We therefore treat light initiation in some detail.

3.3.2.3.1 *Principles of light absorption*

How does light absorption occur, and how may it be used to initiate a reaction and generate a structural signal? Aromatic side chains in proteins and the bases in nucleic acids do not absorb sunlight in the visible region of the spectrum. To do so, naturally light-sensitive proteins bind a small organic molecule known as a chromophore that can directly absorb a photon of visible light. No naturally light-sensitive photoreceptors based on nucleic acids have been discovered – yet (see Text Box 1.2 of (Moffat et al. 2013) and Chapter 9.4.5). However, metabolite-dependent riboswitches that bind FMN are known in certain bacteria (Wickiser et al. 2005; Serganov et al. 2009; Wilt et al. 2020), where they regulate the expression of genes involved in riboflavin biosynthesis and transport. The biological role of similar flavin-based riboswitches could in principle also be dependent on light or redox state.

Representative chromophores are shown in Figure 3.2. The larger the chromophore, the greater the possible spatial extent of light-excited electron delocalization and the longer the wavelength of the spectrum it absorbs. For example, the extended linear tetrapyrrole chromophore, biliverdin, absorbs in the red to far red, from 650 to 750 nm; the smaller FMN chromophore absorbs in the blue, around 450 nm.

Photons in the visible spectrum have energies that are considerably greater than the free energy of stabilization of most proteins, around 40 kJ mol^{-1}. A photon of 400 nm wavelength corresponds to 219 kJ mol^{-1}, 550 nm to 217 kJ mol^{-1}, 700 nm to 170 kJ mol^{-1}, and even in the mid-IR at 1400 nm, to 85 kJ mol^{-1}. Absorption of a visible photon thus constitutes a substantial energetic perturbation, much larger than that typically available from binding of substrate or an inhibitor to an enzyme. Harnessing light energy poses a problem for light-sensitive proteins. They must rapidly direct the energy into a biologically useful pathway and suppress protein unfolding or other undesirable side reactions.

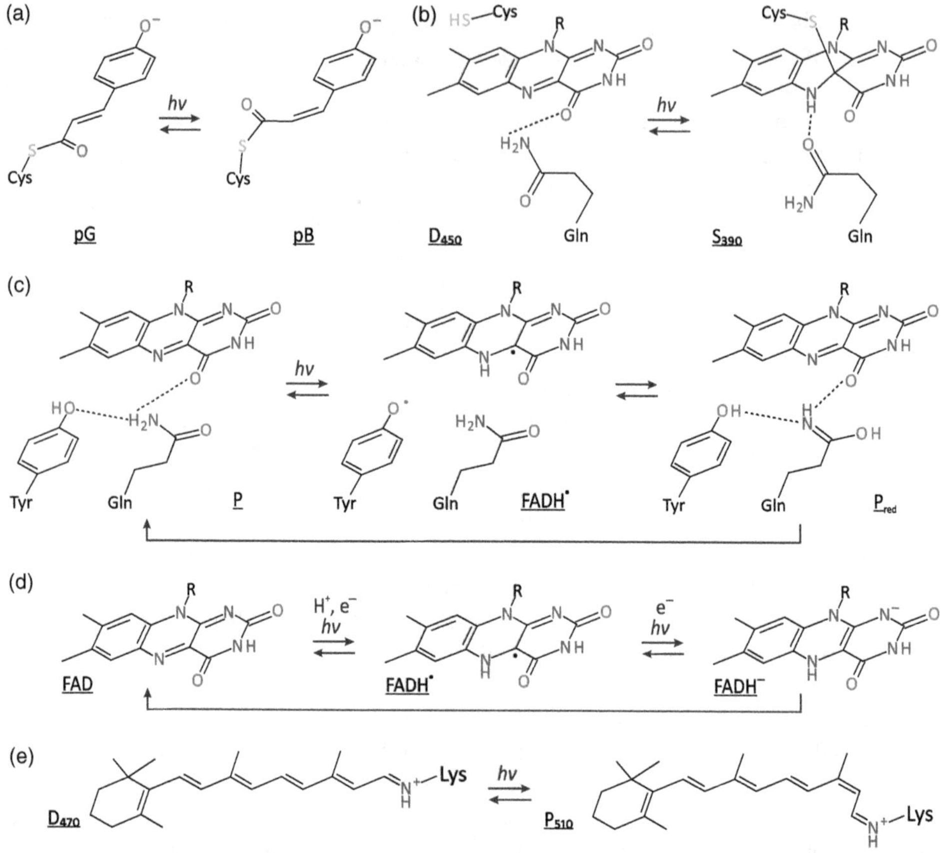

Figure 3.2 **Chromophores of sensory photoreceptors and their principal photochemistry.** The photocycles are shown simplified; short-lived intermediates en route to the signaling state are not included. **Panel a.** Photoactive yellow protein (PYP) covalently binds *p*-coumaric acid as a cysteinyl thioester. Blue light triggers chromophore isomerization from the E isomer in the dark-adapted pG state to the Z isomer in the pB signaling state. **Panel b.** Light-oxygen-voltage (LOV) receptors use flavin nucleotides, mostly flavin mononucleotide (FMN), to absorb blue light which prompts the reversible formation of a covalent thioadduct between the chromophore and a conserved cysteine residue in the signaling state. The resultant flavin protonation at the N5 atom is detected and relayed by a conserved glutamine residue. Intriguingly, neither the cysteine nor the glutamine are strictly required for photoreception. It appears that their roles may be played by water molecules. **Panel c.** As in LOV receptors, BLUF receptors (sensors of blue light using flavin adenine dinucleotide [FAD]) harness flavins for blue-light detection. Light absorption probably drives the tautomerization of a conserved glutamine residue adjacent to the chromophore from its amide form to the imide form in the signaling state. A nearby tyrosine residue is instrumental in mediating this reaction, probably via a radical-pair mechanism. An alternative mechanism proposes that rotation of the glutamine side chain occurs rather than tautomerization. **Panel d.** Cryptochromes employ FAD to detect blue light. In the case of plant cryptochromes (depicted in the figure), absorption of light drives one-electron reduction to the semiquinone radical form which constitutes the signaling state. Absorption of green light drives further reduction to the hydroquinone state. **Panel e.** Rhodopsins rely on retinal chromophores which are covalently bound as a Schiff base (imine) to a conserved lysine residue. Light drives chromophore *cis/trans* isomerization, e.g. from the all-*trans* form to the 13-*cis* signaling state in microbial rhodopsin (shown in the figure). By contrast, animal rhodopsins harness the light-driven conversion of the 11-*cis* to the all-*trans* isomer (not shown).

(f)

Figure 3.2 (Cont'd) **Panel f.** Phytochromes bind a linear tetrapyrrole (a bilin) covalently as a cysteinyl thioether. As shown for a bacterial phytochrome (a bacteriophytochrome) with its biliverdin chromophore, absorption of red or far-red light drives interconversion between the 15Z isomer in the red-absorbing Pr state and the 15E isomer in the far-red-absorbing Pfr state. In conventional phytochromes the Pr state is dark-adapted and the Pfr state formed on absorption of a red light constitutes the signaling state. Conversely, in bathyphytochromes the Pfr state is dark-adapted and the Pr state formed on absorption of a red light constitutes the signaling state. Figure prepared by Andreas Möglich.

Chromophores are embedded in the interior of the protein, to which they may be covalently or non-covalently linked. Upon absorption of a photon, the electronic structure of the chromophore is altered in the as (attosecond) to fs time range. Atoms in the chromophore move and change their interactions with the surrounding, closely packed protein over fs to a few ps. In response, the tertiary structure of the protein changes on a timescale of ps and longer, which extends in some photoreceptors to ms or even s. For example, deeply buried tryptophan side chains flip slowly within proteins with ms time constants (Weininger et al. 2014). The strong interactions between the chromophore and its protein environment lead to substantial differences in photochemistry between an isolated chromophore and the same chromophore embedded in the protein. The protein environment tailors the photochemistry. Conversely, electronic and structural changes in the chromophore upon absorption of a photon such as its isomerization directly perturb the surrounding protein. In the graphic phrase credited to Hans Frauenfelder (Ansari et al. 1985), these changes generate a *protein quake*, a structural signal in which atoms are displaced. The quake emanates from the chromophore epicenter at nearly the velocity of sound, around 2000 m s^{-1} (2.0 nm ps^{-1}). Signal propagation from the chromophore to the surface of the protein has been visualized by ultrafast structural dynamics over the fs to ps time scale in several very different proteins, for example myoglobin (Barends et al. 2015), PYP (Pande et al. 2016), and bacteriorhodopsin (Wickstrand et al. 2015). The affinity of a light-sensitive protein for a downstream partner or partners that bind to its surface is thus altered by the quake. The controlled sequence of structural changes constitutes a light-initiated thermodynamic signal.

Many elementary chemical processes such as breaking of a covalent bond, isomerization, electron transfer, and formation of a covalent bond are initiated by absorption of a photon by the chromophore. The covalent bond is broken between the heme Fe and the CO or O$_2$ ligand in heme proteins. Isomerization occurs about a specific double bond in the long polyene chain of the retinal chromophore in visual receptors, or about the only double bond in the tail of the *p*-cinnamic acid chromophore of PYP. Ultrafast, long range electron transfer occurs between chlorophyll moieties in photosystem II. In small signaling photo-receptors of the LOV domain family, ultrafast intersystem crossing in the FMN chromophore occurs within ns, followed by formation of a covalent bond between the FMN and a nearby cysteine side chain.

The photochemical processes occur very rapidly in the fs to ns time range, and cleanly, without competing chemical reactions such as rupture or formation of a different covalent bond, isomerization about a quite different double bond or electron transfer to an unintended acceptor. Evidently, much of the energy deposited in the chromophore can be specifically directed into the desired chemical and structural path. The remainder is dissipated into other, undesired, nonchemical paths. Immediately upon absorption of one photon, the chromophore becomes electronically and vibrationally excited in a Franck-Condon transition that does not involve nuclear movement (Figure 3.3). It may de-excite by one of several paths: non-radiative vibrational relaxation to the ground electronic state whence it originated; radiative relaxation to the ground state with emission of a fluorescence photon; stimulated emission; or motion of atoms in a structural change. The last is the desired path. Each path is characterized by a quantum yield Q which expresses the probability per photon absorbed that de-excitation will occur by that path. We thus identify, for example, the quantum yields Q_V for vibrational decay, Q_F for fluorescence, Q_{SE} for stimulated emission, or Q_S for structural change. The sum of the quantum yields per photon absorbed for all

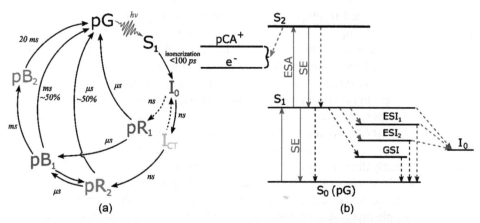

Figure 3.3 Spectroscopic and schematic electronic transitions in photoactive yellow protein.
Spectroscopic transitions are in panel a. In the electronic transitions in panel b, S_0 (pG) denotes the stable, ground state of the chromophore; S_1 denotes the first excited state upon absorption of the first pump photon; S_2 denotes the second excited state upon absorption of a second photon from an intense pump beam by a chromophore already in S_1 (excited state absorption, ESA). Purely electronic transitions (excitation and de-excitation) are shown by vertical arrows; transitions with associated structural changes are shown by diagonal arrows. One route of de-excitation from S_2 is by stimulated emission, SE; other routes that may cause damage are not shown. De-excitation from S_1 may occur directly by electronic and structural changes to the desired, next structural intermediate I_0, or indirectly to I_0 by excited state intermediates ESI_1 or ESI_2; or return to the ground state S_0 by the ground state intermediate GSI. Non-radiative transitions such as vibrational decay are shown by dashed arrows. Radiative transitions such as fluorescence and other non-radiative transitions are not shown. Hutchison et al. 2016/Reproduced with permission from ELSEVIER.

decay paths is unity. In this example, $Q_V + Q_F + Q_{SE} + Q_S = 1.0$. Enhancement of the quantum yield for one decay path must come at the expense of the quantum yield of others. Vibrational decay – heating – is unavoidable but nonproductive. As might be expected, evolutionary pressure has tended to promote higher values of Q_S in signaling proteins and higher values of Q_F in fluorescent proteins. That said, Q_S is generally much less than 1.0. The energy delivered by most of the photons absorbed in the ground state – the first pump photon – dissipates by other undesired but inescapable paths. However, evolutionary pressure has apparently ensured that the products of these undesired paths, derived by absorbing a single pump photon, are not damaging.

The quantum yields for each path are directed by the structural constraints imposed on the chromophore by the surrounding protein. For example, isomerization of the chromophore requires extensive motion of the protein atoms and groups adjacent to the double bond about which isomerization is to occur. Fluorescence typically requires much less motion of atoms. A protein environment that tightly constrains motion of chromophore atoms may therefore favor decay by fluorescence over structural change. In such a protein, the quantum yield for fluorescence may be increased and that for isomerization may be decreased, relative to a protein in which constraints are more relaxed. However, life is often not so simple. The spatial constraints may explicitly favor the motions required by a specific mode of isomerization such as hula-twist which requires a carefully coordinated motion of atoms in the chromophore and the protein, but disfavor the motions required by a different mode such as one-bond flip (Warshel 1976; Liu and Asato 1985). Although one-bond flip does not require as careful coordination, the chromophore atoms in motion sweep out a larger volume which of necessity displaces more protein atoms. Creating a large cavity is likely to traverse a larger energy barrier and thus to disfavor the specific path of isomerization by one-bond flip. The protein environment not merely favors a specific chemical process such as isomerization but directs the exact set of atomic motions and their time course by which that process is executed.

De-excitation processes after light absorption are irreversible, first order and occur over a time range from fs to a few ps. When conducting ultrafast experiments at an XFEL source, their presence is only important in the earliest time points. After electronic excitation, some motion of atoms and their associated electrons may occur both in the excited state prior to its decay and after return to the ground state via a conical intersection (Hosseinizadeh et al. 2021). Undesired structural changes such as those associated with fluorescence, stimulated emission, or vibrational decay will occur at the same time as the desired, biologically relevant structural change. The structural probe of course reveals the superposition of all structural changes present at a given time, both undesired and desired. The presence of undesired structural changes confuses identification of the desired change and its interpretation.

Of more serious consequence, a second pump photon may be promptly absorbed by a chromophore that has already absorbed one photon but remains in its first electronically excited state S_1. It has not yet decayed either by returning to the ground state or by passing through the conical intersection and along the functional trajectory to the next intermediate I_0 (Figure 3.3). Absorption of a second photon in this excited state is likely if the pump pulse is of very high intensity, and if its wavelength is such that the absorption coefficient of the S_1 state is also substantial. The intensities encountered in nature by light-sensitive proteins are much too weak to generate absorption of a second photon. The danger is thus that any structural changes associated with absorption of a second photon by an already excited chromophore are nonbiological. That is, structural changes induced by very high intensities of the pump pulse are likely to be artifactual (Hutchison et al. 2016).

The experimental conditions which might lead to absorption of a second photon remain a point of contention (Grünbein et al. 2020; Brändén and Neutze 2021). What characterizes a pump pulse as too intense? The high peak power (and peak intensity) of the pulse is the most damaging. For the same pulse energy, the peak power and intensity depend on the pulse duration and can become prohibitively high for ultrashort pulses, notably in the fs time range at XFEL sources. The difficulty that tempts experimenters is that the authentic structural signal is enhanced by increasing the intensity of the pump laser from a moderate level. More molecules in the sample are excited until saturation is reached when all chromophores have absorbed one photon. Due to absorption of the pump beam along its path, a laser intensity just sufficient to saturate absorption and maximize the structural signal at the entry face of the sample is too low to saturate at the exit face. Conversely, low intensity reduces the authentic signal, perhaps to the point at which the signal disappears in the noise and becomes undetectable. The absence of signal has consequences to be avoided: no results; no publication; no approval of a subsequent XFEL beamtime application; perhaps no funding of a key research grant, or delay of submission of a doctoral thesis. The clear experimental temptation is to increase the pump laser intensity to maximize crystallographic occupancy.

The first fs X-ray diffraction experiments (Barends et al. 2015; Pande et al. 2016; Nogly et al. 2018) used pump power densities of 350–500 GW cm^{-2}, corresponding at least nominally to many tens of photons absorbed per molecule. A careful review (Brändén and Neutze 2021) of almost all time-resolved studies at XFEL and SR sources assesses the (nominal) number of photons absorbed per molecule from each pump pulse. If real rather than nominal, substantial damage is likely from multiphoton absorption at these peak powers, and that damage should be clearly evident in the structural changes revealed in DED maps derived from such fs pulses. It is relatively straightforward to measure the average intensity, cross section, and pulse duration of the pump laser immediately upstream of the crystal and its environment. However, many uncertainties tend to reduce the real experimental intensities in the crystal itself. These intensities are necessary to estimate the number of photons actually absorbed per molecule throughout the crystal and to assess whether the observed structural changes are authentic or artifactual (Grünbein et al. 2020). True, the physical properties of a crystal such as its dimensions and optical properties such as the absorbance of its ground state and of its first and higher electronic excited states are well known. The problems are that the crystal is often delivered into the X-ray beam in a roughly cylindrical liquid jet (Chapter 4.2). The liquid acts as a lens of uncertain dimensions and shape that vary in time, and which both absorbs and scatters light. An individual crystal embedded in a liquid jet is of uncertain orientation, thickness, and cross-sectional area, and located in a laser beam of nonuniform transverse intensity. The intensity at the entry face of a crystal can only be roughly estimated.

A critical biological question is whether the DED maps show "changes that correlate with functional [biologically relevant] pathways and concur with biologically important structural changes on microsecond and millisecond time scales" ((Brändén and Neutze 2021) and references therein to the original DED maps). Or, do they show artifacts resulting from multiphoton absorption that persist into these time ranges? The authors conclude that the former holds, and that the structural changes observed are biologically relevant.

There is no doubt that absorption of a second pump photon could occur and if so, would be likely to cause damage on the fs time scale. (Such damage artifacts confused the interpretation of ultrafast spectroscopic experiments using the first fs pulsed lasers.) As yet, no features in DED maps have been identified as damage arising from the absorption of a second pump photon, on any system studied by fs time-resolved crystallography. It remains possible that any structural artifacts are – unexpectedly – rapidly reversible. Although generated on the fs time scale, they may decay very rapidly and are no longer detectable in DED maps at longer time points.

However, the problem cannot be ducked. A major goal of ultrafast light-driven experiments at XFELS is to reveal accurately the structural changes on the fs time scale of isomerization and electron transfer. The longer time scales are readily accessible at SR sources. The prudent course to ensure single photon absorption and avoid artifacts at XFELs is experimental. As an essential control experiment, a laser power titration should be carried out *in situ*. The titration should reveal a linear dependence of the X-ray signal on pump laser power from a typical sample, up to a measurable maximum power at which linearity begins to break down. Absorption of a second photon becomes appreciable and the dependence on laser power becomes quadratic or higher. The power titration should be repeated at intervals throughout the experiment to guard against drifts and misalignments which will invariably reduce or worse, eliminate the X-ray signal. This approach is easy to state but in practice unrealistic to execute fully. A statement such as "beamtime is just too precious to devote much of it to control experiments; we need to carry out the *real* experiment" may be heard at the XFEL. A personal example is detailed in Box 3B. Yes, until a referee of your subsequent manuscript claims – with force born of his or her own bitter experience – that most of your hard-won structural signal is an artifact arising from too high a pump intensity leading to absorption of a second pump photon, and that your biological interpretation of the ensuing structural changes is consequently wrong. (We have been both experimenters and referees.)

Box 3B

When Marius Schmidt and colleagues proposed their initial fs time-resolved crystallographic studies of the photocycle of PYP at an XFEL, they insisted on first conducting a control experiment. (The story is previewed in Moffat 2017.) This experiment would compare the nature and magnitude of structural signals at an XFEL (the LCLS) and an SR (the BioCARS beamline at the APS). Success would ensure that comparable data could be collected despite the very different characteristics of X-ray beams from SR and XFEL sources (Chapter 6), and establish an appropriate X-ray data collection design for the subsequent, main LCLS experiment. There was considerable opposition to their insistence from a reviewer and LCLS management, who asserted – correctly – that unlike the main experiment, the control experiment would not reveal anything new about the photocycle of PYP. It was therefore not worthy of precious LCLS beamtime. Vigorous discussion led to the award of beamtime to conduct the control experiment. This experiment convincingly demonstrated that the much higher X-ray laser power at the LCLS was not damaging. DED maps at 10 ns and 1 μs after initiation of the PYP photocycle by a ns pump laser pulse revealed structural signals in tiny single crystals of PYP which

---

(Continued)

Box 3B (Continued)

were qualitatively comparable with, or even quantitatively superior to, those obtained earlier at BioCARS (Figure 3.4; (Tenboer et al. 2014)). The ns pump laser used in both the BioCARS and LCLS experiments obtained a much greater fraction of photoinitiation (~40%) at the LCLS than at BioCARS (~10%) – a promising result. The LCLS crystals were much smaller and hence more optically transparent than BioCARS crystals.

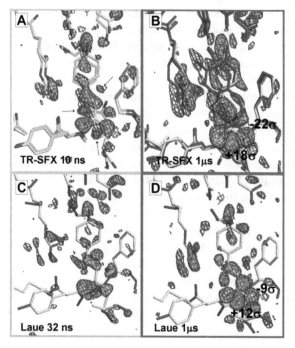

Figure 3.4 Comparison of time-resolved DED maps of the chromophore of PYP and its pocket, obtained at the LCLS/XFEL and BioCARS/APS. DED maps at 1.6 A resolution obtained at the LCLS (panel A at 10 ns, panel B at 1 µs). DED maps obtained at BioCARS (panel C at 32 ns time point, panel D at1 µs). Maps contoured in panels A, B, and D red/white −3 σ/−4σ and blue/cyan +3σ/+4 σ; and in panel C red/white −3σ/−4 σ and blue/cyan +3σ/+7σ. (σ is the rms DED across the map.) Note similarity of features at LCLS with the corresponding time points at BioCARS; compare panel A with C, and panel B with D. However, indicated features at LCLS (+18σ and −22σ) are much more significant than the corresponding features at BioCARS (+12σ and −9σ). Tenboer et al. 2014/Reproduced with permission from American Association for the Advancement of Science.

Emboldened by these encouraging results, the main experiment was then awarded more beamtime at the LCLS to explore the fs to 100 ps structural changes in the photocycle of PYP. Achieving near-fs time resolution required a fs pump laser of much higher and potentially damaging peak power, with probe X-ray pulses of ~150 fs duration. However, the resultant DED maps were again of excellent quality. Highly significant, chemically plausible structural signals in the fs to 3 ps time range arose from isomerization of the p-cinnamic acid chromophore of PYP as it converted from the dark (trans) to the light (cis) configuration (Pande et al. 2016). These signals directly revealed, for the

Box 3B (Continued)

first time, passage through a conical intersection and the structural basis of isomerization. The results could only be obtained with ~150 fs X-ray pulses at an XFEL such as
the LCLS and were completely inaccessible to the much longer, 100 ps X-ray pulses at
BioCARS. Much later, reanalysis of these LCLS data by novel mathematical approaches
(Fung et al. 2016; Hosseinizadeh et al. 2021) greatly improved the time resolution to a
few tens of fs and visualized, again for the first time and with much more structural and
temporal detail, coherent passage through the conical intersection between the electronically excited and ground states. See Chapter 7.7 for discussion of a related example.

3.3.2.3.2 *Practical considerations*

In designing a light-initiated reaction, the user must select several variables: the *wavelength
of the light* that serves as the pump, its intensity, duration and whether this is a single pulse
or a pulse train, the quantum yield per photon absorbed for the desired structural reaction
and undesired side reactions, and the sample size. The choice of wavelength depends on
three main considerations. First, absorbance (the probability that a photon will be absorbed)
is given by the product of the concentration of the sample, its extinction coefficient, and the
expected path length. This product should be less than roughly 0.5 to minimize the longitudinal gradient of intensity as the pump beam traverses the sample. Second, the photochemistry expected at that wavelength must be optimized. The quantum yield of the
desired reaction must be as large as possible. However, values much lower than 1.0 are
both common and acceptable if small differences in structural signals can be confidently
identified. Although the quantum yield is typically independent of wavelength for a single
electronic transition, this is not always the case. For example, it may not be identical at the
blue and red edges of a single absorption peak or at two distinct absorption peaks, even
though the extinction coefficient is the same at these wavelengths. Third, if at all possible
the wavelength must be chosen such that the extinction coefficient of the first electronically excited state resulting from the absorption of one photon is less – preferably much
less – than that of the ground state. If so, the beam bleaches the sample as absorption pro-
gresses during the pump pulse. The mean extinction coefficient of the sample drops as the
pulse proceeds, which minimizes the longitudinal gradient of intensity and the probability
of absorbing a second photon.

The choice of *pump characteristics*: pulse duration, single pulse or pulse train with multiple pulses, and pulse intensity is also important. As discussed above, intensity must be high
enough to generate sufficient signal but low enough to avoid excitation by two X-ray photons. The temporal resolution desired determines whether a single pulse or a pulse train is
used. For XFEL sources, the duration of a single pump pulse is typically in the fs range, with
the exact value depending on the properties of the fs laser available at the XFEL beamline.
Since all experiments are destructive, pump pulse trains are not used at XFELs. For SR
sources where the desired temporal resolution ranges from 100 ps to a few ns, a pump laser
with a single pulse duration in the few ps range is adequate. For temporal resolutions longer
than a few ns, a simpler, 10–20 Hz repetition rate, ns pump laser is adequate, from which a
single pulse can be selected by a fast shutter. In all cases, the routine availability of the laser
at the beamline and its reliability are important. The larger the sample size, the stronger its

X-ray scattering, the larger the absolute change in scattering as the desired structural changes occur, and the greater the possibility (with an SR source) to obtain more than one scattering pattern or scattering curve per sample. However, the smaller the sample, the lower its absorbance and the easier it is to achieve spatial near uniformity in absorption and completeness of the reaction throughout the volume of the crystal. That is, small samples offer a larger occupancy of the structural changes and a larger fractional change in structure amplitudes. Thus, sample size offers a distinct tradeoff which varies from sample to sample. Admittedly limited experience (Tenboer et al. 2014; Pande et al. 2016) suggests that smaller is better. The reduced X-ray signal from a smaller sample is offset by the higher extent of reaction initiation in the entire, smaller sample volume. A higher extent of reaction initiation means that the intensity of each spot contains a larger fraction of the desired, excited state and a lower fraction of the background, dark state. Shape and habit of the sample crystal can also play a role. A crystal may be very optically anisotropic, with an extinction coefficient in plane-polarized light at a given wavelength that varies strongly with the orientation of the pump beam relative to the crystal axes. This can be exploited if the crystals are in a fixed and known orientation with respect to the pump beam. (As an example, Figure 5.6 shows the strongly optically anisotropic absorbance of a PYP crystal (Ng et al 1995)). Crystal habits of plate-like or needle form have one or two short dimensions along which the pump pulse may be directed if control of the crystal orientation is possible. Two or one long dimensions are retained to maintain the volume illuminated by the X-ray probe.

References

Anderson, S., Srajer, V. Pahl, R. et al. (2004). Chromophore conformation and the evolution of tertiary structural changes in photoactive yellow protein. *Structure* 12 (6): 1039–1045.

Ansari, A., Berendzen, J. Bowne, S.F. et al. (1985). Protein states and proteinquakes. *Proceedings of the National Academy of Sciences* 82 (15): 5000–5004.

Barends, T.R., Foucar, L., Ardevol, A. et al. (2015). Direct observation of ultrafast collective motions in CO myoglobin upon ligand dissociation. *Science* 350 (6259): 445–450.

Bar-Even, A., Noor, E., Savir, Y. et al. (2011). The moderately efficient enzyme: evolutionary and physicochemical trends shaping enzyme parameters. *Biochemistry* 50 (21): 4402–4410.

Bower, J.L. and Christensen, C.M. (1995). Disruptive technologies: catching the wave. *The Journal of Product Innovation Management* 1 (13): 75–76.

Bränden, G. and Neutze, R. (2021). Advances and challenges in time-resolved macromolecular crystallography. *Science* 373 (6558): eaba0954.

Chapman, H.N., Fromme, P., Barty, A. et al. (2011). Femtosecond X-ray protein nanocrystallography. *Nature* 470 (7332): 73–77.

Chen, Y., Tokuda, J.M., Topping, T. et al. (2014). Revealing transient structures of nucleosomes as DNA unwinds. *Nucleic Acids Research* 42 (13): 8767–8776.

DePonte, D.P., Weierstall, U., Schmidt, K. et al. (2008). Gas dynamic virtual nozzle for generation of microscopic droplet streams. *Journal of Physics D-Applied Physics* 41 (19).

Donatelli, J.J., Zwart, P.H., and Sethian, J.A. (2015). Iterative phasing for fluctuation X-ray scattering. *Proceedings of the National Academy of Sciences* 112 (33): 10286–10291.

Frank, J. and Ourmazd, A. (2016). Continuous changes in structure mapped by manifold embedding of single-particle data in cryo-EM. *Methods* 100: 61–67.

Fung, R., Hanna, A.M. Vendrell, O. et al. (2016). Dynamics from noisy data with extreme timing uncertainty. *Nature* 532 (7600): 471–475.

Grünbein, M.L., Stricker, M., Kovacs, G.N. et al. (2020). Illumination guidelines for ultrafast pump–probe experiments by serial femtosecond crystallography. *Nature Methods* 17 (7): 681–684.

Halavaty, A.S. and Moffat, K. (2007). N-and C-terminal flanking regions modulate light-induced signal transduction in the LOV2 domain of the blue light sensor phototropin 1 from Avena sativa. *Biochemistry* 46 (49): 14001–14009.

Hekstra, D.R., White, K.I., Socolich, M.A. et al. (2016). Electric-field-stimulated protein mechanics. *Nature* 540 (7633): 400–405.

Hosseinizadeh, A., Breckwoldt, N., Fung, R. et al. (2021). Few-fs resolution of a photoactive protein traversing a conical intersection. *Nature* 599: 697–701.

Hutchison, C.D., Kaucikas, M., Tenboer, J. et al. (2016). Photocycle populations with femtosecond excitation of crystalline photoactive yellow protein. *Chemical Physics Letters* 654: 63–71.

Ihee, H., Rajagopal, S., Šrajer, V. et al. (2005). Visualizing reaction pathways in photoactive yellow protein from nanoseconds to seconds. *Proceedings of the National Academy of Sciences* 102 (20): 7145–7150.

Kam, Z. (1977). Determination of macromolecular structure in solution by spatial correlation of scattering fluctuations. *Macromolecules* 10 (5): 927–934.

Konold, P.E., Arik, E., Weißenborn, J. et al. (2020). Confinement in crystal lattice alters entire photocycle pathway of the photoactive yellow protein. *Nature Communications* 11 (1): 1–12.

Kupitz, C., Olmos, J.L., Jr, Holl, M. et al. (2017). Structural enzymology using X-ray free electron lasers. *Structural Dynamics* 4 (4): 044003.

Kupitz, C. and Schmidt, M. (2018). Towards molecular movies of enzymes. In: *X-ray Free Electron Lasers* (ed. S. Boutet, P. Fromme and M.S. Hunter), 357–376. Springer.

Liu, R. and Asato, A.E. (1985). The primary process of vision and the structure of bathorhodopsin: a mechanism for photoisomerization of polyenes. *Proceedings of the National Academy of Sciences* 82 (2): 259–263.

Martin, J.-L. and Vos, M.H. (1992). Femtosecond biology. *Annual Review of Biophysics and Biomolecular Structure* 21 (1): 199–222.

Meents, A., Wiedorn, M., Srajer, V. et al. (2017). Pink-beam serial crystallography. *Nature Communications* 8 (1): 1–12.

Meier, A., Nelson, R., and Connaughton, V.P. (2018). Color processing in zebrafish retina. *Frontiers in Cellular Neuroscience* 12: 327.

Moffat, K. (2014). Time-resolved crystallography and protein design: signalling photoreceptors and optogenetics. *Philosophical Transactions of the Royal Society B: Biological Sciences* 369 (1647): 20130568.

Moffat, K. (2017). Small crystals, fast dynamics and noisy data are indeed beautiful. *I.U.Cr.J.* 4 (4): 303–305.

Moffat, K., Zhang, F. Hahn, K. et al. (2013). The biophysics and engineering of signaling photoreceptors.In: *Optogenetics*, (P. Hegemann and S. Sigrist), 7–22. Berlin: De Gruyter.

Möglich, A. and Moffat, K. (2010). Engineered photoreceptors as novel optogenetic tools. *Photochemical and Photobiological Sciences* 9 (10): 1286–1300.

Möglich, A., Yang, X. Ayers, R.A. et al. (2010). Structure and function of plant photoreceptors. *Annual Review of Plant Biology* 61: 21–47.

Ng, K., Getzoff, E.D., and Moffat, K. (1995). Optical studies of a bacterial photoreceptor protein, photoactive yellow protein, in single crystals. *Biochemistry* 34 (3): 879–890.

Nogly, P., Weinert, T., James, D. et al. (2018). Retinal isomerization in bacteriorhodopsin captured by a femtosecond x-ray laser. *Science* 361 (6398).

Pande, K., Hutchison, C.D., Groenhof, G. et al. (2016). Femtosecond structural dynamics drives the trans/cis isomerization in photoactive yellow protein. *Science* 352 (6286): 725–729.

Pande, K., Schmidt, M., Schwander, P. et al. (2015). Simulations on time-resolved structure determination of uncrystallized biomolecules in the presence of shot noise. *Structural Dynamics* 2 (2): 024103.

Pandey, S., Calvey, G., Katz, A.M. et al. (2021). Observation of substrate diffusion and ligand binding in enzyme crystals using high-repetition-rate mix-and-inject serial crystallography. *IUCrJ* 8 (6).

Park, H.Y., Qiu, X., Rhoades, E. et al. (2006). Achieving uniform mixing in a microfluidic device: hydrodynamic focusing prior to mixing. *Analytical Chemistry* 78 (13): 4465–4473.

Pearson, A.R. and Mehrabi, P. (2020). Serial synchrotron crystallography for time-resolved structural biology. *Current Opinion in Structural Biology* 65: 168–174.

Perman, B., Srajer, V., Ren, Z. et al. (1998). Energy transduction on the nanosecond time scale: early structural events in a xanthopsin photocycle. *Science* 279 (5358): 1946–1950.

Ren, Z., Bourgeois, D., Helliwell, J.R. et al. (1999). Laue crystallography: coming of age. *Journal of Synchrotron Radiation* 6 (4): 891–917.

Ren, Z., Perman, B., Šrajer, V. et al. (2001). A molecular movie at 1.8 Å resolution displays the photocycle of photoactive yellow protein, a eubacterial blue-light receptor, from nanoseconds to seconds. *Biochemistry* 40 (46): 13788–13801.

Rimmerman, D., Leshchev, D., Hsu, D.J. et al. (2018). Probing cytochrome c folding transitions upon phototriggered environmental perturbations using time-resolved X-ray scattering. *Journal of Physical Chemistry B* 122 (20): 5218–5224.

Roder, H., Maki, K., and Cheng, H. (2006). Early events in protein folding explored by rapid mixing methods. *Chemical Reviews* 106 (5): 1836–1861.

Rupp, B. (2009). *Biomolecular Crystallography: Principles, Practice, and Application to Structural Biology*. Garland Science.

Sakai, K., Tsutsui, K., Yamashita, T. et al. (2017). Drosophila melanogaster rhodopsin Rh7 is a UV-to-visible light sensor with an extraordinarily broad absorption spectrum. *Scientific Reports* 7 (1): 1–11.

Schmidt, M. (2013). Mix and inject: reaction initiation by diffusion for time-resolved macromolecular crystallography. *Advances in Condensed Matter Physics* 2013: 10.

Serganov, A., Huang, L., and Patel, D.J. (2009). Coenzyme recognition and gene regulation by a flavin mononucleotide riboswitch. *Nature* 458 (7235): 233–237.

Spence, J.C. (2017). *Imaging Protein Dynamics by XFELs. X-Ray Free Electron Lasers: Applications in Materials, Chemistry and Biology* (U. Bergman, V. Yachandra, and J. Yano). 47–69. London: Royal Society of Chemistry.

Srajer, V., Teng, T.Y., Ursby, T. et al. (1996). Photolysis of the carbon monoxide complex of myoglobin: nanosecond time-resolved crystallography. *Science* 274 (5293): 1726–1729.

Stagno, J.R., Liu, Y., Bhandari, Y.R. et al. (2017). Structures of riboswitch RNA reaction states by mix-and-inject XFEL serial crystallography. *Nature* 541 (7636): 242–246.

Svergun, D.I., Koch, M.H., Timmins, P.A. et al. (2013). *Small Angle X-ray and Neutron Scattering from Solutions of Biological Macromolecules*. Oxford University Press.

Tenboer, J., Basu, S., Zatsepin, N. et al. (2014). Time-resolved serial crystallography captures high-resolution intermediates of photoactive yellow protein. *Science* 346 (6214): 1242–1246.

Thompson, M.C., Barad, B.A., Wolff, A.M. et al. (2019). Temperature-jump solution X-ray scattering reveals distinct motions in a dynamic enzyme. *Nature Chemistry* 11 (11): 1058–1066.

Toh, K., Stojković, E.A., van Stokkum, I.H. et al. (2010). Proton-transfer and hydrogen-bond interactions determine fluorescence quantum yield and photochemical efficiency of bacteriophytochrome. *Proceedings of the National Academy of Sciences* 107 (20): 9170–9175.

Warshel, A. (1976). Bicycle-pedal model for the first step in the vision process. *Nature* 260 (5553): 679–683.

Weininger, U., Modig, K., and Akke, M. (2014). Ring flips revisited: 13C relaxation dispersion measurements of aromatic side chain dynamics and activation barriers in basic pancreatic trypsin inhibitor. *Biochemistry* 53 (28): 4519–4525.

Wickiser, J.K., Winkler, W.C., Breaker, R.R. et al. (2005). The speed of RNA transcription and metabolite binding kinetics operate an FMN riboswitch. *Molecular Cell* 18 (1): 49–60.

Wickstrand, C., Dods, R., Royant, A. et al. (2015). Bacteriorhodopsin: would the real structural intermediates please stand up? *Biochimica et Biophysica Acta (Bba)-general Subjects* 1850 (3): 536–553.

Wilt, H.M., Yu, P., Tan, K. et al. (2020). FMN riboswitch aptamer symmetry facilitates conformational switching through mutually exclusive coaxial stacking configurations. *Journal of Structural Biology: X* 4: 100035.

Yeremenko, S., Van Stokkum, I.H., Moffat, K. et al. (2006). Influence of the crystalline state on photoinduced dynamics of photoactive yellow protein studied by ultraviolet-visible transient absorption spectroscopy. *Biophysical Journal* 90 (11): 4224–4235.

4

The Sample

4.1 Crystal and Solution Samples

An X-ray-based structural dynamics experiment can be conducted on a sample such as a single crystal, a solution, or a single particle of an isolated protein or nucleic acid, protein-nucleic acid complex or protein complex. The last is known as single particle imaging (SPI). The ability to conduct SPI minimizes the necessity for large-scale purification and avoids the uncertain challenges of crystallization. SPI formed a compelling set of experimental proposals in the scientific case that ultimately led to construction of the LCLS at the Stanford Linear Accelerator Center (Bogan et al. 2008). SPI was envisaged as the "killer app" in structural biology. However, initial SPI experiments at the LCLS proved challenging. Even from very large single viruses, the inevitably weak sample scattering and high background scattering lowered the signal-to-noise and limited the structural resolution (Seibert et al. 2011). Since that date, SPI by cryo-EM has undergone extremely rapid development that at present has far outpaced X-ray-based approaches to SPI. We consider cryo-EM in more detail below (Chapter 8.2), and potential advantages of X-ray-based SPI in Chapter 9.2. We concentrate here on single crystal and solution samples.

For studies in structural dynamics, the essential requirement is that samples must be demonstrably active under the conditions necessary to introduce them into the X-ray beam (Chapter 1). A second requirement is that the small, time-dependent changes in X-ray scattering must not be masked by systematic errors inherent even in time-independent, static experiments (Chapter 3.2.1). Since hard X-rays interact only weakly with electrons (Chapter 4.3.4), the final samples must be in high concentration, often at the few mM or tens of mM level. High purity and the ability to attain high concentration are prerequisites for both crystallization and solution scattering. For solution scattering, gel filtration is normally applied immediately before loading the sample to remove larger aggregates, contaminants, and solvation changes. If present, these would affect the SAXS measurements and their ability to detect changes in oligomerization or large-scale tertiary and quaternary structural changes. WAXS is less sensitive to these contamination effects and offers higher resolution, but its low signal can be difficult to interpret in isolation.

The usual first products of crystallization attempts are microcrystals. Much time and effort are put into enlarging their size to the point where they reproducibly yield diffraction

Dynamics and Kinetics in Structural Biology: Unravelling Function Through Time-Resolved Structural Analysis,
First Edition. Keith Moffat and Eaton E. Lattman.
© 2024 John Wiley & Sons Ltd. Published 2024 by John Wiley & Sons Ltd.

patterns of good quality. That size depends substantially on the X-ray intensity at the crystal, which is determined by the source (SR or XFEL; Chapter 6), the diffraction technique (monochromatic or Laue; Chapters 5.1.1 and 5.1.2), and the properties of the beamline that delivers X-rays from the source to the crystal (Chapter 6). We consider here another important factor, the scattering power of the crystals.

The total energy E reflected by a stationary crystal of total volume V_{cryst} as it generates a Laue reflection of structure amplitude $|F(hkl)|$ with an exposure time Δt_L is

$$E = CI'(k)\lambda^2 \left(V_{cryst}/V^2\right) F\left(hkl\right) \Delta t_L$$

where $I'(k)$ is the incident spectral intensity dI/dk, $k = 1/\lambda$ where λ is the X-ray wavelength that stimulates hkl, V is the unit cell volume, and C is a constant (Kalman 1979). By equating the total energy in a Laue reflection with the corresponding monochromatic energy under comparable conditions, the relative exposure times of Laue and mono experiments may be estimated (Moffat 1997). The value of the intensity $|F(hkl)|^2$ of a typical reflection with Bragg indices hkl may be approximated by $\sum_j f_j^2$, where f_j is the scattering power of the j-th atom and the summation is taken over all atoms in the unit cell. The number of unit cells in the crystal is $N_{cell} = V_{cryst}/V$; for proteins $\sum_j f_j^2 \sim 3.2Mn$, where M is the molecular weight of the protein and n is the number of molecules per unit cell; and the Matthews parameter $V_M = V / Mn; Mn = V/V_M$. Hence

$$E = 3.2 \, C \, I'(k) \text{ x (sample scattering power)},$$

where the sample scattering power S is given by

$$S = \lambda^2 \left(N_{cell}/V_M\right).$$

S is dimensionless and may be used to compare the scattering power of crystals of different proteins, sizes, and Matthews parameters (Moffat et al. 1986).

As a practical matter, the quality of a diffraction pattern depends on the resolution desired and on the scattering power – the signal – and on the background underlying the spots – the noise. Together, these control the accuracy with which individual intensities and structure amplitudes can be measured. For dynamic experiments, the required quality depends on the anticipated magnitude of the time-dependent changes in signal (Holton and Frankel 2010) relative to the magnitude of errors in the structure amplitudes. Finally, when reaction initiation by a pump laser pulse is planned, the optical properties of the crystal must also be considered, such as its absorbance at the laser wavelength and the laser intensity (Chapter 3.3.2.3.2).

Together with the crystal scattering power, these constitute a daunting list of relevant sample parameters, not least because only rough estimates of the sample, X-ray, beamline, and optical factors can be made in advance. Prior to full-length experiments, offline optical experiments and initial small-scale X-ray experiments at a candidate beamline are therefore highly desirable and often essential. In these experiments, crystal size, quality, and limiting resolution are evaluated, the X-ray exposure time is varied, radiation damage is explored, and the appropriate reaction initiation or laser setup is tested. These planning experiments lead to firmer estimates of the parameters critical to subsequent full-length experiments. Realistic estimates greatly enhance the likelihood of later success. Recognizing their benefits, SR and XFEL beamlines routinely support brief beamtime requests to conduct planning experiments, usually under a special access program known as "Protein Crystal (or Solution) Screening." Success helps the user, the beamline staff, and the facility. We cannot overestimate the importance of planning experiments.

One aspect of crystal selection for dynamic experiments is novel. Many, perhaps most, proteins and nucleic acids are polymorphic. Crystals can be obtained under different solution conditions in a wide variety of space groups, size, and external morphologies, with varying ease of reproducible crystallization and extent of biological activity. For *de novo* static structure determination, a major reason for selecting a particular space group is that it yields the highest resolution from readily available, reproducible crystals. Such crystals usually contain only one molecule per asymmetric unit in a space group with relatively small cell dimensions, and thus require the minimum amount of scattering data to the desired resolution. For dynamic experiments, an unusual but significant advantage may be gained by selecting a crystal with more than one molecule per asymmetric unit (always provided that the molecules retain biological activity). That is, the crystal possesses non-crystallographic symmetry. The advantage arises because chemically identical molecules in the asymmetric unit differ in the structural constraints that arise from packing interactions between neighbors (Ren et al. 2012). If the constraints are weak, the overall mechanism is likely to proceed via the same intermediates for each molecule, though the rate coefficients for each elementary step in the mechanism may differ between molecules. That is, the mechanism differs quantitatively for each molecule. If so, each molecule in the asymmetric unit would proceed at a different rate down the reaction coordinate. A consequence is that although a diffraction pattern is recorded at a single physical time point, the pattern arises from different reaction time points, one time point for each molecule in the asymmetric unit. However, each molecule is from the same protein preparation, examined under identical solvent conditions at the same temperature, with the same dose and radiation damage and the same X-ray pulse. Systematic errors from these sources in the time-dependent structural changes can be significant but under these circumstances, such errors are absent. The advantage offered in the quality and interpretability of kinetic data may outweigh the disadvantage of a larger asymmetric unit that requires the collection of more diffraction data. Systematic errors from other sources of course remain. In particular, the packing constraints may be sufficiently large to produce a qualitatively (not merely quantitatively) different mechanism in each molecule. In an extreme case, some molecules in the asymmetric unit may be active, others inactive.

For crystallographic experiments in structural dynamics, the more (active) molecules in the asymmetric unit, the better.

4.2 Introduction of the Sample to the X-ray Beam: Injection and Fixed Targets

At SR sources, techniques to introduce a single crystal or liquid sample to the X-ray beam for dynamic experiments are similar to those for static experiments. Radiation damage (Chapter 4.3) must be taken into account but even at room temperature, damage is generally not severe enough to preclude collection of several excellent scattering patterns from a single sample, and X-ray exposures are usually long enough to include many X-ray pulses (Chapter 6.2 discusses the pulse structure of SR sources). The sample delivery system for dynamic experiments must also incorporate reaction initiation e.g. via a flow mixing device or laser illumination (Chapter 3.3). For example, crystals may be held in a loop at suitable humidity and temperature where they can readily be subjected to a laser pulse (Teng 1990). A liquid may be contained in a flow cell such as a thin-walled quartz capillary, or as a tiny droplet on a conveyor system (Roessler et al. 2013; Martiel et al. 2019). Capillaries may also be used to flow crystals through the X-ray beam (Stellato et al. 2014; Nam 2020b).

At XFEL sources (Chapter 6.3), each tiny crystal or liquid volume yields only a single scattering pattern before it is destroyed by a single, extremely intense fs X-ray pulse (diffraction before destruction, Chapter 4.3.3), and must be immediately replaced by rapidly introducing serial samples to the X-ray beam. Repeated sample replacement is unavoidable and experimentally challenging. The overall experimental approach is known as serial femtosecond crystallography (SFX) or serial solution X-ray scattering (SSX). See (Sierra et al. 2018) for an excellent introduction to SFX sample delivery methods. Serial crystallography (SX) and SSX are also being rapidly adapted to SR sources (Pearson and Mehrabi 2020). Approaches for both XFEL and SR sources have been comprehensively reviewed (Grünbein and Nass Kovacs 2019; Martiel et al. 2019).

Serial approaches at an XFEL source are subject to severe constraints. Research aimed at most nearly satisfying the constraints remains active. Perfection is elusive and may well be unattainable.

Constraints take the following forms.

1) Samples are valuable, limited in amount and must not be wasted. The goal is to illuminate each tiny crystal or liquid sample by one X-ray pulse.
2) XFEL sources are difficult to access and beamtime is precious; it must be efficiently used. XFEL pulses occur at high frequency (120 Hz at the LCLS, and up to MHz at the EuXFEL). The goal is to ensure that most X-ray pulses illuminate a sample. For crystal samples, the fraction of pulses that do so is known as the hit rate.
3) Tiny crystalline and liquid samples scatter X-rays weakly. The diameter of the X-ray beam must match the cross-sectional size of the sample, while allowing for jitter in the position of both. A larger X-ray beam will increase the background from the liquid that surrounds the sample. Other background scattering from the sample mount and from beam paths upstream and downstream of the sample must be minimized to avoid masking the weak scattering from the sample and reducing the effective spatial resolution.
4) The sample must be undamaged when the X-ray pulse arrives. Samples are fragile and prone to artifactual structural changes and disorder. Manipulation prior to and during injection must be minimized. In particular, samples must not be subjected to large hydrodynamic shear forces, unplanned electric fields, shock waves arising from destruction of the prior sample by the preceding X-ray pulse, or excessive dehydration or cooling from injection into a vacuum.
5) All processes must operate reliably over many hours without serious interruption from, for example, clogging of an aperture. This low-tech constraint turns out to be challenging to meet.

Two basic approaches to serial sample introduction have been used: delivering the sample as a liquid jet directed across the X-ray beam (DePonte 2017) or mounting the sample on a fixed target that can be physically scanned across the beam. In a liquid jet, the liquid itself may be the sample as in SSX, or the liquid may act as the carrier for microcrystals as in SFX and SX. Motion of the liquid continuously delivers fresh sample to the X-ray beam. In a fixed target, the sample is preloaded in a known and fixed location on a mount such as a microwell in a 2D array on a chip, or in a 1D array of droplets on a conveyor belt. The mount or belt is then moved across the X-ray beam, to illuminate the crystal in each well or droplet in turn (Grünbein and Nass Kovacs 2019). Alternatively, crystals fall randomly on the chip, on a simple mesh or even on a thin Mylar foil which is then covered with a second foil to form a closed crystal sandwich. A single crystal can be loaded in each microwell but with random crystal distributions, crystals may clump and give rise to superimposed diffraction patterns. Another fixed target approach is designed to minimize crystal handling and improve sample economy (Ren et al. 2020).

A liquid pressurized through an aperture will emerge as a continuous, cylindrical liquid jet which initially has the diameter of the aperture. Although the jet eventually breaks up into tiny droplets, most experiments arrange the X-ray beam to fall on the continuous jet. The diameter of a tightly focused X-ray beam can be as small as 0.1 μm but more typically is 1–5 μm. To deliver crystals of sizes in the range of a few μm, the diameter of the jet must be minimized to reduce the X-ray background to satisfy constraint 3. However, a physical aperture of 1–10 μm will rapidly clog and violate constraint 5. To reduce or with luck avoid clogging, the liquid jet is first constrained by a less clogging-prone physical aperture to about 50 μm diameter, then further reduced to about 5 μm (satisfying constraints 3 and 5) by surrounding the jet with a coaxial gas stream. This assembly forms a gas dynamic virtual nozzle (GDVN), in which the liquid velocity in the constricted jet is accelerated to around 10 m s^{-1} at a flow rate of 10 μL s^{-1}. (See Sierra et al. 2018).

The GDVN has been widely and successfully used for SFX, particularly at the LCLS. However, it has serious limitations. With an X-ray pulse duration of 100 fs and a pulse frequency of 120 Hz at the LCLS, X-rays are only present for 12 ps s^{-1}. No X-rays are present for the vast majority of the time that the jet is flowing. Most crystals are not illuminated by X-rays, which seriously violates constraint 1. Clearly, a higher pulse frequency such as that from the EuXFEL in the MHz range (with a complicated bunch structure within each pulse train; Chapter 6.3) will illuminate more crystals per sec.

At low crystal concentrations in the jet, most X-ray pulses will not hit a crystal, thus violating constraint 2. If the crystal concentration in the jet is raised to increase the hit rate, the probability that two or more crystals will be in the beam simultaneously also increases. However, the resulting superposition of individual diffraction patterns from multiple crystals, each in a random orientation, is more challenging to index and if unsuccessful, the effective hit rate drops. Progress on indexing multiple crystal patterns is being made in systems such as CrystFEL (White 2019) and pinkIndexer (Hadian-Jazi et al. 2021). Thus, fundamental aspects of the GDVN limit its efficiency in using both samples and X-ray pulses and fail to satisfy constraints 1 and 2 by a large margin. Initial, notably successful applications of the GDVN have largely been to proteins such as myoglobin and PYP that can readily be purified and crystallized in very large quantities. Inefficiency can be tolerated in the use of such crystals and X-ray pulses.

Integral membrane proteins such as G-protein coupled receptors (GPCRs) are major drug targets of great biological interest (Stauch and Cherezov 2018) but are challenging to purify, limited in amount, and often crystallized from a lipidic cubic phase (LCP) medium, a membrane mimetic (Zabara et al. 2017). LCPs have a viscosity resembling that of toothpaste, much higher than that of aqueous solutions. Their viscosity greatly reduces the flow rate of a jet in which crystals are suspended (Weierstall et al. 2014). Gas focusing is not required and even with a larger jet diameter of 20–75 μm, flow rates as low as 2 mm s^{-1} can be satisfactory. However, the large diameter increases the background X-ray scattering. Not all samples are compatible with LCPs, and other high viscosity media such as agarose and various greases have been introduced (Conrad et al. 2015; Nam 2020a).

Jet approaches remain under active development but the problem of inefficient use of crystals (constraint 1) will remain. These could be mitigated if introduction of each crystal could be synchronized with an X-ray pulse. Synchronization can be envisaged with a viscous medium with a low flow rate in which the position of each crystal in the jet could in principle be determined prior to injection but is much harder to achieve with a low viscosity medium at high flow rate. The remaining problem of inefficient use of X-ray pulses (constraint 2) seems less serious with the advent of many more XFEL sources of high X-ray pulse frequency. We acknowledge that this viewpoint may simply represent the bias of structural biologists rather than accelerator physicists.

Fixed target injection approaches, largely applicable to crystalline samples, attack constraint 1 offline. By preloading tiny crystals in individual microwells on a chip, adjusting the amount of liquid around the crystal, and then sealing the microwell with X-ray-transparent windows, the position of each crystal in the well relative to fiduciaries in the chip mount can be accurately imaged (Frank et al. 2014; Shelby et al. 2020). The crystals must be attached to a surface of the microwell to ensure their position does not change prior to illumination by the X-ray beam, which is of course delivered horizontally. With their positions relative to the fiduciaries accurately known, the mount can be scanned across the X-ray beam to illuminate the crystal in each well only when it is correctly positioned, thus fully satisfying constraint 1. Chips can be delivered sequentially. Considerable progress is being made in crystallizing proteins *in situ* in an array of microwells that can be mounted on a goniometer and scanned across the X-ray beam (Mueller et al. 2015; Cheng et al. 2020).

A limitation of the sealed microwell approach is that it allows reaction initiation only by illumination with light. Another very promising fixed target approach known as droplet on demand (Orville 2017) overcomes this by preloading tiny droplets of protein solution alone (for solution scattering) or solution plus tiny crystal (for crystallography) at known positions on an X-ray-transparent conveyor tape which delivers them in a controlled environment to the X-ray beam. Prior to illumination, a second droplet of a reactant can be added to the first, to initiate the desired reaction by rapid mixing. Alternatively, droplets on the tape can be illuminated by a laser pulse or pulses at a known time or times before introduction to the X-ray beam, or for very short delay times, as the droplet enters the X-ray beam. The droplets are placed on the conveyor belt by acoustic droplet ejection. Ejection describes the process of droplet formation at a liquid meniscus by an acoustic wave and propagation of the droplet into the gas above or below the liquid. This overall approach is flexible and supports several approaches to reaction initiation. By ensuring that droplets are only formed and delivered to the X-ray beam on demand, the approach also fully satisfies constraint 1 (Fuller et al. 2017).

As structural dynamics develops, the most biologically interesting questions are likely to require samples available only in very limited amounts. This places major emphasis on constraint 1. Sample delivery must be compatible with flexible approaches to reaction initiation. The goal is to obtain an excellent, indexable diffraction pattern from a large majority of crystals, or a good SAXS/WAXS scattering pattern from a concentrated solution. Development of novel sample delivery methods remains a very active research area.

4.3 Radiation Damage

4.3.1 Introduction

X-ray absorption and the generation of radiation damage in macromolecular crystals have been evident from the earliest days of protein crystallography (Blake and Phillips 1962) and continue to have a major impact on the design, conduct, and interpretation of static and dynamic experiments in crystals and solution. The extent of radiation damage varies strongly with temperature, crystal size, X-ray energy, intensity, and pulse length. Its origins and approaches to minimization at SR and XFEL sources are discussed in detailed reviews (Owen et al. 2006; Holton 2009; Holton and Frankel 2010; Chapman et al. 2014; Garman and Weik 2019; de la Mora et al. 2020). In solution scattering, radiation damage manifests rather differently than in crystallography. Modification of surface charges alters interparticle interactions and promotes aggregation, clearly evident in SAXS experiments.

Radiation damage differs in two major aspects in dynamic experiments from static experiments. First, data from the entire dynamic experiment is more extensive since it spans four variables in dynamic crystallography (*hkl, t*) and two variables in dynamic solution scattering (the scattering vector q (Chapter 5.2) and t). With careful design, complete diffraction data in a static experiment can be collected from one crystal at an SR source, with acceptable radiation damage. This is not possible in a dynamic experiment when the complete diffraction data is much larger. A time series with, say, 20 time points spanning a reaction is required, along with complete coverage of the unique volume in reciprocal space with suitable redundancy. To assemble the complete data set, multiple crystals are needed and the data from each must be scaled together.

Second, of critical importance and unavoidably, a dynamic experiment by definition seeks to follow the generation and evolution with time of structural changes. Liquid-like diffusive motions in the protein and the surrounding solvent are suppressed when the temperature is lowered below about 200 K. The result is that protein structural changes essential to activity are also lost (Doster et al. 1989; Cornicchi et al. 2005). Dynamic crystallography is still possible at reduced temperatures, but dynamic cryocrystallography is not possible at cryo temperatures. However, the propagation of radiation damage through the sample also depends on diffusive motions. Retention of these motions at temperatures greater than 200 K unavoidably enhances radiation damage over that present at cryo temperatures. Radiation damage must be considered in planning all dynamic experiments.

4.3.2   Generation of Primary and Secondary Radiation Damage

X-rays interact only weakly with matter. When hard X-ray photons fall on an atom, most (in the range of 95–98%) do not interact in any way with its electrons and are directly transmitted. Those few photons that do interact do so in two ways: elastic scattering and inelastic scattering.

In crystallography, non-Bragg scattering describes all photons not scattered into Bragg peaks by coherent, elastic scattering from a crystalline mosaic block, and includes several different components. The dynamic motion of protein and RNA molecules ensures that at any instant no two unit cells contain exactly the same structure. Diffuse scattering describes coherent scattering from this rapidly varying, nonperiodic portion of the electron density function. Such scattering is elastic and includes the familiar isotropic scattering arising from atomic thermal motions. However, larger-scale coherent motions such as those in α-helices leave much more detailed imprints on diffuse scattering. Ultrashort XFEL pulses eliminate the time averaging that characterizes current measurements at SR sources. The potential for diffuse scattering data to illuminate protein dynamics has received a thoughtful review (Wall et al. 2018), which also discusses strategies for reducing background in scattering experiments.

The remainder of the scattering is inelastic. Of this portion, about 10% comprises inelastic scattering of an X-ray photon from an electron to yield a lower energy X-ray photon, whose energy varies over broad limits. The remainder of the interactions produces photoelectrons; discussed below.

Elastic scattering from the electrons in a crystal generates the X-ray diffraction patterns. By definition, elastic scattering occurs without any change in X-ray energy. Since it deposits no energy in electrons associated with atoms, it causes no damage to the sample. The X-ray exposure must be long enough for sufficient photons to be elastically scattered to yield a diffraction pattern to the desired resolution with good signal-to-noise. In contrast, inelastic scattering occurs with a change in X-ray energy and transfers energy to those electrons and atoms through the process of photoabsorption. Materials science experiments on hard condensed matter can make effective use of inelastic scattering, for example to determine the density of vibrational states. However, for biological samples in aqueous solution, the energy transferred is sufficient to disrupt structure and cause radiation damage.

The probability that an X-ray will undergo elastic or inelastic scattering is expressed as a cross section with units of area denoted barns. The barn originated in high energy physics; 1 barn = 100 fm^2. Elastic and inelastic scattering cross sections vary substantially with X-ray photon energy, the nature of the absorbing atom, and its electronic structure. In inelastic scattering, the energy transferred to the sample per unit mass is known as the dose. The quantitative unit of dose is the Gray; 1 J kg^{-1} = 1 Gray, abbreviated Gy.

The energy of a hard X-ray photon lies in the range of 5–15 keV, which is easily large enough to cause extensive electronic and structural damage to biological samples: ionize atoms, disrupt their electronic structure and interactions with their neighbors, and break covalent bonds (Holton 2009). Dose is directly related to radiation damage but as we shall see, the extent of radiation damage to biological samples from a given dose may be reduced.

Radiation damage has two main classes: primary and secondary. Primary damage is a consequence of the interaction of electromagnetic radiation with atoms. It denotes the direct interaction between X-rays and the electrons in an atom via photoabsorption (Chapman et al. 2014). Primary damage is specific to the absorbing atom and its immediate environment, depends on the number of electrons in the atom and thus increases with atomic number. We emphasize that its origin in atomic physics makes primary damage independent of temperature: its magnitude at cryotemperatures and room temperature is identical. For hard X-rays, the principal form of primary damage is ejection of a photoelectron from an inner core shell of the absorbing atom. The resultant core hole is promptly filled by a valence electron from an outer shell of that atom with ejection of a second, Auger electron from a valence shell. At X-ray energies greater than about 25 keV, Compton scattering becomes significant (Figure 4.1). For atoms of higher atomic number such as transition metals, emission of an X-ray fluorescence photon may also occur.

The outgoing photoelectron generates secondary damage. Its initial energy of many keV is easily sufficient to eject further inner shell photoelectrons when it interacts with other atoms in the protein and solvent along its path. A damage cascade is thus generated. The photoelectron travels a few μm in its 100 fs lifetime and as it loses energy, most damage occurs near the end of its path. Auger electrons have an energy of a few hundred eV, much lower than that of photoelectrons. Ejection of photo and Auger electrons generates an ionized atom whose electrostatic and bonding interaction with its neighboring atoms are perturbed. Nearby atoms and their electrons move, probably on the few fs timescale. For the first few tens of fs after photoabsorption, damage is largely primary and localized to the immediate neighborhood of the absorbing atom. Thereafter, secondary damage dominates. Many more ionizations and damage occur through secondary damage along the path of the photoelectron than from primary damage associated with initial photoabsorption (Chapman et al. 2014). Interaction of the outgoing photoelectrons and Auger

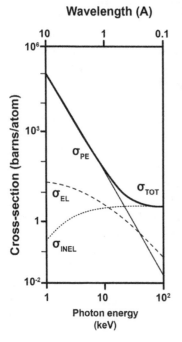

Figure 4.1 Dependence of scattering processes on X-ray energy. The cross sections σ in units of barns on a log scale are shown for elastic scattering (s_E), the photoelectric effect (s_{PE}), inelastic Compton scattering (s_{INEL}), and total for all scattering processes (s_{TOT}), versus X-ray energy. The cross section for the photoelectric effect dominates all scattering processes for X-ray energies up to around 25 keV and in particular, is substantially larger throughout the energy range than the cross section for elastic scattering that generates the crystal diffraction patterns and solution scattering. Henderson 1995/Reproduced with permission from Cambridge University Press.

electrons with the protein and solvent atoms generates highly reactive solvated electrons, hydroxyl radicals, and other radiolytic products. These readily diffuse at room temperature, react chemically, and distribute damage widely throughout the sample. Secondary damage is much more extensive in space and time than primary damage.

Primary damage is inescapable. However, the extent of secondary damage can be lessened by four distinct routes. First, a tiny crystal or liquid sample may have dimensions less than the few μm path length of the outgoing photoelectrons. If photoelectrons escape before depositing all their energy (Nave and Hill 2005), the extent of the radiation damage is reduced. Second, if X-ray fluorescence occurs, X-ray photons may escape from the sample and cause no damage. Third, the highly reactive radiolytic products generated by the energetic photoelectrons and Auger electrons are frozen at cryotemperatures and cannot diffuse over a larger volume; further radiation damage is suppressed.

The first and second routes reduce the extent of radiation damage only in samples from which escape is possible. Since elastic scattering from tiny crystals is also weak, an exceptionally intense X-ray source such as an XFEL is required to obtain good diffraction patterns. At less intense SR sources, a more typical crystal size lies in the tens of μm range, which exceeds the path length of the photoelectrons. The third route accounts for the substantial reduction in radiation damage at temperatures below 200 K compared with room temperature. This reduction has been exploited by conducting almost all static crystallography at cryo temperatures: cryocrystallography. However, as we repeatedly emphasize, cryocrystallography is not applicable to dynamic experiments. This route cannot be exploited to reduce secondary radiation damage.

The fourth route is the most powerful and striking: diffraction before destruction.

4.3.3 Diffraction before Destruction

Radiation damage processes do not occur instantaneously. Damage perturbs electronic and atomic structure and causes structural changes which take time to evolve before they can be detected by crystallography or solution scattering. Structural changes are only recorded in the diffraction patterns if they occur before or during the X-ray pulse. All changes after the pulse, no matter how damaging or extreme, are not detected; they are irrelevant. Thus for very brief, hard X-ray pulses, "diffraction [can occur] before destruction." This proposal was forcefully advanced by (Neutze et al. 2000) and first confirmed by imaging a hard condensed matter sample with soft X-rays at the FLASH XFEL in Hamburg (Chapman et al. 2006). Spectacular experiments at the first hard X-ray FEL, the LCLS at Stanford, on diffraction by protein nanocrystals (Chapman et al. 2011), extended confirmation to biological samples at higher X-ray energies.

The fundamental physics that makes diffraction before destruction possible has been thoroughly reviewed (Chapman et al. 2014). How brief must the required X-ray pulse be? Experimental studies at hard XFEL sources of damage as a function of pulse length are ongoing; shorter is better. Very brief pulses clearly lie in the few fs range, perhaps as low as 1 fs. This is sufficiently short to make irrelevant all secondary radiation damage and some of the damage effects of Auger electrons, even for samples at room temperature. It is unclear to what extent primary damage is present in diffraction patterns obtained with X-ray pulses of a few fs. These patterns are unlikely to be damage-free, as sometimes claimed by enthusiastic proponents of XFEL experiments. However, there is no doubt they are greatly damage-reduced compared with patterns from longer pulses. At the least, ejection of a photoelectron by primary damage reduces the contribution of the absorbing atom by one electron to the X-ray scattering and hence to the final electron density map. It also results in an immediate change to the electrostatic field around the absorbing atom, which propagates at the speed of light, 300 nm fs^{-1}. These issues have been thoughtfully reviewed (Caleman and Martin 2018; Nass 2019).

An important question is whether primary radiation damage leads to prompt, mechanistically significant chemical changes such as sulfur oxidation or changes in metal coordination. The diffraction patterns recorded even with brief X-ray pulses in the few fs range may not be truly free of chemically significant radiation damage. This question is being addressed (Bergmann et al. 2021) by combining measurements sensitive to oxidation state such as X-ray fluorescence (Chapter 8.4) with scattering measurements on the same sample at the same time and dose. For example, significant radiation damage was seen in the iron-sulfur cluster of ferredoxin when data were collected using relatively long 80 fs pulses (Nass et al. 2015).

The ratio of X-ray inelastic to elastic scattering cross sections is large throughout the hard X-ray range, with a value around 30 (Figure 4.1). For 1 photon scattered, 30 are absorbed. It might be thought that this would be very unfavorable for obtaining diffraction patterns arising from elastic scattering with low radiation damage. However, crystals offer a further advantage. If the X-ray intensity is restricted to ensure that no more than one inelastic scattering/photoabsorption event occurs per atom in a crystal, hollow atoms are produced in which one photoelectron has been ejected from the inner shell. If in addition the pulse duration is sufficiently short that little structural change has occurred, most of the elastic scattering occurs from virgin atoms (Chapman et al. 2014).

A further advantage for crystals results from the "Bragg boost" (Spence 2017). Elastic scattering is greatly concentrated at the Bragg peaks by the translational symmetry of crystals, rather than being distributed across all of space as in continuous scattering from a concentrated protein solution or from an individual, large protein complex. The X-ray exposure must be long enough to allow the peak intensity of a few tens of strong spots in the crystal diffraction pattern to be identified with good signal to noise. (We distinguish here the peak intensity of a spot from its integrated intensity.) When these spots are indexed, the orientation of the crystal can be determined. With the cell dimensions and orientation known, the location of all Bragg spots, not just of strong spots, is known. The integrated intensities of all spots can then be determined by combining information from many frames, even if their individual values of peak signal to noise are much less favorable and insufficient for their independent indexing.

The advantage obtained by concentration of the elastic scattering at Bragg peaks is of course absent for single objects such as a large protein complex, which may be studied by single particle imaging (SPI). It has been argued that the structures of these single objects should be studied by electron scattering, where the ratio of inelastic to elastic scattering at typical electron energies of 300 keV is around 3 (Henderson 1995), a ratio much more favorable than for hard X-rays. Cryo-EM studies by SPI of static complexes is proving extremely successful but obviously is conducted at cryo temperatures. Application of cryo-EM to dynamic structural problems has been restricted to complexes where reaction intermediates have been freeze-trapped in the ms time range. See for example (Frank 2017) and Chapter 8.2. Ideas for moving into the ms range have recently been proposed (Voss et al. 2021). We also consider the possibilities for X-ray (Chapter 9.2) and EM (Chapter 9.4) studies of biologically active single molecules or complexes at near-physiological temperatures: dynamic SPI.

4.3.4 Quantitative Studies of Radiation Damage

Radiation damage can be studied in real space or in reciprocal space. Specific structural effects of damage such as rupture of disulfide bonds (Van Den Bedem and Wilson 2019), decarboxylation of acidic side chains, and photoreduction of metal centers are local and readily identified in real space. See for example (Kwon et al. 2021) and references therein; also (Caleman et al. 2020; Nass et al. 2020). Nonspecific, spatially distributed structural effects are usually studied in reciprocal space. Here, damage is evident in the dose-dependent, progressive loss of diffraction intensities. Loss begins at high resolution and extends toward low resolution as the dose increases, with an increase in overall B-factor and swelling of unit cell lengths (Garman and Weik 2017). Inelastic scattering both damages the sample and contributes isotropic non-Bragg scattering and a component that peaks at reciprocal lattice points. These effects are often ignored but should be corrected for in both crystal and solution samples. Reflecting static structure determinations until recently, most studies of radiation damage have used third generation SR and crystals at cryo temperatures ~100 K. The recommended maximum dose at cryo temperatures for "acceptable" radiation damage is 30 MGy (Owen et al. 2006). Whether damage is acceptable depends on the resolution desired and the accuracy required by the specific crystallographic problem. For example, the threshold of acceptability differs for molecular replacement, *de novo* structure determination, or the presence of functionally important metal centers. These centers absorb X-rays more strongly than C, N, and O and are particularly sensitive to radiation damage.

Quantitative studies of damage at room temperature for SR sources are much less numerous (de la Mora et al. 2020). However, these studies suggest that a maximum dose at room temperature should be 30–50 times lower than the better-established dose at cryo temperatures. The suggested maximum dose at room temperature is thus 0.6–1 MGy. For a beamline of known properties at an SR and crystals of a range of sizes and chemical composition, dose can be calculated by programs such as RADDOSE-3D (Zeldin et al. 2013), RADDOSE-XFEL (Dickerson et al. 2020), or RABDAM (Shelley et al. 2018). This information can be combined with the crystal scattering power (Chapter 4.1) to estimate the minimum size of a crystal that will yield one diffraction pattern at a maximum dose of, say, 1 MGy (Hirata et al. 2016; Hadian-Jazi et al. 2021), or a single complete dataset at one time point (Holton and Frankel 2010). For a dynamic experiment requiring 20 time points, one larger crystal of 20 times the volume (and scattering power) would be needed to collect all time points at a single crystal orientation. If 50 orientations would be needed to collect all unique data in scattering space, then 50 larger crystals would be needed to collect all data in time and scattering space.

Estimates of this type may be useful in planning the amount of protein and crystal sizes required for a complete dynamic experiment but are only semiquantitative. For example, the exposure time for one good diffraction pattern is substantially influenced by the X-ray background. Since dynamic experiments are based on accurate measurement of the time-dependent variation in all spot intensities, errors in difference intensities are minimized if the background is time-independent. However, radiation damage is dose-dependent and for a single crystal and a constant X-ray source, must also be time-dependent. Finally, the scattering from damaged molecules will contribute to the diffuse background. If several diffraction patterns are collected at a SR source from one crystal, the increase in dose from time point to time point with radiation damage will be superimposed on the true, time-dependent structural variation. This complicates interpretation.

In spectacular contrast, at XFEL sources even with the sample at room temperature, there is no maximum dose if the X-ray pulse is sufficiently brief to outrun the damage (Chapman et al. 2014). However, one final source of damage remains. Although crystals in an XFEL experiment are at a controlled, near-room temperature in the reservoir and injector, their temperature may drop as they are injected into the vacuum. Conversely, unless heat is conducted away, a 1 MGy maximum X-ray dose to a crystal by a typical XFEL source will eventually heat it by 200 K. The fact that good diffraction patterns have been obtained from tiny, single protein crystals with doses of 1 GigaGy, one thousand times larger, shows that both thermal damage and radiation damage are relatively slow processes that can be almost entirely outrun. Diffraction before destruction really works!

4.4 Optogenetics and Photopharmacology

Initiation of a reaction in a crystal or solution by a visible or IR laser pulse is experimentally convenient and offers essentially the only approach to the structural dynamics of ultrafast reactions on time scales of ns and below (Chapter 3.3.2). However, a relatively limited number of naturally light-sensitive, interesting biological reactions exist. This raises an obvious question: can sensitivity to light be conferred on other interesting reactions, by either a genetic or a chemical approach? A positive answer would greatly increase the range of reactions that can be studied by structural dynamics.

While acknowledging this to be a rapidly developing topic, the answer indeed appears to be positive. Genetic approaches purposefully modify protein sequences to confer sensitivity to light through engineering of the genes that encode those proteins. These approaches are known as optogenetics (Boyden et al. 2005). Chemical approaches are known as photopharmacology and modify small molecules such as enzyme substrates, effectors or inhibitors, or general solution properties such as pH, or the protein itself (Broichhagen et al. 2015; Zhu and Zhou 2018).

We consider first optogenetics. *De novo* protein design is advancing very rapidly (Korendovych and DeGrado 2020; Leman et al. 2020; Woolfson 2021), but has largely been confined to static, foldable structures at a deep minimum of free energy. Design of dynamic, light-sensitive proteins from scratch remains a major and largely unexplored challenge. Rather, applications are based on bio-inspired adaptation of existing proteins. This approach examines systems in which a biological reaction is naturally light-sensitive, seeks general principles for this sensitivity, applies these principles to a new system on which sensitivity to light is to be conferred, then improves the newly introduced, light-dependent properties by cycles of mutation and selection.

Natural photoreceptors in which absorption of a photon alters a biological activity are known as signaling photoreceptors (Kottke et al. 2018). Signaling photoreceptors use light as an information source, distinct from its more common roles in imaging, or as an energy source in photosynthesis. Since light readily traverses biological membranes, most signaling photoreceptors are intracellular, water-soluble, and directly regulate an intracellular activity. They contain a readily bio-available cofactor such as FMN or FAD which serves as their light-absorbing chromophore in the blue region of the visible spectrum, or bilin (a linear tetrapyrrole degradation product of heme) in the red-far red region. Other photoreceptors such as visual rhodopsins and signaling rhodopsins such as channelrhodopsin have been very widely used in optogenetic applications in the neurosciences at the molecular, cellular, and even organismal level (Kandori 2020). Rhodopsins are integral membrane proteins which use retinal as their chromophore. Absorption of a photon leads to rapid isomerization about a specific double bond in the polyene chain of the retinal and generates a structural signal. This signal confers sensitivity to light on an intrinsic property of the membrane where the rhodopsins are located, such as its permeability to certain cations or anions. However, the ability to modify rhodopsins to confer light sensitivity on other systems has been little exploited so far.

The protein structures of soluble signaling photoreceptors are often modular. An input, sensor domain binds the chromophore which absorbs a photon, undergoes an electronic and structural change, and generates a signal which may propagate over very long range, 10 nm or more, to an output, effector domain whose activity is affected by receipt of the signal. The sensor and effector domains may be directly and covalently linked but some are separated by one or more intervening linker domains. Linker domains may belong to the same domain class as the sensor domains, for example the PAS domain family (Stuffle et al. 2021) but linkers lack a chromophore and are not themselves light-sensitive.

The minimum requirement for generation and propagation of a signal is that receipt of the signal changes the affinity between two components of the system. The signal may be absorption of a photon to a photoreceptor or binding of a small molecule to a chemoreceptor. Though a signal is essentially thermodynamic in nature, it is largely evident as a change in structure (Prischi and Filippakopoulos 2021). The affinity changed by the signal can be intramolecular between two regions of the same photoreceptor, e.g. between its chromophore and the surrounding protein, or between the sensor domain, a linker domain and the effector domain. The affinity can be intermolecular between the photoreceptor and another, non-covalently bound protein. Finally, an intermolecular affinity can be between photoreceptor monomers in equilibrium with a dimer or higher polymer. In all cases, absorption of a photon changes the structure of the interface between the regions (measurable in nm as a root mean square deviation (rmsd) in atomic positions) and alters the dissociation equilibrium constant between them (measurable as a free energy in kJ mol^{-1}). Changes in structure are usually presented as an rmsd with the dimensions of length. However, we assert that the inherently thermodynamic nature of a signal with the dimensions of energy is much more significant.

A sensor domain such as the FMN-binding LOV domain is often found covalently linked via an α-helical linker domain to the N-terminus of structurally very different effector domains that display diverse biological activities (Figure 4.2). Conversely, similar effector domains such as a histidine kinase may be linked to sensor domains that

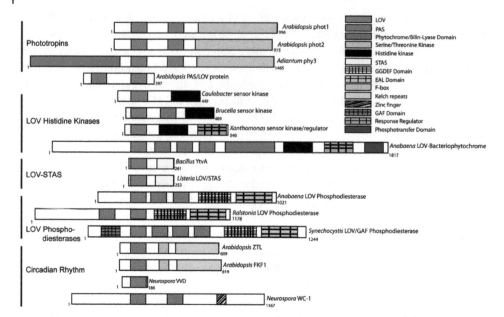

Figure 4.2 Domain diversity of proteins containing a LOV domain. Five major subsets of LOV proteins are represented: phototropins, LOV histidine kinases, LOV-STAS proteins, LOV phosphodiesterases, and proteins involved in the regulation of circadian rhythms. Each domain is shown as a colored box, with LOV domains in blue. Crosson et al. 2003/Reproduced with permission from American Chemical Society.

respond to entirely different signals, such as light or binding of a small molecule such as O_2. These observations suggest a direct structural interaction between the sensor and effector domains is not required. An α-helical linker is a common mode of joining sensor, linker, and effector domains while keeping the cores of sensor and effector domains at a distance. These convenient facts suggest a general principle for conferring sensitivity to a signal. By gene fusion, interchange the sensor domains linked to an effector, or add a sensor domain to the N-terminus of a desired effector. Thus emboldened, Möglich and colleagues (Möglich et al. 2009) applied gene deletion and fusion to replace a chemosensor PAS domain with a photosensor LOV domain. The chemosensor PAS domain responds via O_2 binding to its heme group. Photosensor LOV domains respond to absorption of blue light by their FMN chromophore, and are also members of the more general PAS domain family (Crosson et al. 2003). As hoped, the new construct placed the histidine kinase enzymatic activity of the effector domain – normally completely insensitive to light – under the control of blue light (Möglich et al. 2009). However, control depended in an unusual way on the length of the linker between the sensor and effector domains. Variation of its length by insertion or deletion of amino acids showed a seven amino acid periodicity in the output of the effector kinase. This strongly suggested that the linker had a coiled-coil structure in the dimeric sensor–kinase structure. This suggestion was subsequently confirmed by structural studies (Diensthuber et al. 2013).

An approach by careful gene fusion of a light-sensitive sensor domain to a desired, normally light-inert effector domain appears to be directly applicable to pumping the system by light and probing it by solution-based structural dynamics. However, there are challenges if crystal-based structural dynamics is envisaged. Many signaling photo-receptors are dimeric and undergo a substantial quaternary structural change upon light absorption. As in the classical example of hemoglobin (Chapter 1), this change in quaternary structure is central to signaling. The change may be inhibited or abolished by intermolecular packing constraints in the crystal. Truly monomeric systems may have somewhat different signaling principles that involve direct structural interaction between sensor and effector domains (Strickland et al. 2008). To confer sensitivity to light on such an effector, a sensor domain must be engineered whose structure is specific to that effector. This one-off design process is more challenging and lacks generalizability.

Other approaches take advantage of modular domains and non-covalent interactions. New ways to confer sensitivity to light remain an active research area.

Chemical approaches, now known as photopharmacology (Broichhagen et al. 2015; Shell and Lawrence 2015), have a longer history than optogenetics. The general strategy is to chemically modify an essential component of the reaction and thus cage (or block) activity. A comprehensive review is (Monteiro et al. 2021). The caged component could be a protein or a small molecule reactant such as a cofactor, solution, or environmental component. When the cage is removed photochemically by a laser pulse, the normal biological reaction is initiated and proceeds in the sample. For example, consider an enzyme whose catalytic activity is to be made sensitive to light. Several distinct strategies are possible: cage the enzyme, cage the substrate (or cofactor), or cage a solvent component. In each strategy, the caged compound could involve covalent modification where the primary photochemical reaction is bond cleavage, or non-covalent modification where the primary reaction is isomerization. To cage the enzyme, a photocleavable group could covalently modify an active site residue or an essential cofactor. Classical examples include caged ATP, in which an ortho-nitrobenzoyl group blocks the γ-phosphate (Monteiro et al. 2021), or a caged cysteine side chain in the active site of TEV protease (Nguyen et al. 2014). To cage the substrate, a photocleavable group could modify a substrate moiety essential to binding. To cage the solvent, the pKa of a solvent component (such as a phenolate whose pKa is light-dependent, or caged acetic acid) could be altered by light and thus generate a pH jump. Light-dependent lipids are an example of caging an environmental component (Morstein and Trauner 2020). A caged compound was used to initiate dynamic solution scattering probes of folding transitions in the small protein, cytochrome c (Rimmerman et al. 2018) at the APS SR source.

Covalent modification has limitations. The uncaging reaction may irreversibly break a covalent bond. This is of no consequence for naturally irreversible XFEL experiments but may limit the conduct of SR experiments that could usefully apply repetitive light pulses to a single sample. The initial photochemistry of uncaging often proceeds via highly reactive intermediates such as carbenes and nitrenes, which may modify any nearby group in undesirable side reactions. More commonly, release of the final, desired product such as fully uncaged ATP may be delayed into the μs range or even longer by slow, dark reactions, which limits the time resolution of time-resolved measurements. As a practical limitation, uncaging often requires illumination in the near UV at around 360 nm, where the absorbance of the sample may be unacceptably high.

Non-covalent modifications usually avoid the first three limitations but retain absorption in the near UV. A common modification which also circumvents the first three is based on cis-trans isomerization of an azobenzene analog of a substrate, where one isomer is a substrate and the other is not (Merino and Ribagorda 2012). Re-isomerization to the initial, dark state may occur thermally, or by absorption of a second photon to drive the back reaction and regain the initial state. The concept of isomerization and re-isomerization is based on the natural photochemistry of the bilin and bilin-based chromophores in bacteriophytochromes and phytochromes. These signaling photoreceptors are widely distributed in bacteria, fungi, and plants where they control many developmental and behavioral processes (Möglich et al. 2010). Photochemistry is based on isomerization about a single double bond linking the third and fourth pyrrole rings of their bilin chromophore, by absorption of a photon in the red or far-red region of the visible spectrum.

New light-dependent reactions have recently been discovered (Huang et al. 2020) in a long-studied, flavin-dependent enzyme, Old Yellow Enzyme, that exhibits photocatalysis. Photocatalysis shows promise as an attractive method to accomplish mechanistically novel and synthetically useful transformations (Stephenson et al. 2018). In Old Yellow Enzyme, the enzyme functions as an "ene" reductase to catalyze an unnatural intermolecular cross-coupling reaction. The reaction utilizes FMN as a redox-active center to catalyze the 2-electron reduction of carbonyl-substituted alkenes and achieve intermolecular radical hydroalkylation. In its active site, histidine and asparagine residues are proposed to act as hydrogen bond donors and a nearby tyrosine as a proton donor.

Photocatalysis may emerge as a novel synthetic route. The unexpected consequence is that in new enzyme systems, structural dynamics can usefully complement synthetic organic chemistry.

Flavin cofactors are widespread in biochemistry and exhibit a range of both light-dependent and redox-dependent reactions. The classical FMN-containing signaling photoreceptor is the LOV domain, in which the primary photoreaction upon absorption of a photon in the blue visible spectrum is formation of a covalent adduct between the C4a atom of the FMN and a nearby cysteine side chain, which generates the structural signal. However, when formation of the covalent bond is disabled in the LOV domain of the circadian clock protein Vivid, its FMN can still undergo photoreduction to the neutral semiquinone form, which generates an apparently different signal (Yee et al. 2015). The essential feature in signal generation evidently lies in flavin protonation, which occurs in both the cysteinyl adduct and the neutral semiquinone. The proton was thought to be associated with a hydrogen bond from a nearby asparagine residue, but recent studies propose that a proton from a buried water molecule may suffice (Dietler et al. 2022). Protonation can be achieved by light in photoreceptors or by chemical reduction in chemoreceptors, thus illustrating the flexibility of flavin cofactors in signal generation. Evidently, there are many ways to generate a significant biological signal from a flavin and its immediate environment.

To date, all naturally-occurring signaling photoreceptors are proteins. We consider the possibilities for RNA-based signaling photoreceptors in Chapter 9.4.5.

References

Bergmann, U., Kern, J., Schoenlein, R.W. et al. (2021). Using X-ray free-electron lasers for spectroscopy of molecular catalysts and metalloenzymes. *Nature Reviews Physics* 3 (4): 264–282.

Blake, C. and Phillips, D. (1962). *Effects of X-irradiation on Single Crystals of Myoglobin. Biological Effects of Ionizing Radiation at the Molecular Level*. Vienna: International Atomic Energy Agency 183.

Bogan, M.J., Benner, W.H., Boutet, S. et al. (2008). Single particle X-ray diffractive imaging. *Nano Letters* 8 (1): 310–316.

Boyden, E.S., Zhang, F., Bamberg, E. et al. (2005). Millisecond-timescale, genetically targeted optical control of neural activity. *Nature Neuroscience* 8 (9): 1263–1268.

Broichhagen, J., Frank, J.A., and Trauner, D. (2015). A roadmap to success in photopharmacology. *Accounts of Chemical Research* 48 (7): 1947–1960.

Caleman, C., Jares Junior, F., Grånäs, O. et al. (2020). A perspective on molecular structure and bond-breaking in radiation damage in serial femtosecond crystallography. *Crystals* 10 (7): 585.

Caleman, C. and Martin, A.V. (2018). When diffraction stops and destruction begins. In: *X-ray Free Electron Lasers* (ed. S. Boutet, P. Fromme and M.D. Hunter), 185–207. Springer.

Chapman, H.N., Barty, A., Bogan, M.J. et al. (2006). Femtosecond diffractive imaging with a soft-X-ray free-electron laser. *Nature Physics* 2 (12): 839–843.

Chapman, H.N., Caleman, C., and Timneanu, N. (2014). Diffraction before destruction. *Philosophical Transactions of the Royal Society B: Biological Sciences* 369 (1647): 20130313.

Chapman, H.N., Fromme, P., Barty, A. et al. (2011). Femtosecond X-ray protein nanocrystallography. *Nature* 470 (7332): 73–77.

Cheng, R., Huang, C.Y., Hennig, M. et al. (2020). In situ crystallography as an emerging method for structure solution of membrane proteins: the case of CCR2A. *The FEBS Journal* 287 (5): 866–873.

Conrad, C.E., Basu, S., James, D. et al. (2015). A novel inert crystal delivery medium for serial femtosecond crystallography. *International Union of Crystallography Journal* 2 (4): 421–430.

Cornicchi, E., Onori, G., and Paciaroni, A. (2005). Picosecond-time-scale fluctuations of proteins in glassy matrices: the role of viscosity. *Physical Review Letters* 95 (15): 158104.

Crosson, S., Rajagopal, S., and Moffat, K. (2003). The LOV domain family: photoresponsive signaling modules coupled to diverse output domains. *Biochemistry* 42 (1): 2–10.

de la Mora, E., Coquelle, N., Bury, C.S. et al. (2020). Radiation damage and dose limits in serial synchrotron crystallography at cryo-and room temperatures. *Proceedings of the National Academy of Sciences* 117 (8): 4142–4151.

DePonte, D.P. (2017). Sample delivery methods: liquids and gases at FELs. In: *X-Ray Free Electron Lasers* (ed. V.Y.U. Bergmann and J. Yano), 323–336. Royal Society of Chemistry.

Dickerson, J.L., McCubbin, P.T., and Garman, E.F. (2020). RADDOSE-XFEL: femtosecond time-resolved dose estimates for macromolecular x-ray free-electron laser experiments. *Journal of Applied Crystallography* 53 (2): 549–560.

Diensthuber, R.P., Bommer, M., Gleichmann, T. et al. (2013). Full-length structure of a sensor histidine kinase pinpoints coaxial coiled coils as signal transducers and modulators. *Structure* 21 (7): 1127–1136.

Dietler, J., Gelfert, R., Kaiser, J. et al. (2022). Signal transduction in light-oxygen-voltage receptors lacking the active-site glutamine. *Nature Communications* 13 (1): 1–16.

Doster, W., Cusack, S., and Petry, W. (1989). Dynamical transition of myoglobin revealed by inelastic neutron scattering. *Nature* 337 (6209): 754–756.

Frank, J. (2017). Time-resolved cryo-electron microscopy: recent progress. *Journal of Structural Biology* 200 (3): 303–306.

Frank, M., Carlson, D.B., Hunter, M.S. et al. (2014). Femtosecond X-ray diffraction from two-dimensional protein crystals. *International Union of Crystallography* 1: 95–100.

Fuller, F.D., Gul, S., Chatterjee, R. et al. (2017). Drop-on-demand sample delivery for studying biocatalysts in action at X-ray free-electron lasers. *Nature Methods* 14 (4): 443–449.

Garman, E.F. and Weik, M. (2017). Radiation damage in macromolecular crystallography. *Methods in Molecular Biology* 1607: 467–489. doi: 10.1007/978-1-4939-7000-1_20. PMID: 28573586.

Garman, E.F. and Weik, M. (2019). X-ray radiation damage to biological samples: recent progress. *Journal of Synchrotron Radiation* 26 (4): 907–911.

Grünbein, M.L. and Nass Kovacs, G. (2019). Sample delivery for serial crystallography at free-electron lasers and synchrotrons. *Acta Crystallographica Section D: Structural Biology* 75 (2): 178–191.

Hadian-Jazi, M., Berntsen, P., Marman, H. et al. (2021). Analysis of multi-hit crystals in serial synchrotron crystallography experiments using high-viscosity injectors. *Crystals* 11 (1): 49.

Henderson, R. (1995). The potential and limitations of neutrons, electrons and x-rays for atomic-resolution microscopy of unstained biological molecules. *Quarterly Reviews of Biophysics* 28 (2): 171–193.

Hirata, K., Foadi, J., Evans, G. et al. (2016). Structural Biology with Microfocus Beamlines. In: Senda, T., Maenaka, K. (eds) Advanced Methods in Structural Biology. Springer Protocols Handbooks. Springer, Tokyo. https://doi.org/10.1007/978-4-431-56030-2_14

Holton, J.M. (2009). A beginner's guide to radiation damage. *Journal of Synchrotron Radiation* 16: 133–142.

Holton, J.M. and Frankel, K.A. (2010). The minimum crystal size needed for a complete diffraction data set. *Acta Crystallographica Section D: Biological Crystallography* 66 (4): 393–408.

Huang, X., Wang, B., Wang, Y. et al. (2020). Photoenzymatic enantioselective intermolecular radical hydroalkylation. *Nature* 584 (7819): 69–74.

Kalman, Z. (1979). On the derivation of integrated reflected energy formulae. *Acta Crystallographica Section A: Crystal Physics, Diffraction, Theoretical and General Crystallography* 35 (4): 634–641.

Kandori, H. (2020). Biophysics of rhodopsins and optogenetics. *Biophysical Reviews* 12 (2): 355–361.

Korendovych, I.V. and DeGrado, W.F. (2020). De novo protein design, a retrospective. *Quarterly Reviews of Biophysics* 53, e3.

Kottke, T., Xie, A., Larsen, D.S. et al. (2018). Photoreceptors take charge: emerging principles for light sensing. *Annual Review of Biophysics* 47: 291–313.

Kwon, H., Basran, J., Pathak, C. et al. (2021). XFEL crystal structures of peroxidase compound II. *Angewandte Chemie International Edition* 60.

Leman, J.K., Weitzner, B.D., Lewis, S.M. et al. (2020). Macromolecular modeling and design in Rosetta: recent methods and frameworks. *Nature Methods* 17 (7): 665–680.

Martiel, I., Müller-Werkmeister, H.M., and Cohen, A.E. (2019). Strategies for sample delivery for femtosecond crystallography. *Acta Crystallographica Section D: Structural Biology* 75 (2).

Merino, E. and Ribagorda, M. (2012). Control over molecular motion using the cis–trans photoisomerization of the azo group. *Beilstein Journal of Organic Chemistry* 8 (1): 1071–1090.

Moffat, K. (1997). [22] Laue diffraction. *Methods Enzymol, Elsevier* 277: 433–447.

Moffat, K., Bilderback, D., Schildkamp, W. et al. (1986). Laue diffraction from biological samples. *Nuclear Instruments and Methods in Physics Research Section A: Accelerators, Spectrometers, Detectors and Associated Equipment* 246 (1–3): 627–635.

Möglich, A., Ayers, R.A., and Moffat, K. (2009). Design and signaling mechanism of light-regulated histidine kinases. *Journal of Molecular Biology* 385 (5): 1433–1444.

Möglich, A., Yang, X., Ayers, R.A. et al. (2010). Structure and function of plant photoreceptors. *Annual Review of Plant Biology* 61: 21–47.

Monteiro, D.C., Amoah, E., Rogers, C. et al. (2021). Using photocaging for fast time-resolved structural biology studies. *Acta Crystallographica Section D: Structural Biology* 77 (10).

Morstein, J. and Trauner, D. (2020). Photopharmacological control of lipid function. *Methods Enzymol* 638: 219–232.

Mueller, C., Marx, A., Epp, S.W. et al. (2015). Fixed target matrix for femtosecond time-resolved and in situ serial micro-crystallography. *Structural Dynamics* 2 (5): 054302.

Nam, K.H. (2020a). Polysaccharide-based injection matrix for serial crystallography. *International Journal of Molecular Sciences* 21 (9): 3332.

Nam, K.H. (2020b). Stable sample delivery in viscous media via a capillary for serial crystallography. *Journal of Applied Crystallography* 53 (1): 45–50.

Nass, K. (2019). Radiation damage in protein crystallography at X-ray free-electron lasers. *Acta Crystallographica Section D: Structural Biology* 75 (2): 211–218.

Nass, K., Foucar, L., Barends, T.R. et al. (2015). Indications of radiation damage in ferredoxin microcrystals using high-intensity X-FEL beams. *Journal of Synchrotron Radiation* 22 (2): 225–238.

Nass, K., Gorel, A., Abdullah, M.M. et al. (2020). Structural dynamics in proteins induced by and probed with X-ray free-electron laser pulses. *Nature Communications* 11 (1): 1–9.

Nave, C. and Hill, M.A. (2005). Will reduced radiation damage occur with very small crystals? *Journal of Synchrotron Radiation* 12 (3): 299–303.

Neutze, R., Wouts, R., van der Spoel, D. et al. (2000). Potential for biomolecular imaging with femtosecond X-ray pulses. *Nature* 406 (6797): 752–757.

Nguyen, D.P., Mahesh, M., Elsässer, S.J. et al. (2014). Genetic encoding of photocaged cysteine allows photoactivation of TEV protease in live mammalian cells. *Journal of the American Chemical Society* 136 (6): 2240–2243.

Orville, A.M. (2017). Acoustic methods for on-demand sample injection into XFEL beams. In: *X-Ray Free Electron Lasers: Applications in Materials, Chemistry and Biology* (ed. V.Y.U. Bergmann and J. Yano), 348–364. Springer.

Owen, R.L., Rudiño-Piñera, E., and Garman, E.F. (2006). Experimental determination of the radiation dose limit for cryocooled protein crystals. *Proceedings of the National Academy of Sciences* 103 (13): 4912–4917.

Pearson, A.R. and Mehrabi, P. (2020). Serial synchrotron crystallography for time-resolved structural biology. *Current Opinion in Structural Biology* 65: 168–174.

Prischi, F. and Filippakopoulos, P. (2021). Structural studies of protein complexes in signaling pathways. *Frontiers in Molecular Biosciences* 8.

Ren, Z., Srajer, V., Knapp, J.E. et al. (2012). Cooperative macromolecular device revealed by meta-analysis of static and time-resolved structures. *Proceedings of the National Academy of Sciences of the United States of America* 109 (1): 107–112.

Ren, Z., Wang, C., Shin, H. et al. (2020). An automated platform for in situ serial crystallography at room temperature. *International Union of Crystallography* 7 (6).

Rimmerman, D., Leshchev, D., Hsu, D.J. et al. (2018). Probing cytochrome c folding transitions upon phototriggered environmental perturbations using time-resolved X-ray scattering. *Journal of Physical Chemistry B* 122 (20): 5218–5224.

Roessler, C.G., Kuczewski, A., Stearns, A. et al. (2013). Acoustic methods for high-throughput protein crystal mounting at next-generation macromolecular crystallographic beamlines. *Journal of Synchrotron Radiation* 20 (5): 805–808.

Seibert, M.M., Ekeberg, T., Maia, F.R. et al. (2011). Single mimivirus particles intercepted and imaged with an X-ray laser. *Nature* 470 (7332): 78–81.

Shelby, M., Gilbile, D. et al. (2020). A fixed-target platform for serial femtosecond crystallography in a hydrated environment. *International Union of Crystallography* 7 (1): 30–41.

Shell, T.A. and Lawrence, D.S. (2015). Vitamin B12: a tunable, long wavelength, light-responsive platform for launching therapeutic agents. *Accounts of Chemical Research* 48 (11): 2866–2874.

Shelley, K.L., Dixon, T.P., Brooks-Bartlett, J.C. et al. (2018). RABDAM: quantifying specific radiation damage in individual protein crystal structures. *Journal of Applied Crystallography* 51 (2): 552–559.

Sierra, R.G., Weierstall, U., Oberthuer, D. et al. (2018). Sample delivery techniques for serial crystallography. In: *X-ray Free Electron Lasers* (ed. S. Boutet, P. Fromme and M.D. Hunter), 109–184. Springer.

Spence, J.C.H. (2017). XFELs for structure and dynamics in biology. *International Union of Crystallography* 4: 322–339.

Stauch, B. and Cherezov, V. (2018). Serial femtosecond crystallography of G protein–coupled receptors. *Annual Review of Biophysics* 47: 377–397.

Stellato, F., Oberthür, D., Liang, M. et al. (2014). Room-temperature macromolecular serial crystallography using synchrotron radiation. *International Union of Crystallography* 1 (4): 204–212.

Stephenson, C.R., Yoon, T.P., and MacMillan, D.W. (2018). *Visible Light Photocatalysis in Organic Chemistry*. John Wiley & Sons.

Strickland, D., Moffat, K., and Sosnick, T.R. (2008). Light-activated DNA binding in a designed allosteric protein. *Proceedings of the National Academy of Sciences* 105 (31): 10709–10714.

Stuffle, E.C., Johnson, M.S., and Watts, K.J. (2021). PAS domains in bacterial signal transduction. *Current Opinion in Microbiology* 61: 8–15.

Teng, T.-Y. (1990). Mounting of crystals for macromolecular crystallography in a free-standing thin film. *Journal of Applied Crystallography* 23 (5): 387–391.

Van Den Bedem, H. and Wilson, M.A. (2019). Shining light on cysteine modification: connecting protein conformational dynamics to catalysis and regulation. *Journal of Synchrotron Radiation* 26 (4): 958–966.

Voss, J.M., Harder, O.F., Olshin, P.K. et al. (2021). Microsecond melting and revitrification of cryo samples. *Structural Dynamics* 8 (5): 054302.

Wall, M.E., Wolff, A.M., and Fraser, J.S. (2018). Bringing diffuse X-ray scattering into focus. *Current Opinion in Structural Biology* 50: 109–116.

Weierstall, U., James, D., Wang, C. et al. (2014). Lipidic cubic phase injector facilitates membrane protein serial femtosecond crystallography. *Nature Communications* 5 (1): 1–6.

White, T.A. (2019). Processing serial crystallography data with CrystFEL: a step-by-step guide. *Acta Crystallographica Section D* 75 (2): 219–233.

Woolfson, D.N. (2021). A brief history of de novo protein design: minimal, rational, and computational. *Journal of Molecular Biology* 433 (20): 167160.

Yee, E.F., Diensthuber, R.P., Vaidya, A.T. et al. (2015). Signal transduction in light–oxygen–voltage receptors lacking the adduct-forming cysteine residue. *Nature Communications* 6 (1): 1–10.

Zabara, A., Meikle, T.G., Trenker, R. et al. (2017). Lipidic cubic phase-induced membrane protein crystallization: interplay between lipid molecular structure, mesophase structure and properties, and crystallogenesis. *Crystal Growth & Design* 17 (11): 5667–5674.

Zeldin, O.B., Gerstel, M., and Garman, E.F. (2013). RADDOSE-3D: time-and space-resolved modelling of dose in macromolecular crystallography. *Journal of Applied Crystallography* 46 (4): 1225–1230.

Zhu, M. and Zhou, H. (2018). Azobenzene-based small molecular photoswitches for protein modulation. *Organic & Biomolecular Chemistry* 16 (44): 8434–8445.

5

Time-resolved Crystallography, Solution Scattering, and Molecular Dynamics

5.1 Time-resolved Crystallography

A summary of the fundamentals of conventional, static crystallography is in Appendix A. A comprehensive textbook covering all aspects of biological crystallography is (Rupp 2010).

Chapter 1.5 presents a brief overview of the components of time-resolved crystallography experiments. Here we discuss the X-ray aspects of the probe experiment itself and summarize the main considerations that influence time-resolved scattering experiments.

1) Reaction initiation can be synchronized to establish time zero for the reaction.
2) Short-lived intermediates in a reaction are metastable, thermodynamic states. Each intermediate has a distinct, time-independent structure, to be determined.
3) Intermediates are populated for different lengths of time – their lifetime. This can range from very short (fs–ps) to very long (ms, s, or even longer). The lifetime determines the experimental time resolution needed to determine the structure of molecules in that state.
4) Kinetics describes how the population of molecules in each state varies with time.
5) The kinetics, the reactants, intermediates, and products constitute an ordered array of states, a chemical kinetic mechanism.
6) Each intermediate is associated with a different energy minimum in a highly multidimensional conformational space.
7) Intermediate states adjacent in the overall reaction are separated by an energy barrier. Molecules cross the barrier very rapidly, at random times. The highest point of the barrier is denoted the transition state.
8) How individual molecules cross the barrier is an active research topic. Although most barrier crossings are unsynchronized, crossings may remain synchronized during the initial, few fs to ps time scale.
9) An experiment that probes a statistically large number of molecules (such as crystallography, solution scattering, or optical spectroscopies) does not result in a molecular movie of continuously-changing molecular structure. Rather, the time dependence in such experiments arises from kinetics, the variation with time of the populations of structurally distinct reactant, intermediates, and product molecules.

Dynamics and Kinetics in Structural Biology: Unravelling Function Through Time-Resolved Structural Analysis, First Edition. Keith Moffat and Eaton E. Lattman.
© 2024 John Wiley & Sons Ltd. Published 2024 by John Wiley & Sons Ltd.

10) In contrast, any experimental or computational experiment that probes the structure of one molecule at a time (such as single particle X-ray scattering, cryo-EM, or molecular dynamics) can generate an authentic movie of continuously changing molecular structure that arises from dynamics.

Time-resolved experiments have been reviewed from the earliest, tentative developments to the present (Moffat 1989, 2001; Cruickshank et al. 1992; Schlichting et al. 1994; Hajdu 2002). More recent reviews (Moffat 2017; Šrajer and Schmidt 2017; Spence 2018) provide experimental details, cite papers in which these originate and are developed, and identify successful applications (Colletier et al. 2018; Orville 2020; Pearson and Mehrabi 2020; Malla and Schmidt 2022). The details are briskly evolving. We concentrate here on the underlying experimental principles, likely to remain relevant.

At $t=0$ the crystal is in thermal equilibrium. Rapid mixing of reactants, or illumination by a pulse from a visible laser, IR laser, or an electric field, drives the system from equilibrium (Chapter 3). We reemphasize that the populations of the reactant, intermediate, and product states of the molecules in the crystal rise and fall as the reaction proceeds and the system ultimately returns to equilibrium. The structures and structure amplitudes differ from state to state but are themselves time-independent. The critical experimental requirement at both SR and FEL sources is to obtain accurate measurements of the time-dependent, X-ray integrated intensities $I(hkl,t)$, from which the structure amplitudes $|F(hkl,t)|$ can be directly obtained. (The meaning of "integrated" intensities is discussed below.) The 4D data space hkl,t is large and defined by three considerations. The values of hkl must span the unique volume in reciprocal space with high completeness to the desired resolution; the delay times t between the pump and probe must cover reaction initiation ($t=0$) to completion ($t=t_{max}$); and the values of $I(hkl,t)$ must be measured with redundancy sufficient to achieve the required accuracy in both $|F(hkl,t)|$ and $\Delta F(hkl,t) = \{|F(hkl,t)| - |F(hkl,0)|\}$ for all values of hkl,t. The time-dependent structural changes are always small. As a rule of thumb, the mean value of $\Delta F(hkl,t)/|F(hkl,0)|$ lies in the range of 2–5%, but varies with the system and with time t. Since all intensities are subject to radiation damage (Chapter 4.3), complete coverage of the data space requires that diffraction patterns from many crystals be measured and scaled together to achieve accurate values of the mean value of $|F(hkl,t)|$ and its variance for all values of hkl and t. Accuracy requires minimization or elimination of systematic errors in $|F(hkl,t)|$ and $|F(hkl,0)|$ (Chapter 3.2.1).

The evolution with time of $|F(hkl,t)|$ can be followed by either monochromatic X-ray diffraction or Laue diffraction.

5.1.1 Time-resolved Monochromatic Diffraction

X-rays may be characterized by their wavelength λ or energy E where $E = hc/\lambda$, h denotes Planck's constant, and c is the velocity of light. The energy of an X-ray photon of 0.1 nm (1.0 Å) wavelength is 12.4 keV. The bandwidth (also known as bandpass) of an X-ray spectrum is $\Delta\lambda/\lambda = \Delta E/E$, where $\Delta\lambda = \lambda_{max} - \lambda_{min}$.

The generation of diffraction patterns by monochromatic and polychromatic X-ray beams is most clearly illustrated by the Ewald construction in reciprocal space (Appendix A and (Rupp 2010)). See also the powerful video tutorial XrayView created by George Phillips, accessible at https://xray.utmb.edu/tutorials/index.html. We consider first the more familiar monochromatic X-ray probe beam of wavelength λ, derived for example at a polychromatic SR source by including a narrow bandwidth monochromator in the beam-line between the source and the crystal. During the X-ray exposure, the crystal must rotate through an angle large enough to span the effective angular rocking width of all spots in the diffraction pattern. Each spot is derived from a single reciprocal lattice point (RLP) as it traverses the Ewald sphere of radius $1/\lambda$ (Figure A4 in Appendix A). Rotation ensures that all volumes within the crystal diffract X-rays from all areas within the source. This condition is necessary to generate the integrated intensities from which structure ampli-tudes can be directly extracted. In monochromatic diffraction, integration takes place over angle as the crystal is rotated. See Appendix A.

One component of the rocking width arises from crystal disorder, modeled by tiny mosaic blocks within the macroscopic crystal whose angular orientation and cell dimensions differ slightly (Rupp 2010). The existence of a rocking width implies that each RLP is no longer a mathematical point as it would be for a perfect, disorder-free crystal of infinite size. Rather, each RLP is associated with a small volume whose shape and dimensions depend on the crystal disorder (Figure 5.1). If the rotation angle is less than the rocking width (or if the RLP lies at the edge of the full rotation range), only a fraction of the volume associated with the RLP will contribute to the observed intensity. Such spots will be recorded as only a partial intensity at that RLP. The extent of partiality must be estimated for each spot and applied to the observed intensities to obtain the integrated intensities. Accurate estimation can be challenging, which emphasizes the desirability of directly measuring integrated intensities for all RLPs.

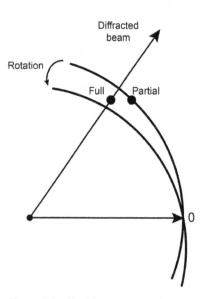

A second component enlarges the experimen-tal rocking width and arises from the nonparallel nature of the X-ray beam. Angular X-ray cross fire is introduced by tightly focusing the beam on a tiny crystal. Focusing is modeled by replacing the single Ewald sphere that represents a paral-lel, truly monochromatic beam by a nest of Ewald spheres, displaced in angle from each other. Each sphere passes through the origin of recipro-cal space, and each arises from a slightly differ-ent direction with which the X-ray beam is incident on the crystal. To visualize cross fire in practice, a single direction for the incident beam is retained, represented by a single Ewald sphere. The angular cross fire is modeled by combining it with angular crystal disorder to increase the angular range spanned by the mosaic blocks and thus slightly enlarge the volume associated with each RLP.

Figure 5.1 Ewald construction for monochromatic diffraction with crystal rotation. RLPs (filled circles) are either full if the entire volume of the RLP is traversed by the rotation of the crystal (and the reciprocal lattice), or partial if the entire volume is not traversed.

During crystal rotation, each RLP traverses the fixed Ewald sphere and generates a diffraction spot at a distinct, readily identifiable time. Rotation thus automatically introduces time dependence into the X-ray intensities and structure amplitudes. This simple approach has no special detector requirements. Even an integrating detector (Chapter 6.4) that lacks any time resolution will be adequate, since the time dimension is introduced solely by the crystal rotation. If the crystal will withstand fast rotation and the detector offers fast readout, then higher time resolution can be achieved by following the time dependence of the intensities of individual spots as the volume of each RLP traverses the Ewald sphere. However, each RLP diffracts only during the time that this volume traverses the Ewald sphere. This time is generally a small fraction of the total X-ray exposure time. Achieving stronger scattering would require slower rotation. Although conceptually very simple, this monochromatic approach to time-resolved crystallography is limited by these factors to time resolutions of ms and greater, and thus to probing only slower reactions in this range. Though straightforward and feasible, the approach has not been widely explored.

5.1.2 Time-resolved Laue Diffraction

Laue diffraction is named after von Laue, who with his students Friedrich and Knipping obtained the first X-ray diffraction patterns from a single crystal (Friedrich et al. 1912). Two characteristic features distinguish Laue diffraction from monochromatic diffraction: the crystal is stationary during the exposure and the X-ray beam is polychromatic, spanning a significant range of wavelengths. Although the *polychromatic* characteristic is generally regarded as distinctive, the *stationary* characteristic turns out to be more important.

The X-ray pulses emitted by SR sources have a duration around 100 ps. Those from XFEL sources are three or four decades in time shorter, with durations in the 10–100 fs range (Chapter 6.2 and 6.3). These times are much too short for any significant crystal rotation to occur as a single SR or XFEL X-ray pulse illuminates a crystal; crystals are effectively stationary. It follows that ALL single crystal diffraction experiments at SR and FEL sources that use single X-ray pulses (or pulse trains whose total duration is short enough to ensure no significant rotation) generate Laue diffraction patterns. The advent of intense single pulse SR sources and particularly, of XFEL sources has rejuvenated the Laue approach. Under these widely used experimental conditions, the Laue approach is inescapable.

A concise description of the crystallographic principles underlying Laue diffraction is given by (Moffat 1997; Ren et al. 1999). We give a brief summary here.

A stationary crystal may be illuminated by an X-ray beam of wide bandwidth ($\Delta\lambda/\lambda = 0.5 - 1.0$; informally known as a white beam), medium bandwidth ($\Delta\lambda/\lambda = 0.01 - 0.5$; informally known as a pink beam), or narrow bandwidth ($\Delta\lambda/\lambda = 10^{-2} - 10^{-4}$; a quasi-monochromatic beam). Experimental beams with these properties are common. For example, a rotating anode laboratory source offers a wide bandwidth beam in which intense CuKα and CuKβ emission lines, each of quasi-monochromatic bandwidth, are accompanied by a broad X-ray background of wide bandwidth. (In the pre-synchrotron days when the authors were research students, diffraction data were largely collected on film by precession photography, using a laboratory source. The process of orienting a newly mounted crystal in order to take a precession photograph of a specific reciprocal lattice plane (such as *hk0* or *h2l*) began by recording the diffraction from a stationary crystal, yielding a wide bandwidth Laue pattern.) The full spectrum from a wiggler or bending magnet source of synchrotron radiation is also of wide bandwidth (Chapter 6.2). The first harmonic spectrum of an undulator (Chapter 6.2.3 and 6.2.4), or the somewhat narrower spectrum obtained after an undulator beam traverses a multilayer monochromator, is an example of a medium bandwidth beam. An XFEL beam derived from self-amplified spontaneous emission (SASE; Chapter 6.3) or the beam from an SR after traversing a single crystal monochromator is an example of a narrow bandwidth beam.

The fact that diffraction patterns from a stationary crystal, irrespective of X-ray bandwidth, are Laue patterns is experimentally important. All diffraction patterns map part of the 3D reciprocal lattice on to a 2D detector. The mapping differs fundamentally between quasi-monochromatic, rotating crystal diffraction patterns and stationary crystal, Laue diffraction patterns. A monochromatic diffraction pattern is essentially a spatial pattern in reciprocal space, in which each spot on the detector maps an individual RLP. In a Laue diffraction pattern this mapping is quite different and is essentially an angular pattern in reciprocal space. A single Laue spot on the detector maps a central line in reciprocal space. (A central line is radial, passing from the origin *000* of reciprocal space through the RLPs *hkl, 2h2k2l, 3h3k3l, ...nhnknl....*) Each central line thus contains between one and several RLPs and is mapped into a single Laue spot (Figure 5.2).

Monochromatic and Laue diffraction are distinguished by four other features. First, Laue diffraction generates integrated intensities by integrating over X-ray wavelength, not by integrating over angle as the crystal is rotated in a monochromatic beam. However, for X-ray beams of quasi-monochromatic bandwidth such as those emitted by an XFEL or by an SR source with monochromator, the wavelength range may not be sufficient to give full integrated intensities. Each Laue spot is then partial (Figure 5.3). The extent of partiality depends on the bandwidth, the nature of crystal disorder, the angular cross fire, and the resolution of the RLPs contained within that spot.

Second, all Laue spots diffract for the entire duration of the X-ray exposure. This has both advantages and disadvantages. For a given integrated intensity, Laue spots typically require shorter exposures than monochromatic spots (Moffat 1997), which has the advantage for dynamic studies of enhancing the time resolution. Since all Laue spots are recorded simultaneously, systematic errors arising from intensity fluctuations in the X-ray beam during the exposure are minimized – an advantage. Errors arising from irregularities in rotation are absent. However, if the background X-ray scattering around each Laue spot is derived from a larger volume in reciprocal space than in a monochromatic rotation pattern, the background is increased – a disadvantage.

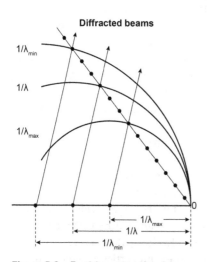

Figure 5.2 Ewald construction for Laue diffraction: stationary crystal, wide bandwidth. The crystal is illuminated by an X-ray beam of wide bandwidth from λ_{min} to λ_{max}, with an intermediate wavelength L. Diffracted X-ray beams from all RLPs along the central line in reciprocal space from the origin O are parallel. The diffracted beams are exactly superimposed. RLPs along each central line produce a single Laue spot on the detector.

Third, a Laue pattern automatically records all RLPs that lie between three spheres in reciprocal space: the limiting Ewald spheres corresponding to $1/\lambda_{min}$ and $1\lambda_{max}$, and the sphere centered on the origin of reciprocal space of radius d^*_{max}, corresponding to the maximum resolution to which the crystal diffracts (Figure 5.4). For a beam of medium or wide bandwidth, the volume between these three spheres is substantial and contains a large number N_L of RLPs to a resolution d^*, where

$$N_L = \pi(d^*)^4 (\lambda_{max} - \lambda_{min})/4V^*$$

and V^* is the volume of the reciprocal unit cell. Fewer Laue patterns are necessary to record the complete, unique volume – an advantage. However, there is a tradeoff: as the bandwidth increases, so does the background around each spot – a disadvantage. The dependence of N_L on $(d^*)^4$ means that RLPs at higher resolution whose structure amplitudes tend to be lower are measured with higher multiplicity – an advantage.

The fourth feature is the most significant. The mapping of each central line to a spot in a Laue diffraction pattern means that an individual Laue spot may contain contributions from several orders of an RLP. For example, *3h3k3l, 4h4k4l, 5h5k5l ... nhnknl ...* are the third, fourth, fifth, and nth orders (also denoted harmonics) of the first-order RLP *hkl*, in which *h*, *k*, and *l* are co-prime i.e. they do not have a common factor other than unity. Each set of planes in the crystal has a different d spacing and acts as its own monochromator. Thus, each order within an individual Laue spot automatically selects its own, different wavelength λ to satisfy Bragg's Law, and gives rise separately to an integrated intensity. However, the diffraction from all orders of an RLP along a central

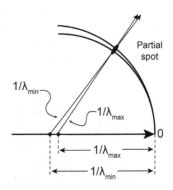

Figure 5.3 Ewald construction for Laue diffraction: stationary crystal, narrow bandwidth. The crystal is illuminated by an X-ray beam of narrow bandwidth (exaggerated for clarity) from λ_{min} to λ_{max}. Crystal mosaicity enlarges the volume associated with all RLPs; this volume for one RLP is shown as a small ellipse. Since some of its volume lies outside the Ewald spheres for λ_{min} and λ_{max}, this spot is partial.

Figure 5.4 Ewald construction for Laue diffraction: volume stimulated. All RLPs lying within the three spheres of radius $1/\lambda_{min}$, $1/\lambda_{max}$, and the limiting crystal resolution d^*_{max} are in diffracting position and contribute to a Laue spot. The spheres are obtained by rotation of the three circles shown around the horizontal, incident X-ray beam.

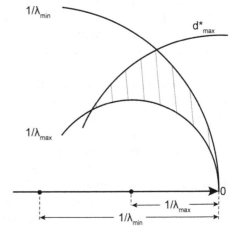

line in reciprocal space overlaps exactly to form a single Laue spot on the detector (Figure 5.2). Although each Laue spot contains an array of wavelengths, large area detectors lack sensitivity to X-ray wavelength. Detectors can record only the total intensity, the sum of the intensities from each order contained in that spot. This is known as the *overlapping orders problem*. However, quantitative crystallography requires the individual structure amplitudes for all orders within each spot. This problem is potentially serious. If it cannot be overcome, the application of Laue diffraction to all forms of quantitative crystallography including its dynamic versions is limited – a serious disadvantage. Indeed, although Laue diffraction was widely and successfully used in the earliest days of crystallography (Moffat 2019), quantitative limitations led to its slow replacement in the 1930s by quasi-monochromatic, rotating crystal approaches.

The advent 50 years later of naturally polychromatic, medium to wide bandwidth SR sources led to reexamination of the principles of Laue diffraction and its suitability for structure determination (Moffat et al. 1984). Does the overlapping orders problem affect all, many, or only a few RLPs? How many and which RLPs contribute to each Laue spot? The answers depend on the X-ray bandwidth (Cruickshank et al. 1987). Even in the extreme case of an infinite bandwidth, 72.8% of RLPs occur as the single order in a Laue spot, from which it is straightforward to extract structure amplitudes. For experimentally realistic pink beams of medium bandwidth, this value increases to 83–85%. Evidently, the overlapping orders problem affects only a minority of RLPs, 15–17%. Even when the bandwidth is further increased in a white beam to stimulate a larger volume of reciprocal space, the intensities of the individual orders of each RLP in the minority of Laue spots with multiple orders could be experimentally and analytically isolated (Ren and Moffat 1995a). Concomitantly, it proved possible to determine directly from the diffraction patterns the wavelength-dependent correction factors to the raw intensities of each Laue spot such as the shape of the incident X-ray spectrum. This allowed the structure amplitudes of each individual order to be obtained from all spots in Laue diffraction patterns. The accuracy of these structure amplitudes is entirely comparable to those obtained from conventional, monochromatic rotation patterns (Ren and Moffat 1995b). If accurate quantitation did not exist, static and dynamic structure determination by serial ns or serial fs crystallography at SR and FEL sources would be at best seriously hindered and at worst, impossible.

Initial complexities in indexing Laue patterns from SR and XFEL sources have largely been overcome. Several excellent software packages for pre-indexing, indexing, and extraction of accurate structure amplitudes from XFEL data are available such as CrystFEL (White 2019); nxds (Kabsch 2014); Cheetah (Barty et al. 2014); DIALS (Winter et al. 2018); and pinkindexer (Gevorkov et al. 2020). The very narrow bandwidth from XFEL sources does not require wavelength normalization or resolution of the intensity contribution of each order. In contrast, the wider bandwidth from SR sources requires these additional features to obtain accurate structure amplitudes from all orders within a Laue spot. Precognition/Epinorm, widely used under its developer Zhong Ren at BioCARS (Ren et al. 2020), has not been separately published but has been commercially developed by Renz Research Inc. Its predecessor package Laueview, also developed by Ren, Srajer, and colleagues at BioCARS (Ren and Moffat 1995b), continues to be widely used. CrystFEL has been modified by the Meents group to process medium bandwidth Laue data from SR sources (Stellato et al. 2014). Rather than determining the X-ray spectrum directly from the

Laue data, their approach uses a separately-measured X-ray spectrum to conduct wavelength normalization, and includes pinkIndexer to index these Laue patterns.

A fast shutter train was developed (LeGrand et al. 1989) that could cleanly isolate single, useful X-ray pulses from a prototype hard X-ray undulator at CHESS (Szebenyi et al. 1988, 1992). Successors to this prototype became the workhorse undulator A at the APS. By the mid-1990s, undulator sources at the ESRF and the APS were intense enough to allow single 100 ps X-ray pulses to generate excellent Laue diffraction patterns from strongly diffracting, stationary protein crystals (Bourgeois et al. 1996). When single pulses are used, the time resolution is enhanced from the duration of a short train of X-ray pulses (say, 100 μs) to the duration of that single pulse (say, 100 ps) – a very large factor of 10^6. Structure determination by ultrafast crystallography on the 100 ps to ms time scale becomes feasible. Indeed, the first example of dynamic time-resolved crystallography with single X-ray pulses was conducted at the ESRF. CO rebinding to myoglobin after photolysis could be followed with ns time resolution (Srajer et al. 1996).

The advent of the first hard XFEL source, the LCLS (Chapman et al. 2011), posed new problems to quantitative analysis. The quasi-monochromatic X-ray beam meant that in a single Laue pattern, most spots yielded only partial intensities. The mean X-ray energy jittered from pulse to pulse over about 20 eV at 10 keV. The fundamental lasing process based on self-amplified spontaneous emission (SASE) was derived from the amplification of noise in the spatial distribution of electrons in each bunch (Chapter 6.3). Lasing generated X-ray pulses that, despite their narrow bandwidth and quasi-monochromatic nature, were extremely spiky in X-ray energy, in time and in intensity. Furthermore, there was no correlation in these properties from pulse to pulse. The extreme brilliance of each tightly focused pulse severely damaged or even destroyed samples; all experiments were irreversible.

In short, most spots were partial, their intensities were subject to several inherently serious systematic errors, and the set of individual intensities from repeated measurements on different samples was extremely noisy. It was hard to imagine a more unpromising set of source characteristics with which to measure small, error-prone, dynamic differences in Laue spot intensities.

Small differences underlie not merely dynamic but also all static crystallographic measurements such as *de novo* phase determination by isomorphous replacement and anomalous scattering, and exploration of inhibitor and small molecule binding. Indeed, it could be argued that *crystallography is almost entirely based on the accurate measurement of small differences in X-ray diffraction*. This raised a fundamental question. Could quantitative conventional or novel time-resolved crystallography be conducted at XFEL sources? The prospective advantages of XFELs were so compelling that the challenges to accurate measurement of structure amplitudes and the small differences between them were directly attacked by many groups and progressively overcome.

First, the very welcome result was that diffraction before destruction proposed in (Neutze et al. 2000) was shown experimentally to be valid (Chapman et al. 2006, 2011). Diffraction data could indeed be obtained before the crystal was destroyed. Second, by recording many Laue diffraction patterns and averaging over the resultant large number of measurements of the raw intensity of each Laue spot, sufficiently accurate structure amplitudes could be acquired for each RLP (Kirian et al. 2011). In essence, averaging over a very large number of highly imprecise intensity measurements at an XFEL, each from a different tiny crystal, achieved the same result as relatively few, accurate measurements at an SR, each from a much larger crystal. Today, diffraction patterns can be recorded at XFELs under near-native conditions from tiny crystals, tiny liquid volumes and in favorable cases, even from single, strongly scattering objects – X-ray-based single particle imaging (SPI). Despite the fact that in nanocrystals, a nontrivial fraction of the molecules lie on the surface and may scatter differently from molecules in the interior, quantitative crystallography using the approaches developed for much larger crystals may indeed be conducted. Challenges of course remain; this is by no means a mature area.

5.1.3   Time-dependent Substitutional Disorder and Loss of Crystal Symmetry

In a static, non-reacting system in equilibrium at near-physiological temperature, all molecules in a crystal possess the same mean structure, and the crystal exhibits translational symmetry. However, the structures of individual molecules fluctuate thermally around that mean. Strictly speaking, all symmetry elements of crystals, including translational symmetry, are statistical in nature. As the reaction begins in a dynamic experiment, each molecule changes its structure as it proceeds through the states along the reaction coordinate. The population of molecules in the reactant, intermediate, and product states varies with time (Chapter 1.5). That of the reactant decays, that of each intermediate builds up and decays, and that of the product ultimately increases. A consequence of the changes in structure and population is that the translational symmetry characteristic of crystals is at least partly lost during the reaction. This process is known crystallographically as time-dependent substitutional disorder. The adjective *substitutional* describes the fact that the structure of an individual molecule originally in the reactant state is sequentially replaced – substituted – by its rather different structures in the intermediate states and finally in the product state.

An artifactual contribution to loss of translational symmetry is produced when the pump process does not produce spatially uniform reaction initiation throughout the crystal volume illuminated by the probe X-ray beam. One common example occurs in pumping by a laser pulse (Chapter 3.3.2.3) when the optical absorption of the crystal is high. The intensity of the pump beam diminishes markedly along its path, and hence the extent of reaction initiation also diminishes. Loss of translational symmetry also occurs when a reaction is initiated by diffusion after rapid mixing. The reaction is initiated first in the surface layers of the crystal and then at deeper layers as the reagent diffuses into the crystal: an onion model. This model is well illustrated in (Pandey et al. 2021) who studied substrate diffusion and ligand binding in enzyme crystals using high-repetition-rate, mix-and-inject serial crystallography.

Smaller crystals or crystals with a plate-like habit with one much smaller dimension will minimize the effects of absorption and if appropriate, of diffusion. The advantage of low absorption is exemplified by efficient initiation of the photocycle of PYP by a blue light pump pulse in tiny crystals (~5 μm) used at the LCLS, in contrast with larger crystals (~50 μm) at an SR source, BioCARS. Compare the DED maps in panels A and B of Figure 3.5 from the LCLS with the DED maps in panels C and D from BioCARS.

The effect of time-dependent substitutional disorder on X-ray diffraction depends on whether the molecules in the crystal behave independently of each other. Almost all experimental evidence to date in time-resolved crystallography suggests that this is so. This does not mean that molecules in the crystal behave exactly as if they were in solution (Chapter 1.2). A counterexample to independent structural changes might involve formation of an intermediate in one molecule promoting – or hindering – that formation in an immediately adjacent molecule. Examples of this behavior are found in hard crystals of certain minerals and inorganic molecules where their hard nature arises from very strong interatomic interactions and the absence of liquid water. However, examples are almost entirely absent in soft and squishy crystals of macromolecules with only weak intermolecular interactions and substantial liquid water. A recent, possible exception (Ramakrishnan et al. 2021) identified an ordered phase transition in a crystal of the riboO RNA aptamer. Cell parameters and the RNA structure change substantially but very slowly, spanning several hundred s. The changes associated with the phase transition across this macroscopic crystal are evidently synchronous on this lengthy time scale, which permits structure determination throughout the transition. Whether synchrony holds on the much shorter time scales of ns to s characteristic of tertiary structural changes is not known. We expect this fascinating observation to be explored further.

When molecules behave independently of each other, then the various intermediates appear and disappear spatially at random in the crystal lattice. The time-dependent structure factors over the volume of the crystal illuminated by the X-ray beam are simply the average over the structure factors of the different structures present, weighted by their populations i.e. their occupancies (Appendix A, Equation 10). The kinetics of the time dependence takes the form of a sum of exponentials, as they do in dilute solution (Figure 5.5).

Time-dependent substitutional disorder leads to transient diminution in crystal symmetry or even its complete loss. Not merely is exact translational symmetry lost but rotational

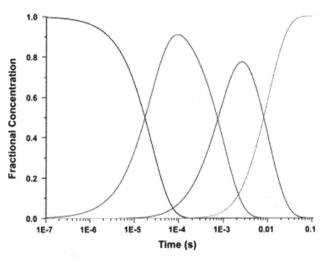

Figure 5.5 The kinetics of time dependence in a simple chemical kinetic mechanism. Example of a sequential, irreversible, chemical kinetic mechanism with four states: I0 -> I1 -> I2 -> I3 . The time-dependent fractional concentrations of the four states are plotted against log time. The three first-order rate coefficients are $k1 = 4 \times 10^4$ s^{-1}, $k2 = 10^3$ s^{-1}, and $k3 = 10^2$ s^{-1}. The rise and fall of the concentration of each state is a simple exponential. Rajagopal et al. 2004/Reproduced with permission from ELSEVIER.

and screw axes and centers of symmetry may no longer be exact. Molecules that prior to reaction initiation are related by the exact symmetry operators in the space group may progress along the reaction coordinate to a different extent. As they do so, they acquire differences in structure. A previously exact rotation axis becomes a time-dependent pseudo-rotation axis; previously exact screw axes or centers of symmetry become inexact. An interesting consequence is that diffraction spots, initially of zero intensity due to space-group-specific systematic absences, acquire time-dependent, nonzero intensity. This effect does not appear to have been looked for experimentally.

A diminution in crystal symmetry can also be generated by the nature of the pump beam itself. Chromophores in light-sensitive photoreceptors, either natural or produced by optogenetic approaches, are often planar and strongly optically anisotropic. If a pump light pulse is plane-polarized, then the probability of absorbing a photon and hence of initiating a reaction depends strongly on the orientation of the plane of polarization of the pump pulse with respect to the crystallographic axes (Figure 5.6). If for example a dimeric molecule lies on a twofold symmetry axis of the space group, one monomer may have a high

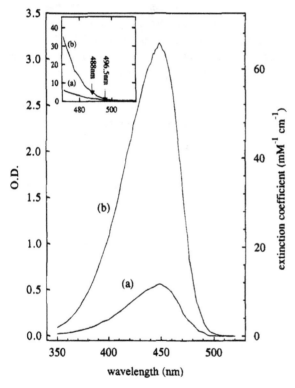

Figure 5.6 Plane polarized absorption spectra of a single PYP crystal. Absorption spectra at room temperature with (a) the plane of polarization parallel to the crystal long axis, which contains the sixfold crystallographic axis of symmetry, and (b) the plane perpendicular to the crystal long axis. The ratio of the optical density (O.D.) parallel/perpendicular is constant at 0.19 ± 0.01 over the wavelength range from 400 to 470 nm, which indicates that the optical transition moment can be characterized by a single polarization direction. Ng et al. 1995/Reproduced with permission from American Chemical Society.

probability of absorption and its twofold symmetry mate, a low probability: the crystal symmetry is lowered by absorption.

Not all molecules absorb visible photons, but all are sensitive to an electric field (Chapter 3.3) and the direction of its application with respect to the crystallographic axes. Unless the crystal is triclinic, applying an electric field pulse to a crystal lowers its symmetry. However, if the field is exactly aligned along a rotation axis in a (non-triclinic) space group, that rotational symmetry is retained.

Molecules that are structurally identical by space group symmetry may become nonidentical by one or more of the effects of substitutional disorder or the pump beam itself. The reduction in symmetry can be exploited to distinguish the desired authentic, biologically associated structural changes from noise such as structural changes produced by an artifactual temperature jump, which does not affect the crystal symmetry.

Finally, disorder in the crystal, whether translational, rotational, screw, or center-of-symmetry, results in an increase in the non-Bragg, diffuse scattering between the diffraction spots. In general, such scattering arises from coherent motions or static disorder of substructures within the crystal (Moore 2009). The diffuse scattering must also be time-dependent and, although much weaker than the Bragg scattering, is rich in structural information. If based purely on translational disorder, it may even extend to higher resolution than the Bragg spots due to its independence from the temperature factor inherent in Bragg scattering (Ayyer et al. 2016). The existence and nature of time-dependent diffuse scattering is an active area of research (Meisburger et al. 2020) that needs additional investigation.

5.1.4 Are My Crystals Active? What Time Points Should Be Collected?

It seems obvious: no X-ray-based dynamic experiment should be considered until it is established that the molecules in the intended sample (crystals or solution) are indeed active under the conditions planned for the experiment (Chapter 1.5). The requirement for demonstration of activity is met by an easily conducted, semiquantitative or better yet, quantitative spectroscopy experiment where a property such as fluorescence or absorbance varies as the reaction proceeds in the sample. A literature search aided by your friendly spectroscopists, biophysicists, and biochemists will show how activity in your system is assayed in dilute solution. Can you adapt this assay to the crystalline state? What time scales are accessible? Are there time periods during which no spectroscopic changes occur? What sample concentrations and amounts are required? The goal is not a publication-worthy kinetic study, though that is certainly possible. Rather, it is to demonstrate that your crystals are indeed active under specified conditions, both to your own group and to prospective funders and reviewers of SR and XFEL beamtime proposals. Initial, offline spectroscopy experiments may be conducted on the thin film obtained by crushing a larger crystal between coverslips and sliding them repeatedly over each other. Although crushing makes crystallographers wince, it's in an excellent cause. Sliding makes the film more uniform, achieves the desired thickness with low optical density and removes any effects of optical anisotropy in single crystals by randomizing the molecular orientations. It may then be possible to show that the kinetic behavior and overall mechanism in the crystal film at least qualitatively resembles that in the favorite conditions of biochemists and biophysicists: a dilute solution under not too extreme conditions of pH, ionic strength, and temperature. Even if your initially desired crystal form turns out to be marginally active or worse yet, inactive, other crystals with quite different molecular packing and constraints on activity may be obtained under different solution conditions, or of the homologous molecule from different species. For example, crystal forms of hemoglobin were eventually obtained in which crystalline order is retained throughout the large, functionally important change in quaternary structure on ligand binding and release.

Ideally, three time points per decade of time are required to scan a reaction in the absence of any prior knowledge of the time course in the sample. It is very unlikely that sufficient beamtime will be awarded at the SR or XFEL source to enable this entire time range to be explored, at least initially. Reviewers of your beamtime proposals – particularly if these proposals request access to experimental stations in high demand – are likely to be highly skeptical if the proposals lack any prior data on the reactivity and dynamics of your sample. Old literature references to an uncertain stage of the reaction in a related species in solution are simply inadequate.

Fortunately, good prior spectroscopic data enable informed selection of a few time points to be explored initially. For example, spectroscopy on crushed crystals at room temperature may reveal a time period, often of quite long duration over one or two decades in time when no spectroscopic changes are evident. These may well define a period when no structural changes are occurring. If only one structure is present throughout this period, that structure is time-independent, homogeneous and if need be, readily refinable by standard, static techniques from data at a single time point within that period. In short, X-ray data from one or only a few carefully chosen time points are likely to be readily analyzed, functionally important, and publishable. Such data make a strong case for further beamtime to explore the entire time course of the reaction. Caution is however needed. It is also – most inconveniently – possible that these are periods when the electronic structural changes that generate the spectra are indeed time-independent but uncoupled from the time-dependent changes in molecular structure that crystallography will reveal (Figure 5.7). In this example from the photocycle of PYP, two intermediate structures denoted pR_{CW} and pR_{E46Q} are simultaneously present for an extended time range. They happen to exhibit the same red-shifted absorption spectrum (hence the pR notation). This circumstance is more likely if the spectroscopy probes only local structure, for example the absorption (as here) or fluorescence of a chromophore, tryptophan or label; but much less likely if the spectroscopy probes overall structure, for example IR or CD absorption (Chapter 1.3).

5.2 Time-resolved X-ray Solution Scattering

X-ray solution scattering (XSS) has progressed dramatically in recent years as elaborated in recent publications (Svergun et al. 2013; Lattman et al. 2018; Brosey and Tainer 2019).

5.2.1 Basics of the Experiment

XSS has three major advantages: the uncertain process of crystallization is not required, molecules are unhindered by crystal lattice contacts, and molecules can be readily studied under conditions where they retain full biological activity.

(A)

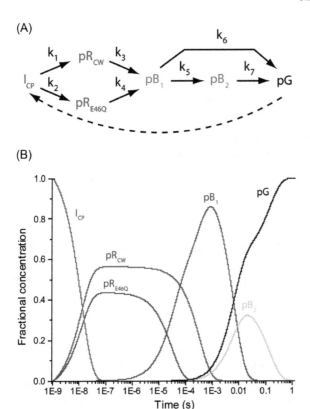

(B)

Figure 5.7   **The time dependence of the photocycle of wild type PYP. Panel A** The chemical kinetic mechanism derived from time-resolved crystallographic data for the photocycle of wild type PYP, with six intermediates. The dashed arrow indicates the light-driven reaction from the reactant (dark) state pG to the first intermediate state, I_{CP}. **Panel B** Predicted fractional concentrations of intermediates I_{CP} (red), pR_{CW} (magenta), pR_{E46Q} (purple), pB_1 (blue), pB_2 (cyan), and pG (black). In the time range from 10^{-7} to 10^{-5} s, intermediates pB_1 and pB_2 are both present and their concentrations are nearly constant in time. At all other times, the concentrations of intermediates vary sharply with time. The values of the seven first-order rate coefficients derived are given in the legend to Figure 4 of the original publication. Ihee et al. 2005/Reproduced with permission from Proceedings of the National Academy of Science.

XSS experiments are simple in outline. A solution of the protein or other sample is illuminated with a parallel, monochromatic, or narrow-bandwidth polychromatic X-ray beam and the scattered radiation is collected on a detector. A typical experimental setup is shown in Figure 5.8. The diffraction pattern is circularly symmetric about the axis defined by the X-ray beam, and the experimental data are recorded as a radial trace through this pattern, conventionally termed $I(q)$ where $q = 4\pi \sin\theta/\lambda$. The circular symmetry arises from two forms of averaging. First, many particles lie in the scattering volume, in random orientations. Second, angular Brownian motion of molecules during the exposure can spherically average diffraction information, but only when their rotational correlation time is short with respect to the duration of the X-ray exposure. Both forms of averaging are reduced or

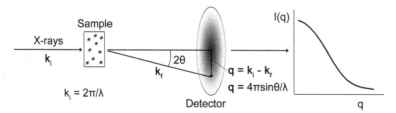

Figure 5.8   SAXS experimental setup. Incident X-rays corresponding to wave vector k_I with magnitude $2\pi/\lambda$ pass through the sample chamber and scatter at an angle of 2θ. The isotropic detector image is integrated and circularly averaged to produce a 1D scattering profile of intensity *I* versus the momentum transfer *q*. Lattman et al. 2018/Reproduced with permission from Oxford University Press.

eliminated when XSS data are collected with an intense and very brief, fs XFEL pulse. The intensity generates a good diffraction pattern with relatively few particles in the scattering volume, and the brief exposure time freezes their rotational motion. When the number of particles illuminated is small enough, important traces of the diffraction patterns of individual particles remain. The field of fluctuational X-ray scattering has developed to make use of this kind of information (Kam 1977; Donatelli et al. 2015).

XSS experiments may be roughly divided into two classes. SAXS experiments collect data for which the maximum value of q is at most a few times $1/D$, where D is the particle diameter. These classical experiments were developed by such leaders as Guinier and Kratky (Guinier et al. 1955; Glatter and Kratky 1982). Before SR sources became available, XSS experiments were confined to laboratory X-ray sources and were very challenging since they required great instrumental skill and long exposures. WAXS experiments extend data collection to much larger values of q where the structural signal becomes very weak, and have only become feasible at much more intense SR or XFEL sources. Though weak, WAXS curves contain much more structural information than SAXS curves. Figure 5.9 presents a schematic plot of $I(q)$ versus q, identifies the WAXS and SAXS ranges, and notes the levels of structural detail revealed at different values of q. Recent reviews provide critical approaches to the interpretation of XSS data (Hub 2018; Gräwert and Svergun 2020).

5.2.2   The *I(q)* Curve

The $I(q)$ curve represents an idealized function arrived at by considerable processing of the raw experimental data, outlined below. $I(q)$ represents the spherical average of the diffraction pattern of a single molecule, $F^2(\mathbf{q})$. If we carry out a *gedanken* XSS experiment on a single, immortal particle, each diffraction snapshot records a section through $F^2(\mathbf{q})$ sampled by an Ewald sphere (or by a narrow range of Ewald spheres in the case of a polychromatic, pink X-ray beam) arising from the current orientation of the single particle. As the particle rotates by diffusion through all angles, these snapshots are averaged over all orientations to produce the spherical average *of $F^2(\mathbf{q})$*, denoted $I(q)$. $I(q)$ is thus a function in three angular dimensions that can be completely specified by a single radial curve in one dimension q, also confusingly denoted $I(q)$. The dimensionality of $I(q)$ must be considered

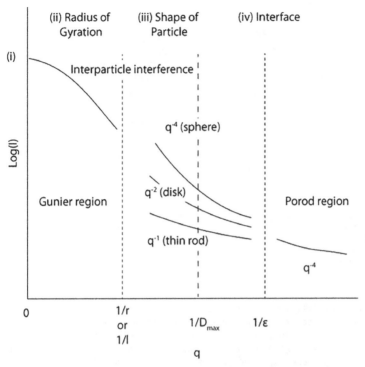

Figure 5.9 *I(q)* **plot with resolution ranges. Schematic of a typical XSS experimental scattering curve.** Four regions of interest are illustrated: (i) the intensity at zero angle; (ii) scattering primarily influenced by the radius of gyration; (iii) scattering associated with shape; and (iv) scattering associated with the interface of the particle and solution. D_{max} is defined in the text; R and L are radius (sphere and disc) and length (rod) and ε is the thickness of a disk or the diameter of the thin rod depending on particle shape. Lattman et al. 2018/Reproduced with permission from Oxford University Press.

in mathematical manipulations such as integration. The connection between these functions is clarified in Figure 5.10.

Raw XSS data are usually acquired as 2D-frames on a position-sensitive area detector. Considerable processing is required to obtain an approximation to the $I(q)$ curve, which ideally represents the scattering from the dissolved molecules alone. The raw 2D data must first be rotationally averaged to generate a preliminary 1D radial curve. This step includes corrections for detector sensitivity which may vary across its area, and accurate determination of the X-ray beam axis about which the averaging occurs. The next step is subtraction of the scattering from the buffer in which the sample is dissolved, typically recorded in a separate experiment. Subtraction of buffer scattering also removes extraneous scattering from slits and other optical elements in the beamline, common to both the sample and buffer datasets. Incoherent scattering is substantial and must also be subtracted. Most XSS experiments are carried out over a series of sample dilutions to remove any effects from sample aggregation at higher concentrations. The dilution series is extrapolated to estimate the $I(q)$ curve that would be obtained from an infinitely dilute sample (but at nonzero

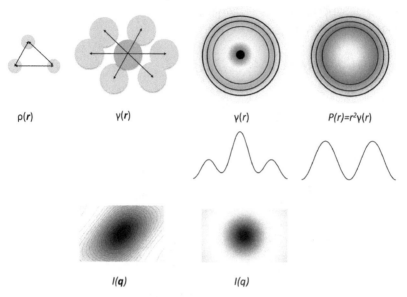

$\rho(r)$ $\gamma(r)$ $\gamma(r)$ $P(r)=r^2\gamma(r)$

$I(q)$ $I(q)$

Figure 5.10 The averaging process in XSS. In two dimensions, how the pair distance distribution function (Chapter 5.2.2.2) and the $I(q)$ profile arise from a single scattering object. In the top row $\rho(r)$ represents scattering density of a simple three-particle object, and $\gamma(r)$ is the Patterson function of ρ. $\gamma(r)$ – note the scalar r – is the rotational average of $\gamma(r)$. Directly below $\gamma(r)$ is a radial trace through it. Finally, $P(r)=r^2\gamma(r)$ depicts the pair distance distribution function telling how many vectors there are of length r, independent of the orientation of the vector. Just below P is a radial trace through it. In the bottom row the diffraction patterns from $\gamma(r)$ and $\gamma(r)$ are shown. Lattman et al. 2018/Reproduced with permission from Oxford University Press.

concentration!). Shot-to-shot variations in beam intensity also must be accounted for. This extensive data processing finally yields an $I(q)$ curve useful for XSS calculations.

Much structural information is unavoidably lost in the spherical averaging process. To compensate for the information loss, XSS is often applied in conjunction with other methods. An interesting example is given by (Kim et al. 2016), who combine optical spectroscopy with XSS to monitor the linkage between local and global structural changes in photoactive yellow protein. Additional examples are discussed below.

Experimental details are given by (Svergun et al. 2013; Lattman et al. 2018) and we emphasize only major points here. XSS experiments are conducted at a low sample concentration in the range of 1% w/w, to prevent aggregation of or interactions between solute molecules. Low concentration leads to weak scattering signal and requires relatively long X-ray exposures, though these are still much shorter at SR sources than at laboratory sources. Even at low concentration, XSS experiments need substantial amounts of pure sample. The high intensity beams at SRs and XFELs have transformed XSS by dramatically reducing the exposure time needed to accurately measure the weak scattering signal, but inevitably introduce radiation damage (Chapter 4.3). They have also enabled new classes of studies that effectively use difference XSS measurements to explore dynamic, time-dependent changes in $I(q)$ as a reaction proceeds, or static, time-independent ligand binding at equilibrium. XSS experiments are intensity-driven rather than brilliance-driven. The relatively low resolution of the scattering data allows both a larger transverse beam size and the use of a pink X-ray beam of medium bandwidth. Compared with a monochromatic beam, a pink beam gives stronger scattering and shorter exposures with better time resolution or alternatively, allows lower sample concentration.

Of even greater importance, the increased intensity at SR and XFEL sources has made WAXS experiments practical. Signal-to-noise issues remain very important. Kim and colleagues (Kim et al. 2020) have provided helpful estimates of the signal and noise in time-resolved X-ray solution scattering data at SR and XFEL sources. In particular, they provide code for significantly improving the accuracy of difference XSS spectra including the basic step of buffer subtraction, especially as applied to TR experiments. They emphasize the separable terms that arise in difference spectra: the solute-only term, the solute–solvent cross term (solvent cage), the solvent-only term, and noise.

5.2.2.1 Calculation of I(q)

We take a semi-intuitive approach to calculation of $I(q)$ based on crystallography. If $\rho(\mathbf{r})$ is the electron density of a particle and $\mathbf{F}(\mathbf{q})$ is its Fourier transform, then $F^2(\mathbf{q})$ is the single particle diffraction pattern that we could in principle observe. The Fourier transform (FT) of this diffraction pattern is simply the Patterson function of the particle density $\rho(\mathbf{r})$. Recall that the Patterson function gives the centrosymmetric distribution of all interatomic vectors in the particle; see Chapter 8 in (Glusker and Trueblood 2010). Thus $I(q)$, the rotational average of $F^2(\mathbf{q})$, is the FT of the rotational average of the Patterson function of the particle, known as $\gamma(r)$. Let us deconstruct $\gamma(r)$ into contributions from single pairs of scatterers. Any two scattering elements in the particle located at $\mathbf{r}_i$ and $\mathbf{r}_j$ contribute two non-origin vectors to the Patterson function, given by $\mathbf{r}_i - \mathbf{r}_j$ and $\mathbf{r}_j - \mathbf{r}_i$. These vectors have the identical magnitude termed r_{ij}. When rotationally averaged, each pair of scatterers contributes to $\gamma(r)$ a fuzzy spherical shell of radius r_{ij}. Thus, the complete distribution of $\gamma(r)$ can be thought of as a sum of such spherical shells, one for each r_{ij}. The FT of a spherical shell of radius r is of the form $sin(qr)/qr$ (Champeney 1973). Each pair of scattering elements adds a weighted $sin(qr)/qr$ ripple into $I(q)$. Summing over all pairs of scattering elements leads directly to the Debye Equation 5.1 for $I(q)$:

$$I(q) = \sum_{ij} f_i f_j \frac{sin\left(qr_{ij}\right)}{qr_{ij}} \tag{5.1}$$

where r_{ij} is the distance between the centers of scattering elements i and j; and f_i and f_j are the scattering form factors. In the case of atoms, f_i and f_j are the atomic scattering factors.

5.2.2.2 Pair Distance Distribution Function

The rotationally averaged Patterson function $\gamma(r)$ is defined in three dimensions but is spherically symmetric, having the same profile along any radius. It is convenient to project $\gamma(r)$ into a 1D function that will tell us the number of vectors of length r that are present in the particle. Such a plot is called the Pair Distance Distribution Function, often denoted $P(r)$, and is given by $r^2\gamma(r)$, where $r^2\gamma(r)dr$ is proportional to the number of vectors of lengths between r and $r + dr$ in $\gamma(r)$. (The conventional factor of 4π is normally omitted in the SAXS literature, appearing elsewhere in an integral.) This leads to the continuous version of the Debye Equation, Equation 5.2:

$$I(q) = \int_0^{D_{max}} 4\pi r^2 \gamma(r) \frac{sin(qr)}{qr} dr. \tag{5.2}$$

Here D_{max} is the maximum diameter of the particle. A standard source such as (Lattman et al. 2018) gives the derivation of the inverse FT relation between $I(q)$ and $P(r)$:

$$P(r) = \frac{r}{2\pi^2} \int_0^\infty qI(q)sin(qr)dq. \tag{5.3}$$

The maximum particle dimension D_{max} can be identified with the value r at which $P(r)$ falls to zero. However, determination of D_{max} lacks accuracy. The value of $I(q)$ is not measured to $q = \infty$, as required by the upper limit of integration in the equation for $P(r)$. This limitation introduces ripples into $P(r)$ and creates uncertainty about where it falls to zero. The method of indirect FTs has been introduced (Glatter 1977) to alleviate this problem. Flexible molecules may have values of D_{max} that fluctuate in time; specifying a unique value may be misleading.

5.2.2.3 Invariants

Even without modeling, analysis of an $I(q)$ curve can yield a number of model-independent quantities termed invariants. Any model being developed must agree with these invariants, of which D_{max} is one example. Perhaps the most important invariant is the radius of gyration R_g. It is obtained from the eponymous Guinier plot of log $I(q)$ versus q^2. For small q this plot has a linear region whose slope is $-R_g^2/3$ (Lattman et al. 2018). Other useful invariants described there and in the detailed sources include the particle mass, the particle volume, and the particle surface area.

The Kratky plot illustrated in Figure 5.11 provides a simple tool to estimate how compact the sample particle is, distinguishing, for example, fully folded proteins from molten globules and expanded structures. A graph of $q^2I(q)$ versus q falls to zero at high q for compact particles, plateaus for somewhat expanded particles, and continues to rise for flexible chain structures.

5.2.3 Dynamic Applications of TR SAXS/WAXS

5.2.3.1 Introduction

Intense, short X-ray pulses from SR or XFEL sources make possible the measurement of $I(q,t)$ curves at a series of time points t after a structural reaction has been initiated in the system, for example by a light pulse (Chapter 3.3.2.3). Families of curves can be collectively analyzed by methods such as singular value decomposition, SVD (Chapter 7.4). SVD reveals the significant orthonormal scattering curves (the left singular vectors), the number of such curves (significant singular values) that represent a component of the scattering to within a given noise level, and the time dependence of each curve (the right singular vectors). Initiation by light, T-jump, electron injection, and rapid mixing have all been applied (Plumridge et al. 2018; Rimmerman et al. 2019; Thompson et al. 2019; Kim et al. 2020). If

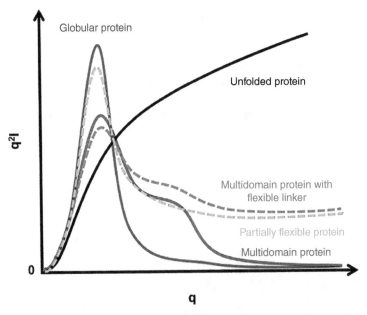

Figure 5.11 Kratky plot. Kratky plots offer a rapid characterization of the specimen. The figure shows such plots for a range of idealized samples including a globular protein, an unfolded protein, multidomain proteins, and samples containing flexible regions. Lattman et al. 2018/ Reproduced with permission from Oxford University Press.

the atomic structure of the ground-state molecule is known, MD simulations or related methods can be used to generate time-dependent ensembles of structures to compare with experiment. In favorable cases this analysis results in sets of atomic models that track the dynamics of the system and agree with the $I(q,t)$ curves. A comparison of conformational states generated in MD simulations and suggested by WAXS data appears in (Volmar et al. 2020). Changes in structural fluctuations at equilibrium, such as those that result from ligand binding, can also be analyzed.

We present several examples of XSS to illustrate different types of problems that are being attacked. Although information in the SAXS/WAXS data itself is somewhat limited, it is effectively complemented by static structures, spectroscopic data, and MD and other computational analyses. In favorable cases these lead to confident structural and mechanistic interpretation.

5.2.3.2 Example 1: Protein Quake

The bacterial photosynthetic reaction center (PRC) contains a set of antenna-like bacteriochlorophyll cofactors that absorb photons of green light and pass the energy on through the protein to initiate photosynthesis. The PRC molecule faces extreme issues of energy management. The energy in a single photon of green light is sufficient to unfold the protein. Although the PRC may absorb many photons over a short period of time, it remains intact in both structure and function. The energetics of protein folding is often monitored by calorimetry, in which the energy to unfold the protein comes in the form of heat. The protein thus remains close to equilibrium at all stages of the unfolding experiment. In contrast, absorption of an energetic photon pushes the system far from equilibrium, which requires novel mechanisms for the redistribution of the absorbed energy. Among these are the "protein quake," a general concept to describe how light energy is deposited, harnessed, and transmitted through a specific system such as the PRC (Ansari et al. 1985; Xie et al. 2001; Miyashita et al. 2003). The Neutze group provided experimental confirmation of this concept through a time-resolved WAXS study of the PRC over the fs to ps time range at the LCLS (Arnlund et al. 2014). Membrane fractions enriched in PRC were solubilized in the detergent lauryldimethylamine-N-oxide, and further purified by gel filtration chromatography. The PRC was pumped using an optical laser at 800 nm in the near IR with pulses of 50 fs duration. The ensuing changes in scattering were probed at 41 time points with time delays from 0.5 ps to 280 ps, each generating an $I(q,t)$ scattering curve. From these were subtracted the $I(q,0)$ curve at zero time, prior to the optical laser pulse, to generate a set of time-dependent difference curves. A subset is shown in panel b of Figure 5.12. A

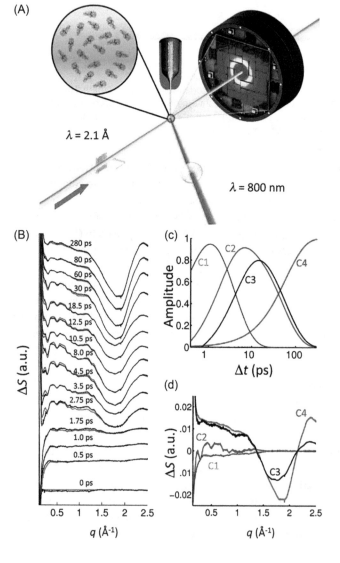

Figure 5.12   Protein quake studied by solution scattering. As described in the text, (A) shows the experimental setup, including the X-ray beam (yellow), the optical pump beam (red), the sample injector, and the detector. (B) shows a subset of the difference XSS curves with delay times. (C) shows the time-dependent weights of the four major components of the difference spectra panel (D). Arnlund et al. 2014/Reproduced with permission from Springer Nature.

combination of MD simulations and SVD analysis of the scattering curves identified four major contributors to (i.e. components of) the set of difference spectra. These were sufficient to account for the family of curves very well and provided a strong basis for a detailed understanding of energy propagation in the system. The four contributors (the left singular vectors) are shown in panel d of Figure 5.12; their amplitudes as a function of time (the right singular vectors) are shown in panel c. A picture of heat flow emerged, beginning in the bacteriochlorophyll antennae which had absorbed a photon from the 50 fs laser pulse and transiently heated to several thousand K. Absorption was followed by nonequilibrium heating of the protein, followed by heat flow over the ps time range into the surrounding protein, micelle, and solvent. MD simulations of the functional trajectory (Chapter 5.3) provided sets of structural coordinates whose calculated scattering curves agreed well with experiment. This work convincingly shows how atomic level information representative of a functional trajectory can be obtained from time-resolved WAXS studies at an XFEL.

5.2.3.3 Example 2: Folding of Cytochrome C from Heterogeneous Unfolded States
The folding of cytochrome c was monitored using time-resolved solution scattering carried out at the APS and ESRF SR sources (Kim et al. 2020). The experiments took advantage of the observation that the oxidation state of the heme cofactor is coupled with the folding

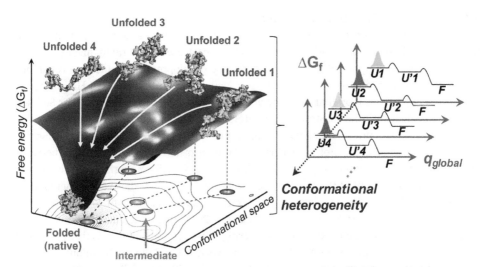

Figure 5.13   Cytochrome c folding dynamics landscape. Schematics of the folding dynamics in a funnel-like free-energy landscape. Upon the photoreduction of cyt-c, the heterogeneously populated initial unfolded state (red dots in the projected plane) undergoes spontaneous folding along the multiple parallel pathways (white arrows). Based on the EOM analysis with one MD trajectory over 1-μs duration, four conformations were obtained to represent the unfolded state. The number of representative unfolded conformations somewhat depends on the trajectory identity. The rugged features around the top of the free-energy funnel may be associated with the existence of intermediates (yellow dots in the projected plane) that are visited in the early phase of folding. The observed stretched exponential folding kinetics is the result of the superposition of the multiple pathways from the unfolded substrates (Ui, i = 1 to N) to the folded state (F) via intermediate substrates (U'i, i = 1 to N), indicating the conformational heterogeneity in the cyt-c folding. Kim et al. 2020/Reproduced with permission from National Academy of Science.

free energy. Conditions exist where the protein is fully unfolded in one oxidation state but fully folded in the other. The change in the oxidation state thus resculpts the free energy landscape (Chapter 8.3) and brings about a strong bias toward the native, folded state. The folding reaction was triggered by illuminating NADH in the solution of cytochrome c with a ns laser pulse at 355 nm, which allows NADH to act a light-dependent, external electron source. A reactive species with a solvated electron was generated in the µs time range which reduced the heme cofactor in unfolded, oxidized cytochrome c. Its rapid refolding could then be followed by SAX/WAXS.

Time-resolved difference scattering curves were measured from 31.6 µs to 316 ms. The difference scattering curves showed both a significant amplitude in the SAXS region ($q <$ 0.4 Å^{-1}) arising from a global conformational change, and oscillatory features in the WAXS region (0.4 Å$^{-1} < q <$ 2.2 Å^{-1}). The latter could be attributed to subtle structural changes such as the rearrangement of the secondary structure. Heat flows from the protein to the surrounding solvent between snapshots. $\Delta I(q,t)$ curves from adjacent time points thus contain contributions from this solvent effect, which must be considered in any modeling. As shown in Figure 5.13, the group analyzed the data using SVD and related methods to identify the number of distinguishable scattering species and made extensive use of MD to model both the kinetics of folding and the structures that represent major clusters in the unfolded ensemble.

5.2.3.4 Example 3: Modulation of HIV Protease Flexibility

The dynamics of target proteins plays a major role in drug design and development. A protease from the human immunodeficiency virus (HIV protease) is central to infectivity and a principal target for important anti-AIDS drugs. The development of multidrug resistance makes continued development of new drugs or lead compounds critical. In HIV protease, Thr 80 forms part of a loop that lines the active site and is highly conserved. Thr 80 itself makes no contact with the substrate or inhibitors, but its replacement by any other residue dramatically reduces protease activity. The mutation T80N is completely inactive. However, conventional spectroscopic probes such as circular dichroism reveal no differences between wild type (WT) and mutants at position 80. A revealing WAXS comparison of this mutant with WT was carried out at the BioCAT facility at the APS. When these experimental data were combined with MD simulations, they demonstrated a significant loss of flexibility across all length scales (Zhou et al. 2015; Wang et al. 2017) including decreased mobility in residues comprising the active site. Their scattering analysis of flexibility effectively rules out more specific, static conformational differences between WT and the T80N mutant. As determined by SAXS, the radius of gyration of T80N was only 0.03 nm smaller than that of WT.

WAXS data on the T80N mutant showed much more pronounced oscillations and more detail than corresponding data on the WT protease. The full analysis included MD simulations based on static crystal structure coordinates, and application of a vector convolution method to introduce flexibility into the calculation of the WAXS spectra in a natural way (Makowski et al. 2011). This study makes clear how nonspecific changes in the flexibility of a protein can dramatically affect function. From the point of view of energy landscape analysis (Chapter 8.3), a single mutation can raise the energy barrier between the ensemble of solution conformations and the set of conformations that are available for substrate binding and catalysis.

5.2.3.5 Example 4: XSS as a Probe with Stopped-Flow Kinetics

Stopped-flow is one of the most versatile and productive reaction initiation techniques in chemical kinetics. Aliquots of two or more solutions are driven by rapid, motorized syringes, first into a mixing chamber and then into an observation cell where flow is stopped. The evolution of the reaction in this stationary sample may be probed by any convenient technique such as optical spectroscopy or XSS. The combination of XSS and stopped flow has a long history in studies of protein folding both in equilibrium and TR studies, and has been the subject of excellent reviews (Doniach 2001; Chaudhuri 2015).

A recent XSS study extends the range of stopped flow approaches to reaction outcomes that involve large conformational changes (Josts et al. 2020). The bacterial membrane protein MsbA is a transporter, a member of the ABC superfamily of ATPases. MsbA plays an important role in multidrug resistance by transporting a broad range of drugs out of the bacterium. One of its normal roles is to transport lipid A to the outermost membrane leaflet of gram-negative bacteria, which delimits the periplasmic space. Since lipid A forms an essential component of the membrane leaflet, its transporter is a potential drug target. The study probed the transient dimerization of the nucleotide binding domain (NBD) of MsbA, an essential step in its functional cycle. A model (Mi et al. 2017) proposes that MsbA flips between periplasm-facing and cytoplasm-facing conformations during the functional cycle. The periplasm-facing conformation favors binding of ATP to the NBD, leading to transient dimerization of the NBD. Subsequent reorganization of the transmembrane domains creates a transport channel to the periplasm.

In the initial experiment, NBD solutions were rapidly mixed with substrate, ATP and Mg^{++} and XSS data collected at time points from 2 ms to 120 s. The evolution of R_g with time showed an initial increase ($\sim$1 s) corresponding to dimerization, followed by a plateau and then a slower return to the monomer profile as the substrate was exhausted (by 120 s). Careful analysis confirmed the presence of only two scattering species whose values of R_g are fully consistent with the known, static crystal structures. This was followed up by a more adventurous experiment in which full length MsbA was reconstituted in lipid nanodiscs. The authors carried out XSS measurements on the apoprotein and on the protein in the presence of a product analog, ADP-vanadate. The latter experiments were presented in detail (Josts et al. 2020). Uncertainties in the calculation of XSS curves for nanodisc-embedded MsbA molecules limited their interpretation. The capacity to perform these calculations with confidence will surely develop when additional data of excellent quality are acquired on this and related systems.

5.2.3.6 XSS and RNA Dynamics

Over the past decade, awareness of the vast repertoire of cellular RNA forms has increased dramatically (Fritz et al. 2016). A key aspect of RNA function is the process by which it folds within the cell. RNA folding processes are far different from those of proteins. In RNA folding, secondary structures comprising base-paired stretches form first. These elements then rearrange in a stepwise manner to achieve the final structural state. Intermediate states are often describable by a set of flexibly-linked secondary structural elements, which suggests that an ensemble approach is needed to describe the folding process (Woodson 2010). Divalent Mg^{++} is an important cofactor in folding, linking together secondary structural elements. A key feature of XSS from nucleic acids is the dominant contribution of the strongly scattering phosphorus atoms, which offer high contrast with the weaker scattering from the C, N, and O atoms in the backbone. When nucleic acid–protein complexes are being studied, this allows the powerful method of contrast matching to be applied. In contrast matching, the XSS contribution from either the protein or the nucleic acid can be masked by tuning the electron density of the solvent to match that of the protein or the nucleic acid, respectively. Solvent matching approaches have been extensively reviewed (Brosey and Tainer 2019; Ganser et al. 2019; Ponce-Salvatierra et al. 2019).

A technically excellent and biologically interesting study from the Pollack group (Chen et al. 2019) applies static and TR SAXS and ensemble modeling to explore the Mg^{++}-induced folding of a small RNA domain, the tP5abc three–helix junction. This moiety is a 56-nucleotide truncated subdomain of a ribozyme from *Tetrahymena* which provides a conveniently sized unit for the scattering studies. Static SAXS experiments on tP5abc were carried out in the presence of varying amounts of KCl or $MgCl_2$. KCl provides generalized electrostatic screening that encourages nonspecific collapse of the RNA, as revealed by a steady but noncooperative decrease in R_g as the concentration of KCl is increased. $MgCl_2$ is far more efficient at collapsing the molecule, since divalent Mg^{++} participates in global tertiary interactions that stabilize specific structures. To interpret the scattering curves at values of q beyond the Guinier region, the group used an insightful simulation procedure. This began with the generation of a very large pool of possible structures in which MD simulations perturbed both the NMR-based structure of the expanded KCl state and the tP5abc structure excised from the crystal structure of the intact ribozyme. This procedure is termed the ensemble optimization method. Intermediates along the putative pathway between the two structures were identified by simulation. SAXS curves from this large pool were calculated for all structures along a folding pathway and compared with experimental curves to find subsets of structures that best fit the SAXS data. These sets of structures allow the construction of an energy landscape for the folding process between the two structures.

A prescient account of TR SAXS in the study of RNA is given by (Pollack 2011). An excellent review of current static and TR SAXS applications is provided by (Brosey and Tainer 2019).

5.3   Molecular Dynamics Simulations

All-atom simulations of intramolecular protein motions have become increasingly powerful and sophisticated in the last decade and represent a tool in some ways orthogonal to the crystallographic and solution scattering experiments discussed above. Indeed, dynamic simulations play a major role in extracting atomic information from various classes of time-resolved studies. They also act as hypothesis-generating engines to suggest what kind of motions experimentalists might look for under various circumstances. We mentioned above (Chapter 5.2.3.2) the role of MD simulations in analysis of experimental data on a protein quake by SAXS (Arnlund et al. 2014). We present the general principles behind the most commonly used MD simulation technique, mention applications of MD to important studies, and provide references at various levels of sophistication.

Two classic works lay out the basics with clarity and vision: a slender book (Brooks et al. 1988) and a massively cited review article (Karplus and McCammon 2002). A more recent review provides a highly readable, nonmathematical, and practical introduction to understanding and carrying out MD simulations (Hollingsworth and Dror 2018).

The National Center for Biotechnology Information maintains a library of MD simulations of thousands of proteins, perhaps including your favorite. Access to this resource is described in (Meyer et al. 2010). The growing use of MD in drug discovery is well reviewed in (Salo-Ahen et al. 2021). And finally, the use of machine learning in advancing MD force fields (see below) is described in (Befort et al. 2021). We encourage readers to watch MD simulations. They help to diffuse the mental pictures we develop by looking at countless illustrations of proteins built of shiny balls, rigidly packed.

Sidebar

In the earliest days of protein crystallography, atomic structures were elaborately constructed of Kendrew models on a scale of 2 cm = 1 Angstrom. Covalent bonds were represented by brass wires, atomic positions by the intersection of these wires, and the whole was bolted down tight by tiny barrel connectors joining the wires. This gave rise to the "brass model mentality" (the phrase is only partly in jest) in which the model of a protein structure was inherently rigid. The models had to be examined with errant sleeves rolled up to avoid snagging surface side chains, which would introduce an artifactual change in structure.

5.3.1 The MD Algorithm

MD typically starts from a static structure such as a set of crystal coordinates and uses parameters such as bond lengths and angles derived from other sources such as precise crystal structures of small molecules. The simulations are straightforward in concept: apply Newton's laws to each of the atoms in a molecule, small or large, in order to follow their motion as a function of time. Simulations proceed by embedding the desired molecule in a large box containing a realistic representation of solvent. Thus, the MD simulation operates on a molecule-solvent system. If perfect (or at least very good) MD existed, experimental structural dynamics might not be needed. Such exactness still seems to be a long way off, but the range and accuracy of MD simulations continue to improve. Range encompasses both the time interval which a simulation can span, and the extent of the molecular potential energy surface that it can explore. Clearly, surmounting a high energy barrier will take more computer cycles than exploring flatter portions of the energy surface or landscape. We now describe the fundamental computational cycle that forms the heart of an MD simulation, and then identify modifications made to extend its range.

It is simplest to start the description in the middle. At any time t during the simulation the program knows:

1) The coordinates $\boldsymbol{x}_k = x_k$, y_k, z_k of each atom k included in the simulation, including solvent.
2) The velocity $\boldsymbol{v}_k = \mathrm{d}\boldsymbol{x}_k / \mathrm{d}t$ of each atom included in the simulation

3) The potential energy U of the molecule as a function of the atomic coordinates x_k. As elaborated below, U contains a number of terms, each describing the energy for a particular type of interatomic interaction.

4) The acceleration $a_k = dv_k / dt$ of each atom included in the simulation. The gradient derivative $-dU / dx_k = -(dU / x_k, dU / y_k, dU / z_k)$ gives the force F_k exerted on atom k by its neighbors. Through Newton's second law, the acceleration $a_k = F_k / m_k$, where m_k is the mass of atom k.

The simulation updates the quantities above after a short time step Δt, as follows:

$$x_k(t + \Delta t) = x_k(t) + v_k(t)\Delta t$$
$$v_k(t + \Delta t) = v_k(t) + a_k(t)\Delta t$$
$$F_k(t + \Delta t) = -dU / x_k(t + \Delta t).$$

A typical time step Δt is 2 fs, so many steps are required to cover appreciable times.

Although the simulation typically starts with a set of experimentally determined atomic coordinates, coordinates from modeling can also be used. Each atom is initially assigned a random velocity, with the average velocity chosen to set the temperature of the molecule. Each atom in a molecule has average energy $(3/2)k_B T$, the same as that of a small gas molecule flying freely at the same temperature. The instantaneous temperature T is related to the kinetic energy of the constituent atoms via an expression such as

$$T = \frac{1}{k_B(3N - 3)} \sum_j m_k \, | v_k^2 \, |.$$

The instantaneous temperature has to be averaged over some time period to be meaningful and can be altered by rescaling the velocities. Very high temperatures are used in the refinement of protein structures, to allow the molecules to move out of false local minima.

Typically, many simulation cycles are run in preparation for an experimental run, to allow the distribution of atomic velocities to reach equilibrium values.

The potential energy function U, often called a *force field*, lies at the heart of MD implementations. It synthesizes experimental data, theoretical development, and physical insights to provide the best possible summary of protein energetics. Each atom is acted on by forces that arise from its interaction with other atoms. These include the stretching and bending of chemical bonds, electrostatic and van der Waals interactions, and many other terms. In addition to atoms within the macromolecule, simulations also include interactions involving solvent molecules such as water or lipid, and buffer components such as ions. Each interaction is represented by a term in the expression for U. Thus in a prototypical representation:

$$U = \sum_{\substack{atom \\ pairs}} \frac{1}{2} K_b (b - b_0)^2 + \sum_{\substack{bond \\ angles}} \frac{1}{2} K_\theta (\theta - \theta_0)^2 +$$

$$\sum_{\substack{dihedral \\ angle}} K_\phi \left[1 + \cos(n\phi - \delta) \right] + \text{additional terms.}$$

In term one, b_0 represents the equilibrium bond length for the atom pair involved, say C–C, obtained from high resolution crystal structures, and K_b is the spring constant for the stretching of this bond, obtained from spectroscopy. Term two parameterizes the angle θ between two bonds originating on the same atom, $/^C_{\backslash}$. Term three describes the energetics of an n-fold rotation about a bond. Additional terms (not specified here) include electrostatic and van der Waals interactions, hydrogen bonds, interactions between water molecules and potential neighbors, and application-specific interactions with the atoms in ligands such as a drug. Finding an appropriate treatment for water molecules has been a key concern. Models such at TIP4 (Jorgensen et al. 1983) have long been workhorses, with continuing adjustments. Recently, attention has been given to the dynamics of water confined to tiny spaces, an idea relevant to protein crevices (Zaragoza et al. 2019). Other terms may be added to preserve sidechain geometry. For example, force fields in their simplest form have difficulty maintaining the planarity of ringlike groups such as a tyrosine side chain. One trick to retain planarity introduces a fictitious mast atom that lies above the plane of the ring, and adds a term to U that tightly constrains the distance between the mast atom and the real atoms of the ring. The mast atom is programmed not to take part in any other interactions such as van der Waals.

The simulation is carried out using a rectangular parallelepiped or cell to house the complete set of protein and solvent atoms being studied. Periodic boundary conditions are applied to this cell to ensure that if a solvent molecule drifts out of the defined cell, a translationally related solvent molecule appears on the other side of the cell.

5.3.2 The Range of MD Simulations

One of the greatest practical limitations on MD is the huge number of processor cycles required to simulate the system over times of biochemical or physiological interest, which may extend from fs to ms or even s. Long-time dynamics is intrinsic to biochemical systems that involve many atoms. In an ideal world, a goal might be to begin with a static molecule of a photosensor at equilibrium, allow it to absorb a visible photon, and model by MD the evolution of its structure. Even for a small PYP-sized molecule, this is a very tall order since the simulation would have to run for ~1 s to cover the photocycle. A second complication is that since the photon is energetic, the system is driven far from equilibrium and nonequilibrium thermodynamics is required to describe the early steps of the photocycle. But suppose this modeling were constrained by short-lived experimental structures on the few fs, or ps, or ns time scale. We then want to drive/morph the molecule along the reaction coordinate in a physically realistic way, from one experimental structure to the next.

Many techniques have been introduced to extend the range of the MD simulation. *Umbrella sampling* is a general method of helping a system overcome an energy barrier by adding an energy term that gently encourages the system to cross that barrier. For example, the unbinding of a ligand can be a very slow process on the MD time scale. By introducing a weak, nonspecific term to repel the ligand from its binding site, the process of unbinding can be speeded up without altering the structural details of how unbinding takes place. Another method for speeding up simulations is the intuitively named *coarse-grained sampling* in which chemical groups comprising several atoms are treated as a single entity (Jewett et al. 2021). *Gaussian networks* treat macromolecules as an elastic network of masses and Hookean springs that can be analyzed to examine important components of long-time dynamics.

Algorithms that generalize and expand upon umbrella sampling form part of what has become known as *accelerated molecular dynamics*. Their recent application to the folding of α-helical proteins illustrates the method very well (Duan et al. 2019). Another useful approach is replica exchange, in which parallel simulations are carried out at different temperatures, with the trajectories from these simulations being periodically swapped (Qi et al. 2018).

5.3.3   Simulation of Structures Not at Equilibrium

Specialized issues arise when, as with PYP, time-resolved experiments yield a set of intermediate structures that are not at equilibrium and may (particularly for short-lived intermediates) be strained, with unusual parameter values. Such issues were faced in a WAXS study of a protein quake in the PRC, introduced in Chapter 5.2.3.2 (Arnlund et al. 2014). In their simulation, they raised the temperature of the light-absorbing antennae in PRC to that calculated for the *in vitro* molecule after the absorption of a visible light photon. The procedure was validated by the agreement between experimental WAXS curves obtained at time *t* and those calculated from the MD model at the same time point. It is often overlooked that the agreement between the experimental and calculated curves provides confirmation for the force field employed during the simulation.

Many issues remain to be explored. How are MD simulations in the middle of a reaction coordinate conducted, i.e. simulations that begin with an intermediate structure? How can the initial equilibration described above be carried out? How can the reaction be driven forward, progressing further along the reaction coordinate, and not allowing it to go backward? The last is almost certainly not a computational problem when the reaction is very downhill energetically, such as after a photosensor absorbs a photon. But it could be a problem when the system remains near thermodynamic equilibrium as in reactions initiated by ligand binding and in most catalysis steps. Clearly, methods related to umbrella sampling will prove critical here.

5.3.4   Example: The Role of Lipids in Rhodopsin Function

An instructive example of the capacity of MD to illuminate biology comes from the Grossfield group (Salas-Estrada et al. 2018) on the mechanisms by which lipids alter the function of bovine rhodopsin. Lipids can interact with proteins in processes that exhibit the character of specific ligand binding, or as nonspecific solvent molecules where other characteristics such as overall lipid chain conformation are important. The group generated an ensemble of multi-ms, all-atom simulations of rhodopsin along the five principal steps in its photocycle. (Space does not allow us to describe them in full.) The molecules were embedded in realistic model lipid membranes. Each simulation was carried on six independent replicas of the rhodopsins. The critical light-activation step, isomerization about the specific C10–C11=C12–C13 double bond of the retinal chromophore in rhodopsin, was accelerated by the use of an external potential – an example of umbrella sampling. The resulting MD trajectories suggest that membrane modulation of rhodopsin activity is due to both ligand-like and solvent-like properties of lipids and that these effects are state-dependent. In the early stages of the photocycle, the stearoyl side chains of the lipid become more ordered. The paper contains significant discussion of lipid conformation and interactions around the whole photocycle. Lipid conformations sampled during the simulation are consistent with ^{2}H quadrupolar splitting observed in NMR experiments on the same lipids. The work illustrates the application of MD to a complex, activated system that requires sophisticated conformational analysis along the resultant trajectories to provide quantitative conclusions.

References

Ansari, A., Berendzen, J., Bowne, S.F. et al. (1985). Protein states and proteinquakes. *Proceedings of the National Academy of Sciences* 82 (15): 5000–5004.

Arnlund, D., Johansson, L.C., Wickstrand, C. et al. (2014). Visualizing a protein quake with time-resolved X-ray scattering at a free-electron laser. *Nature Methods* 11 (9): 923–926.

Ayyer, K., Yefanov, O.M., Oberthür, D. et al. (2016). Macromolecular diffractive imaging using imperfect crystals. *Nature* 530 (7589): 202–206.

Barty, A., Kirian, R.A., Maia, F.R. et al. (2014). Cheetah: software for high-throughput reduction and analysis of serial femtosecond X-ray diffraction data. *Journal of Applied Crystallography* 47 (3): 1118–1131.

Befort, B.J., DeFever, R.S., Tow, G.M. et al. (2021). Machine learning directed optimization of classical molecular modeling force fields. *Journal of Chemical Information and Modeling* 61 (9): 4400–4414.

Bourgeois, D., Ursby, T., Wulff, M. et al. (1996). Feasibility and realization of single-pulse Laue diffraction on macromolecular crystals at ESRF. *Journal of Synchrotron Radiation* 3 (2): 65–74.

Brooks, C., Karplus, M., and Pettitt, B. (1988). *Proteins: A Theoretical Perspective of Dynamics, Structure, and Thermodynamics'*. NY: John Wiley & Sons.

Brosey, C.A. and Tainer, J.A. (2019). Evolving SAXS versatility: solution X-ray scattering for macromolecular architecture, functional landscapes, and integrative structural biology. *Current Opinion in Structural Biology* 58: 197–213.

Champeney, D.C. (1973). *Fourier Transforms and Their Physical Applications*. New York: Academic Press.

Chapman, H.N., Barty, A., Bogan, M.J. et al. (2006). Femtosecond diffractive imaging with a soft-X-ray free-electron laser. *Nature Physics* 2 (12): 839–843.

Chapman, H.N., Fromme, P., Barty, A. et al. (2011). Femtosecond X-ray protein nanocrystallography. *Nature* 470 (7332): 73–77.

Chaudhuri, B.N. (2015). Emerging applications of small angle solution scattering in structural biology. *Protein Science* 24 (3): 267–276.

Chen, Y.-L., Lee, T., Elber, R. et al. (2019). Conformations of an RNA helix-junction-helix construct revealed by SAXS refinement of MD simulations. *Biophysical Journal* 116 (1): 19–30.

Colletier, J.-P., Schirò, G., and Weik, M. (2018). Time-resolved serial femtosecond crystallography, towards molecular. In: *X-ray Free Electron Lasers: A Revolution in Structural Biology* (ed. P.F.S. Boutet and M.S. Hunter), 331. Springer.

Cruickshank, D.W., Helliwell, J.R., and Johnson, L.N. (1992). *Time-resolved Macromolecular Crystallography*. Proceedings of a Royal Society Discussion Meeting, Held on 29 and 30 January 1992. Oxford [England]; New York: Oxford University Press.

Cruickshank, D.W.J., Helliwell, J.R., and Moffat, K. (1987). Multiplicity distribution of reflections in Laue diffraction. *Acta Crystallographica Section A* 43: 656–674.

Donatelli, J.J., Zwart, P.H., and Sethian, J.A. (2015). Iterative phasing for fluctuation X-ray scattering. *Proceedings of the National Academy of Sciences* 112 (33): 10286–10291.

Doniach, S. (2001). Changes in biomolecular conformation seen by small angle X-ray scattering. *Chemical Reviews* 101 (6): 1763–1778.

Duan, L., Guo, X., Cong, Y. et al. (2019). Accelerated molecular dynamics simulation for helical proteins folding in explicit water. *Frontiers in Chemistry* 7.

Friedrich, W., Knipping, P., and Laue, M. (1912). *Interferenz-Erscheinungen bei Röntgenstrahlen*. Sitzungsberichte der Kgl. Bayer. Akad. der Wiss. 303–322.

Fritz, J.V., Heintz-Buschart, A., Ghosal, A. et al. (2016). Sources and functions of extracellular small RNAs in human circulation. *Annual Review of Nutrition* 36: 301–336.

Ganser, L.R., Kelly, M.L., Herschlag, D. et al. (2019). The roles of structural dynamics in the cellular functions of RNAs. *Nature Reviews Molecular Cell Biology* 20 (8): 474–489.

Gevorkov, Y., Barty, A., Brehm, W. et al. (2020). Pinkindexer–a universal indexer for pink-beam X-ray and electron diffraction snapshots. *Acta Crystallographica Section A: Foundations and Advances* 76 (2): 121–131.

Glatter, O. (1977). Data evaluation in small angle scattering: calculation of the radial electron density distribution by means of indirect Fourier transformation. *Acta Physica Austriaca* 47 (1–2): 83–102.

Glatter, O. and Kratky, O. (1982). *Small-Angle Scattering of X-Rays*. New York: Academic Press.

Glusker, J.P. and Trueblood, K.N. (2010). *Crystal Structure Analysis: A Primer*. Oxford University Press.

Gräwert, T.W. and Svergun, D.I. (2020). Structural modeling using solution small-angle X-ray scattering (SAXS). *Journal of Molecular Biology* 432 (9): 3078–3092.

Guinier, A., Fournet, G., and Yudowitch, K.L. (1955). *Small-angle scattering of X-rays*.

Hajdu, J. (2002). The challenge offered by X-ray lasers. *Nature* 417 (6884): 15-15.

Hollingsworth, S.A. and Dror, R.O. (2018). Molecular dynamics simulation for all. *Neuron* 99 (6): 1129–1143.

Hub, J.S. (2018). Interpreting solution X-ray scattering data using molecular simulations. *Current Opinion in Structural Biology* 49: 18–26.

Ihee, H., Rajagopal, S., Šrajer, V. et al. (2005). Visualizing reaction pathways in photoactive yellow protein from nanoseconds to seconds. *Proceedings of the National Academy of Sciences* 102 (20): 7145–7150.

Jewett, A.I., Stelter, D., Lambert, J. et al. (2021). Moltemplate: a tool for coarse-grained modeling of complex biological matter and soft condensed matter physics. *Journal of Molecular Biology* 433 (11): 166841.

Jorgensen, W.L., Chandrasekhar, J., Madura, J.D. et al. (1983). Comparison of simple potential functions for simulating liquid water. *The Journal of Chemical Physics* 79 (2): 926–935.

Josts, I., Gao, Y., Monteiro, D.C. et al. (2020). Structural kinetics of MsbA investigated by stopped-flow time-resolved small-angle X-ray scattering. *Structure* 28 (3): 348–354. e343.

Kabsch, W. (2014). Processing of X-ray snapshots from crystals in random orientations. *Acta Crystallographica Section D: Biological Crystallography* 70 (8): 2204–2216.

Kam, Z. (1977). Determination of macromolecular structure in solution by spatial correlation of scattering fluctuations. *Macromolecules* 10 (5): 927–934.

Karplus, M. and McCammon, J.A. (2002). Molecular dynamics simulations of biomolecules. *Nature Structural Biology* 9 (9): 646–652.

Kim, J., Kim, J.G., Ki, H. et al. (2020). Estimating signal and noise of time-resolved X-ray solution scattering data at synchrotrons and XFELs. *Journal of Synchrotron Radiation* 27 (3).

Kim, T.W., Lee, S.J., Jo, J. et al. (2020). Protein folding from heterogeneous unfolded state revealed by time-resolved X-ray solution scattering. *Proceedings of the National Academy of Sciences* 117 (26): 14996–15005.

Kim, T.W., Yang, C., Kim, Y. et al. (2016). Combined probes of X-ray scattering and optical spectroscopy reveal how global conformational change is temporally and spatially linked to local structural perturbation in photoactive yellow protein. *Physical Chemistry Chemical Physics* 18 (13): 8911–8919.

Kirian, R.A., White, T.A., Holton, J.M. et al. (2011). Structure-factor analysis of femtosecond micro-diffraction patterns from protein nanocrystals. *Acta Crystallographica Section A* 67: 131–140.

Lattman, E.E., Grant, T.D., and Snell, E.H. (2018). *Biological Small Angle Scattering: Theory and Practice*. Oxford University Press.

LeGrand, A.D., Schildkamp, W., and Blank, B. (1989). An ultrafast mechanical shutter for x-rays. *Nuclear Instruments and Methods in Physics Research Section A: Accelerators, Spectrometers, Detectors and Associated Equipment* 275 (2): 442–446.

Makowski, L., Gore, D., Mandava, S. et al. (2011). X-ray solution scattering studies of the structural diversity intrinsic to protein ensembles. *Biopolymers* 95 (8): 531–542.

Malla, T.N. and Schmidt, M. (2022). Transient state measurements on proteins by time-resolved crystallography. *Current Opinion in Structural Biology* 74: 102376.

Matsuoka, H. (1999). An introduction to small-angle scattering. *Nihon Kessho Gakkaishi* 41 (4): 213–226.

Meisburger, S.P., Case, D.A., and Ando, N. (2020). Diffuse X-ray scattering from correlated motions in a protein crystal. *Nature Communications* 11 (1): 1–13.

Meyer, T., D'Abramo, M., Hospital, A. et al. (2010). MoDEL (molecular dynamics extended library): a database of atomistic molecular dynamics trajectories. *Structure* 18 (11): 1399–1409.

Mi, W., Li, Y.,Yoon, S.H. et al. (2017). Structural basis of MsbA-mediated lipopolysaccharide transport. *Nature* 549 (7671): 233–237.

Miyashita, O., Onuchic, J.N., and Wolynes, P.G. (2003). Nonlinear elasticity, proteinquakes, and the energy landscapes of functional transitions in proteins. *Proceedings of the National Academy of Sciences* 100 (22): 12570–12575.

Moffat, K. (1989). Time-resolved macromolecular crystallography. *Annual Review of Biophysics and Biophysical Chemistry* 18: 309–332.

Moffat, K. (1997). [22] Laue diffraction. *Methods Enzymol, Elsevier* 277: 433–447.

Moffat, K. (2001). Time-resolved biochemical crystallography: a mechanistic perspective. *Chemical Reviews* 101 (6): 1569–1582.

Moffat, K. (2017). Small crystals, fast dynamics and noisy data are indeed beautiful. *International Union of Crystallography Journal* 4 (4): 303–305.

Moffat, K. (2019). Laue diffraction and time-resolved crystallography: a personal history. *Philosophical Transactions of the Royal Society A* 377 (2147): 20180243.

Moffat, K., Szebenyi, D., and Bilderback, D. (1984). X-ray Laue diffraction from protein crystals. *Science* 223 (4643): 1423–1425.

Moore, P.B. (2009). On the relationship between diffraction patterns and motions in macromolecular crystals. *Structure* 17 (10): 1307–1315.

Neutze, R., Wouts, R., van der Spoel, D. et al. (2000). Potential for biomolecular imaging with femtosecond X-ray pulses. *Nature* 406 (6797): 752–757.

Ng, K., Getzoff, E.D., and Moffat, K. (1995). Optical studies of a bacterial photoreceptor protein, photoactive yellow protein, in single crystals. *Biochemistry* 34 (3): 879–890.

Orville, A.M. (2020). Recent results in time resolved serial femtosecond crystallography at XFELs. *Current Opinion in Structural Biology* 65: 193–208.

Pandey, S., Calvey, G., Katz, A.M. et al. (2021). Observation of substrate diffusion and ligand binding in enzyme crystals using high-repetition-rate mix-and-inject serial crystallography. *IUCrJ* 8 (6).

Pearson, A.R. and Mehrabi, P. (2020). Serial synchrotron crystallography for time-resolved structural biology. *Current Opinion in Structural Biology* 65: 168–174.

Plumridge, A., Katz, A.M., Calvey, G.D. et al. (2018). Revealing the distinct folding phases of an RNA three-helix junction. *Nucleic Acids Research* 46 (14): 7354–7365.

Pollack, L. (2011). Time resolved SAXS and RNA folding. *Biopolymers* 95 (8): 543–549.

Ponce-Salvatierra, A., Astha, Merdas, K. et al. (2019). Computational modeling of RNA 3D structure based on experimental data. *Bioscience Reports* 39.

Qi, R., Wei, G., Ma, B. et al. (2018). Replica exchange molecular dynamics: a practical application protocol with solutions to common problems and a peptide aggregation and self-assembly example. *Methods in Molecular Biology (Clifton, N.J.)* 1777: 101–119.

Rajagopal, S., Kostov, K.S., and Moffat, K. (2004). Analytical trapping: extraction of time-independent structures from time-dependent crystallographic data. *Journal of Structural Biology* 147 (3): 211–222.

Ramakrishnan, S., Stagno, J.R., Conrad, C.E. et al. (2021). Synchronous RNA conformational changes trigger ordered phase transitions in crystals. *Nature Communications* 12 (1): 1–10.

Ren, Z., Bourgeois, D., Helliwell, J.R. et al. (1999). Laue crystallography: coming of age. *Journal of Synchrotron Radiation* 6 (4): 891–917.

Ren, Z. and Moffat, K. (1995a). Deconvolution of energy overlaps in Laue diffraction. *Journal of Applied Crystallography* 28 (5): 482–494.

Ren, Z. and Moffat, K. (1995b). Quantitative analysis of synchrotron Laue diffraction patterns in macromolecular crystallography. *Journal of Applied Crystallography* 28 (5): 461–481.

Ren, Z., Wang, C., Shin, H. et al. (2020). An automated platform for in situ serial crystallography at room temperature. *IUCrJ* 7 (6): 1009–1018.

Rimmerman, D., Leshchev, D., Hsu, D.J. et al. (2019). Revealing fast structural dynamics in pH-responsive peptides with time-resolved X-ray scattering. *The Journal of Physical Chemistry B* 123 (9): 2016–2021.

Rupp, B. (2010). *Biomolecular Crystallography: Principles, Practice, and Application to Structural Biology*. Garland Science.

Salas-Estrada, L.A., Leioatts, N., Romo, T.D. et al. (2018). Lipids alter rhodopsin function via ligand-like and solvent-like interactions. *Biophysical Journal* 114 (2): 355–367.

Salo-Ahen, O.M., Alanko, I., Bhadane, R. et al. (2021). Molecular dynamics simulations in drug discovery and pharmaceutical development. *Processes* 9 (1): 71.

Schlichting, I., Berendzen, J., Phillips, G.N. et al. (1994). Crystal-structure of photolyzed carbonmonoxy-myoglobin. *Nature* 371 (6500): 808–812.

Spence, J.C. (2018). X-ray lasers for structure and dynamics in biology. *Iucrj* 5 (Pt 3): 236.

Šrajer, V. and Schmidt, M. (2017). Watching proteins function with time-resolved x-ray crystallography. *Journal of Physics D: Applied Physics* 50 (37): 373001.

Srajer, V., Teng, T.Y., Ursby, T. et al. (1996). Photolysis of the carbon monoxide complex of myoglobin: nanosecond time-resolved crystallography. *Science* 274 (5293): 1726–1729.

Stellato, F., Oberthür, D., Liang, M. et al. (2014). Room-temperature macromolecular serial crystallography using synchrotron radiation. *IUCrJ* 1 (4): 204–212.

Svergun, D.I., Koch, M.H., Timmins, P.A. et al. (2013). *Small Angle X-ray and Neutron Scattering from Solutions of Biological Macromolecules*. Oxford University Press.

Szebenyi, D., Bilderback, D., LeGrand, A. et al. (1988). 120 picosecond Laue diffraction using an undulator X-ray source. *Transactions of the American Crystallographic Association* 24: 167–172.

Szebenyi, D.M., Bilderback, D.H., LeGrand, A. et al. (1992). Quantitative analysis of Laue diffraction patterns recorded with a 120 ps exposure from an X-ray undulator. *Journal of Applied Crystallography* 25 (3): 414–423.

Thompson, M.C., Barad, B.A., Wolff, A.M. et al. (2019). Temperature-jump solution X-ray scattering reveals distinct motions in a dynamic enzyme. *Nature Chemistry* 11 (11): 1058–1066.

Volmar, A.Y., Guterres, H., Zhou, H. et al. (2020). Ras structure, dynamics and conformational states assessed by MD simulations and solution wide angle X-ray scattering. *The FASEB Journal* 34 (S1): 1-1.

Wang, Y., Zhou, H., Onuk, Emre. et al. (2017). What can we learn from wide-angle solution scattering? In: *Biological Small Angle Scattering: Techniques, Strategies and Tips* (ed. B. Chaudhuri, I.G. Muñoz, S. Qian, and V.S. Urban), 131–147. Springer.

White, T.A. (2019). Processing serial crystallography data with CrystFEL: a step-by-step guide. *Acta Crystallographica Section D* 75 (2): 219–233.

Winter, G., Waterman, D.G., Parkhurst, J.M. et al. (2018). DIALS: implementation and evaluation of a new integration package. *Acta Crystallographica Section D* 74 (2): 85–97.

Woodson, S.A. (2010). Compact intermediates in RNA folding. *Annual Review of Biophysics* 39: 61–77.

Xie, A., Kelemen, L., Hendriks, J. et al. (2001). Formation of a new buried charge drives a large-amplitude protein quake in photoreceptor activation. *Biochemistry* 40 (6): 1510–1517.

Zaragoza, A., González, M.A., Joly, L. et al. (2019). Molecular dynamics study of nanoconfined TIP4P/2005 water: how confinement and temperature affect diffusion and viscosity. *Physical Chemistry Chemical Physics* 21 (25): 13653–13667.

Zhou, H., Li, S., Badger, J. et al. (2015). Modulation of HIV protease flexibility by the T80N mutation. *Proteins: Structure, Function, and Bioinformatics* 83 (11): 1929–1939.

6

X-ray Sources, Detectors, and Beamlines

6.1 Introduction

Experiments in structural dynamics are complex and depend on a wide range of hardware including pulsed X-ray sources, beamlines, detectors, and other instrumentation. The source and beamline are not just components that are "out of sight, out of mind" but increasingly form an integral part of users' experiments. Taking advantage of the hardware's powerful and often unusual properties is essential to successful experiments.

We present general aspects of the sources, beamlines, and hardware most relevant to the design and conduct of experiments. Our treatment cannot be comprehensive; the detailed aspects of instrumentation such as detectors are rapidly evolving. We cite books and reviews that offer more information.

One important point should not be overlooked. In developing new hardware and software, beamline and facility staff consult as a matter of course with users such as structural biologists. They ascertain user views on the present limitations and how they might be overcome. User views are more valuable if based on the principles behind the existing hardware. Indeed, users often develop research and development ideas themselves.

Biological samples place severe demands on the beamline hardware. The samples are small in cross section, weakly scattering, high in water content, and subject to radiation damage (Chapter 4.3). To yield good scattering from such samples, many photons per unit time must be delivered in a stable, tightly focused, pulsed X-ray beam. These requirements impose strong constraints on the X-ray source. The pulses need to arrive at a suitable repetition rate, shaped and controlled by high precision X-ray optical devices such as monochromators and focusing mirrors, and tunable over a wide range of energy that includes hard X-rays, typically defined as >8 keV. Tunability of a monochromatic beam is critical to measuring anomalous scattering at the absorption edges of elements such as bromine and selenium, and also for various forms of X-ray spectroscopy (Chapter 8.4).

A polychromatic beam is required for Laue diffraction and many SAXS applications. A readily adjustable X-ray bandwidth providing a pink X-ray beam (Chapter 5.1) offers important advantages. In SAXS experiments a dramatic increase in flux can be obtained by increasing $\Delta E/E$ to perhaps 10^{-2}, much larger that the value of 10^{-4} used for crystallographic work. Comparable bandwidths are used for Laue experiments. Many of the benefits of pulsed radiation depend on the ability to isolate single pulses using a fast X-ray shutter train.

For ultrafast dynamic experiments, the time difference between the arrival of a pump light pulse and each probe X-ray pulse must be known for each diffraction pattern (but see Chapter 7.3.4 on powerful, computational approaches to timing uncertainty). Most scattering experiments are of low signal-to-noise and greatly aided by low noise. Background scattering from slits and other beamline elements must therefore be minimized. Finally, the X-ray beam at the sample must be stable, which requires stability in the source itself and all beamline components. For example, components must not suffer from vibration or drift that arises from heating by the X-ray beam.

Almost contradictory demands are placed upon detectors. For single particles, whether isolated or in solution, the scattering intensity $I(q)$ falls off roughly as q^{-4}. See Chapter 3 in (Lattman et al. 2018). This steep power dependence requires a detector that can accurately measure both the strong, low angle (small q; SAXS) and the weak, high angle (large q; WAXS) scattering data on the same sample at the same time, thus minimizing any q-dependent systematic errors (Chapter 5.2). This capability is more formally known as a large dynamic range, which in turn depends on the properties of the detector. Detectors whose inherent design is based on photon-counting or integration have quite different dynamic ranges. Photon-counting detectors typically work well at SR sources, which have much lower peak brilliance than XFELs. For XFELs, all the photons arrive within a time window measured in femtoseconds and integrating detectors are the only plausible technology today.

The detector must have a large area to capture the full wide angle scattering patterns. However, individual pixels must be small in area to resolve fine details in the patterns, highly uniform in their properties from pixel to pixel, and those properties must be stable. There are tradeoffs on pixel size. If pixels are too small ($<50\ \mu m$), charge bleeding between them occurs. If they are too large ($>300\ \mu m$), the sample to detector distance must be large to resolve the angular details. The very large number of pixels which results from large detector area and small pixel size generates data at a high rate, which must be evaluated, transmitted, and stored by a fast data processing chain to take full advantage of a high pulse rate from the X-ray source. Finally, many experiments, particularly at XFEL sources, place great demands on efficient sample delivery. The growth of robotics and automation in areas such as sample delivery, beamline control, and data processing is critical to making effective use of the large amounts of data generated at XFELs with a very high repetition rate such as the EuXFEL.

6.2 Sources of Synchrotron Radiation

6.2.1 Introduction

The advent of SR sources around 1980 triggered a rapid transformation of structural biology. These sources provided pulsed X-rays with a dramatic increase in performance over the rotating anode sources than available in the home laboratory. The most common and useful metric for the effective strength of an X-ray beam is the *brilliance* (Als-Nielsen and McMorrow 2011).

$$Brilliance = \frac{Photons}{Second \cdot mrad^2 \cdot mm^2 \cdot 0.1\% bandwidth}. \qquad (6.1)$$

Brilliance captures the concept of an intense beam, photons s^{-1}, and of a narrow, parallel beam, mm^{-2} and $mrad^{-2}$. A brilliant X-ray beam thus has both a small cross-sectional area (essential to illuminate a small sample such as a tiny crystal or droplet of solution) and a small angular divergence (essential to ensure that X-ray diffraction spots remain tightly focused and that scattered beams do not diverge significantly on their path to the detector). The *bandwidth* is a measure of monochromaticity. The inclusion of bandwidth in Equation 6.1 ensures that white or pink beams with a significant wavelength range – a substantial bandwidth – are compared on an equal footing with more monochromatic beams of narrow bandwidth.

Brilliance is a conserved quantity: if the X-ray beam is focused, its brilliance is unchanged. The decrease in spot size corresponds to an increase in divergence, keeping the brilliance the same.

There is considerable variability in the literature in this terminology. The term *brightness* is often used interchangeably with *brilliance*. The term brilliance was in use before the advent of pulsed sources, and *de facto* refers to *average* brilliance. The important related term *peak brilliance* or *pulse brilliance* describes more accurately the dramatic difference between continuous sources such as rotating anode generators and pulsed sources such as SRs and XFELs. For pulsed sources, peak brilliance is defined as the brilliance over the duration of a single pulse. Thus, for an XFEL source producing 10^{12} photons in a 20 fs pulse, the photons per second factor in Equation 6.1 would be 5×10^{25}. If the source produced 100 pulses per second, the average brilliance would be much lower: the photons per second factor would be $100 \times 10^{12} = 10^{14}$. The peak brilliance of LCLS at 10 keV is perhaps 10^7 times greater than that of SR sources (Boutet and Yabashi 2018). Because SR sources have a higher pulse rate, the average brilliance of the LCLS is only about a hundred times greater. The brilliance of the LCLS-II upgrade under way will increase this further.

X-ray beam brilliance has a counterpart called *emittance* (ϵ) in the electron beam that gives rise to the radiation. Emittance is often expressed in units such as millimeters-milliradians. Low emittance describes an electron beam that has small spatial extent and a narrow spread in momentum. That is, it has low angular divergence. For SR sources, the emittance is often quite different in the two transverse directions x and y (normal to the longitudinal direction z of the electron beam) and is often denoted by ϵ_x and ϵ_y. Low transverse emittance of the electron beam is essential to deliver high brilliance in the X-ray beam. XFELs try to keep the beam round, as do the high-brilliance upgrades of SR sources such as the ESRF and APS.

Both static and dynamic experiments in structural biology depend on a brilliant X-ray beam. Such experiments are said to be *brilliance-driven*. Enhanced brilliance benefits users in almost every discipline. The brilliance offered by SR and now XFEL sources has dramatically increased over time as the sources and the characteristics of the X-ray beams they emit have been progressively optimized (Figure 6.1). SR and XFEL sources now dominate X-ray-based experiments in structural biology generally and notably in structural dynamics.

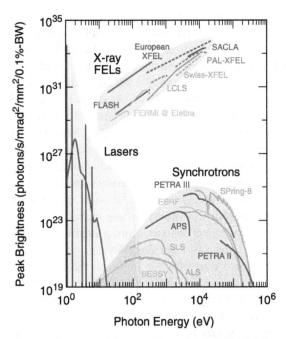

Figure 6.1 **Peak brightness.** Comparison of peak brightness as a function of photon energy between conventional lasers and higher harmonic generation sources, storage ring sources, and X-ray free electron lasers. Boutet and Yabashi 2018/Reproduced with permission from Springer.

Static experiments at SR sources are now routinely conducted remotely. The scientific team ships its samples in advance to the source staff, stays home and to conduct their experiment, logs on to the beamline from home. However, experiments in structural dynamics remain sufficiently complex that an expedition by the scientific team to the beamline is essential.

Synchrotron radiation sources can be roughly classed into so-called generations marking their evolution. The first-generation sources in the 1960s and 1970s used the synchrotron radiation at hard X-ray wavelengths emitted as electrons traversed the bending magnets in accelerators known as electron synchrotrons. These accelerators (such as DESY at Hamburg and NINA at Daresbury) were designed for fixed target experiments in high-energy physics. To the high-energy physicists, synchrotron radiation was an unavoidable nuisance that destabilized the electron beam, degraded its current, and compromised their experiments. To physicists more generally, materials scientists and biologists, the radiation offered promising new probes of the structure of matter (Watson and Perlman 1973). (In the eyes of some high-energy physicists, the synchrotron scientists themselves became the nuisance!) However, the electron energy increased substantially during the acceleration phase of electron synchrotrons, which caused the X-ray spectrum of the radiation to vary markedly with time. This variation greatly complicated X-ray experiments, reduced any advantages of synchrotrons over laboratory sources, and raised doubts whether synchrotrons were indeed useful sources.

Fortunately, the interests of the high-energy physicists and the would-be users of synchrotron radiation coincided. Both benefited from constant electron energy with lower emittance. In the mid-1970s, new storage ring (SR) accelerators such as SPEAR at Stanford and CESR at Cornell were constructed for high-energy physics experiments, in which counter-propagating beams of electrons and positrons at constant energy collided head-on. The low emittance of the electron beam and its constant electron energy for the first time provided stable X-ray beams of high brilliance, though the circulating electron current diminished over a few hours as electrons were lost. Around 1980, second-generation sources such as the National Synchrotron Light Source (NSLS) at Brookhaven National Laboratory and the Photon Factory (PF) at Tsukuba were the first purpose-built sources of synchrotron radiation, independent of high-energy physics experiments. These sources harnessed the synchrotron radiation emitted as electron beams of constant energy passed through their bending magnets. Third-generation sources still used radiation from their bending magnets but placed specialized magnetic arrays, insertion devices known as wigglers and undulators (discussed below), into the straight sections between the bending magnets as their primary hard X-ray radiation sources. Undulators in particular offered a substantial increase in brilliance over bending magnet sources. Third generation sources such as the ESRF at Grenoble, the APS at Argonne National Laboratory, Spring8 at Hyōgo Prefecture, Japan, and Diamond outside Oxford were also SRs in which the circulating electron current was maintained by constant replenishment (top-up), which further stabilized the characteristics of their X-ray beams. Today's fourth generation sources use an arrangement of the bending magnets that differs from third generation sources, and incorporate a low electron emittance photocathode source to further enhance the brilliance of their X-ray beams.

The exceptional stability of the X-ray beams from third and fourth generation SR sources in many properties such as brilliance, spatial and angular position of the source point, X-ray spectrum and timing of pulse arrival at the sample distinguishes them from all predecessor sources. Stability, particularly in time, is central to all dynamic experiments, which is why we have referred throughout to these sources as *storage ring (SR)* sources rather than the more common – but less revealing – usage of *synchrotron* sources.

6.2.2 Generation and Properties of Synchrotron Radiation

SR sources create X-rays using a large, donut-shaped ring at high vacuum in which circulating electrons are constrained to a roughly circular orbit by magnetic devices. This is illustrated by a cartoon in Figure 6.2a. Figure 6.2b shows a more realistic depiction of an SR source, discussed below. When electrons follow a curved trajectory e.g. along the arc of a circle, they are accelerated. Acceleration causes them to emit a spectrum of electromagnetic radiation known as synchrotron radiation. The peak of the spectrum depends on the magnitude of the acceleration and the energy of the electrons. If this energy is several GeV and the acceleration is achieved by passing the electrons through typical bending magnets, the peak lies in the hard X-ray range. In practical units the critical (peak) energy E_c is given by (Attwood and Sakdinawat 2017)

$$E_c(\text{keV}) = 0.6650 E_e^2 (GeV)(B_0),$$

where E_e^2 is the square of the electron beam energy in GeV, and B_0 is the magnetic field strength in Tesla.

Figure 6.2a   Cartoon of storage ring, insertion devices, and beamlines.

1. Electron pulse injection
2. RF system tops up pulse energy to compensate for radiation energy loss
3. Storage Ring/Bending Magnets
4. Beamlines
5. Experiments
6. Wigglers
7. Undulators

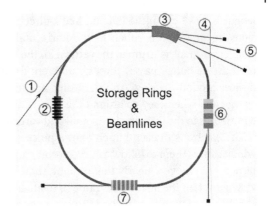

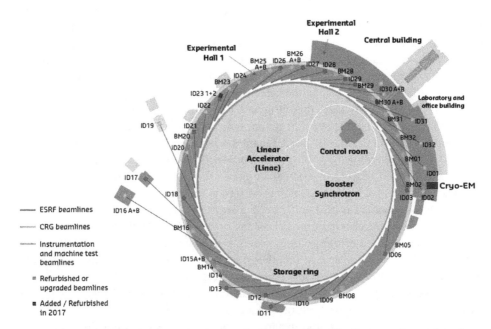

Figure 6.2b   Plan of the European Synchrotron Radiation Facility. The linear accelerator launches the electron pulses that are further accelerated in the booster synchrotron before being injected into the storage ring. Beamlines are shown on the periphery. The circumference of the ring is about 840 meters. Reproduced from the public ESRF website.

Electrons are not continuously distributed around the ring. Rather, they form bunches that occupy *buckets* generated by radiofrequency (RF) cavities that flank part of the ring and maintain the electron energy. RF cavities provide resonant taps to the energy of the electrons in the bunch through irradiation at the same frequency as that of the electron pulse train in its synchrotron orbit, in the MHz range. Accelerator physics effects limit the electron energy spread $\Delta E/E$ in the bunch to about 10^{-3}. At electron energies > 6 GeV, the electron bunch length is 0.5–3 cm in the longitudinal z-direction, along their orbit. The transverse dimensions of the electron bunch in the horizontal and vertical *x*- and *y*-directions are much smaller, as small as 10 μm in the vertical for the upgraded APS, and vary with position in their orbit as the bunch passes through different focusing, quadrupole and sextupole magnetic devices around the ring. An interesting side note is that as the electrons radiate, they lose energy and when they go around the next bend, they move inward. This is one factor that contributes to the differing horizontal and vertical emittances of synchrotron radiation.

The electrons in each bunch emit a pulse of X-rays as they traverse a magnetic device which causes them to follow a curved orbit and be accelerated. Electrons of > 6 GeV energy have a velocity very nearly that of light; they are said to be relativistic. From a laboratory viewpoint ("in the laboratory frame"), relativistic considerations tightly constrain the synchrotron radiation to a forward direction, tangential to the instantaneous path of the electrons. Emission of an X-ray photon of synchrotron radiation causes an electron to lose energy and change from its stable orbit in the ring to an unstable orbit, with the result that some electrons are continuously lost from each bunch. From time to time, top-up of the current in each bunch by injection of further electrons replaces those lost. The resultant intensity of each X-ray pulse is nearly constant in time. This quality is highly desirable for almost all experiments and most decidedly so for dynamic experiments, which depend on the accurate measurement of the time dependence of X-ray scattering.

The X-ray pulses from a storage ring are around 100 ps in duration, determined by the ~3 cm length of the electron bunch; bunches are normally separated in time by about 100 ns. The distribution of bunches and the exact separation between them is set by the operating mode of the SR, determined by how the buckets are filled with electrons. In some modes, all buckets are filled with electrons; in other modes, only some buckets. The mode with all buckets filled gives the highest circulating current of electrons and hence the highest flux and high brilliance of X-rays at every beamline around the ring. This mode is preferred by the relatively small number of users with samples of large cross-sectional area, conducting what are known as flux-driven experiments. In contrast, biological samples are generally tiny in area; high flux in a large beam offers no advantage. As alluded to above, high flux X-ray beams that originate in a low emittance electron source can however be tightly focused on tiny samples. These beams have high X-ray brilliance, well suited to the brilliance-driven experiments in structural biology.

In structural biology, static and dynamic experiments that require a time resolution > 100 ps can utilize the radiation from several sequential X-ray pulses in a short pulse train. Ultrafast experiments in structural dynamics that require a time resolution < 100 ps utilize the X-ray pulse from a single bunch. Such experiments seek an operating mode in which a single electron bucket is topped up with the largest number of electrons possible – a *fat bunch* whose X-ray pulse will be used. That bucket is usually flanked by several buckets that lack any electrons. These buckets are deliberately kept empty to minimize background arising from X-rays that their residual electrons might generate, as the fast shutter system that isolates single X-ray pulses may not be able to eliminate this background. All other buckets in the ring are filled with electrons in the normal way and provide a total circulating current that is high, but not quite the highest. This hybrid operating mode, called the *camshaft mode*, thus satisfies the needs of most flux-driven and brilliance-driven experiments and the specialized, ultrafast experiments.

The ring in a storage ring is in fact a polygon (Figure 6.2b). For example, the third generation APS has 34 straight sides. Dipole bending magnets at the vertices deflect the electron beam path from one edge of the polygon to the next. This centripetal acceleration in a bending magnet produces an X-ray beam with a wide bandwidth, white X-ray spectrum, emitted tangentially to the electron orbit. The bending magnets are designed to tightly focus the X-ray beam in the vertical direction (high brilliance) but provide a fan-shaped beam horizontally (much lower brilliance), as used in first generation sources. As we discuss below, X-ray beams may be monochromatized, focused, collimated, and otherwise processed using X-ray optical elements specialized for each experimental application.

Because the electron bunch traversing a dipole magnet is accelerated in the horizontal plane, the X-ray pulses are polarized in the horizontal plane. This property has been exploited in certain physics applications but has so far had little impact in biology. In crystallography, the polarization correction to spot intensities is significant and must take account of the polarization direction. Rotation cameras at SR sources have their rotation axis horizontal to record the stronger scattering in the vertical direction.

Laboratory X-ray sources are not tunable; the energy is fixed by the characteristic X-ray lines emitted by the element forming the anode. Since the X-ray spectrum emitted by bending magnet, wiggler, and undulator sources is of broad bandwidth, their X-ray beams are easily tunable over a wide range. They can readily be monochromatized by placing an Si or Ge single crystal monochromator in the beamline, which reduces the bandwidth $\Delta E/E$ to 10^{-3} or less. For X-ray solution scattering or other Laue experiments that do not require a narrow energy range, the bandwidth can be increased to 10^{-1} by replacing the single crystal monochromator with a multilayer monochromator to deliver a pink beam. This increases the flux by a highly advantageous factor as large as 10^2. Tuning is readily accomplished by adjusting the monochromator or multilayer monochromator, depending on the source of the radiation and the needs of the experiment (Szebenyi et al. 1992; Gembicky et al. 2005).

6.2.3 Third Generation Storage Ring Sources

Third generation sources such as the ESRF, Spring8, and APS are characterized by the use of wigglers and undulators to generate X-rays, inserted in the straight sections of the ring polygon and hence called insertion devices. Bending magnets at the vertices of the polygon are mostly relegated to beam steering and focusing roles. The insertion devices are optimized to produce X-ray beams of high brilliance.

Wigglers make use of a linear array of pairs of magnets, in which the polarity of the north and south poles alternates from pair to pair:

$$\frac{NSNS...}{SNSN...}$$

In its vacuum chamber, the electron bunch passes through the gap between the poles of the magnets. The alternating magnetic field drives the electrons along a trajectory that is nearly sinusoidal in the horizontal plane but undergoes no net displacement. Each pair of magnets in the wiggler produces an acceleration comparable to that in the single pair in a bending magnet. The X-ray spectrum emitted by a wiggler thus resembles that from a bending magnet. Since the radiation from each pair adds incoherently, the radiation from a wiggler is stronger than that from a comparable bending magnet by a factor equal to the number of pairs. A pinhole placed on the axis of a wiggler source provides a smooth, polychromatic source suitable for Laue diffraction with a large bandwidth. However, the wiggler beam is not uniform. It is tightly constrained vertically and of high brilliance in the vertical plane, but extended horizontally in a fan of low brilliance, as from a bending magnet. X-ray focal spots thus tend to be very elongated in the horizontal direction, which is less desirable.

Undulators also have a linear array of pairs of magnets similar to wigglers. However, they have a different magnetic structure of shorter period and greater overall length than wigglers, with many more poles. Since their magnetic field strength is also much lower, the near-sinusoidal trajectory of the electrons has a much smaller amplitude than in a wiggler. The electrons *undulate* rather than *wiggle* about their mean direction. As the electron beam passes through the undulator, interference effects between the electrons and the X-rays emitted upstream modify the characteristics of the X-ray beam.

The undulator X-ray beam from an SR source is concentrated into an angular cone of width $1/(\gamma N_u^{1/2})$ of small solid angle, which results in much greater brilliance. In this relation, the key relativistic factor γ, where $\gamma = (1 - (v^2/c^2))^{-1/2}$, encodes the energy E of the electron beam through the relation

$$E = \gamma m_e c^2,\tag{6.2}$$

where N_u is the number of periods in the undulator; m_e is the rest mass of the electron; c is the velocity of light; and v is the velocity of the electron pulse in the laboratory frame – very close to c. For example, the forthcoming high-energy upgrade to the LCLS – LCLS-II HE – will accelerate electrons to $E = 6$ GeV while the APS – a representative SR source – operates at 7 GeV.

The path of the electron bunch through the undulator magnets is not quite sinusoidal but contains higher Fourier components. These give rise to harmonics of the fundamental frequency in the emitted X-ray beam. The amplitude of these harmonics is sensitive to the undulator strength parameter K (defined below): the stronger the magnetic field, the greater the amplitude of the harmonics. Symmetry considerations require that on axis only odd harmonics appear, while off axis both odd and even harmonics are present. Since the first harmonic has a natural bandwidth of roughly 5×10^{-2}, the entire peak can be used for pink beam Laue diffraction or SAXS; no monochromator is needed. Alternatively, a monochromator can be inserted to select any desired wavelength. Other even harmonics emerge at distinct but small angles to the optical axis, in a complicated angular distribution. The energy of the fundamental is largely governed by the energy of the electron beam and by the strength B_0 of the undulator field on its axis. The value of B_0 and hence the X-ray energy of the fundamental can be tuned by adjusting the gap between the magnetic arrays of the undulator. This alters the magnetic field on the undulator axis and shifts the fundamental and harmonics in energy. B_0 typically is incorporated in the undulator strength parameter $K = eB_0\lambda_u / 2\pi m_e c$. $K = 1$ is a representative value for SR undulators.

An example typical of the leading third generation sources is the APS (now being upgraded to a fourth-generation source, as described below). A linear accelerator initially accelerates electrons and then transfers them to a synchrotron which further accelerates them to an energy of 7 GeV and injects them into the storage ring. In normal operating mode, the APS ring circulates 324 electron bunches each of typical duration 22 ps, spaced by 11 ns. These generate X-ray pulses of the same duration and spacing. The average circulating current is about 100 mA. A hybrid mode of operation in which a single fat bunch has a larger number of electrons is also available for a limited number of weeks per year.

Each beamline has its own undulator and X-ray optics to suit the experimental needs of its users. For example, the BioCARS Collaborative Access Team (CAT) at the APS operates beamline 14-ID for pump-probe experiments in the structural dynamics of biological systems (Graber et al. 2011). The source is unusual in having two collinear undulators with different magnetic periods. Their undulator gaps can be adjusted separately to alter the energy of the harmonics, which allows considerable user-controlled flexibility in the properties of the X-ray beam. The gaps can be adjusted such that the first harmonics of the two undulators coincide in X-ray energy, giving maximum intensity at the sample; or their energies can be separated to give a broader spectrum. Both configurations provide a pink beam for Laue crystallography and other experiments of wider bandwidth with $\Delta E/E = 5$ to 8×10^{-2}. For ultrafast dynamic experiments, a chopper is phase-locked to the accelerator ring clock and together with a series of shutters, cleanly isolates a single 100 ps X-ray pulse containing about 3×10^{10} photons in both 324 bunch and hybrid modes. Bendable mirrors focus the X-ray beam on to the sample, in a focal spot of 90×30 μm (Graber et al. 2011). In contrast, the BioSync facility at the APS specializes in single crystal experiments suitable for SAD or MAD phasing. It provides about 1.3×10^{12} photons per second at the sample, and a focal spot of 100×20 μm.

6.2.4 Fourth Generation Storage Ring Sources

Fourth generation, diffraction limited SRs are being introduced, either as new sources such as MaxIV in Lund or as upgrades of existing third generation sources such as the ESRF, SPring 8 and APS (Martensson and Eriksson 2018). They make use of novel arrays of bending magnets called multi-bend achromats (MBAs) at each vertex of the ring polygon. The MBAs reduce the emittance of the electron beam and enhance the horizontal X-ray brilliance, resulting in smaller, low divergence X-ray beams. The MBAs increase the brilliance and coherent flux of harder X-rays (>20 keV) above today's capabilities for both static, serial crystallography and ultrafast dynamic experiments by one to two orders of magnitude beyond the already excellent capabilities of third generation sources. At the upgraded APS, a further goal is to increase the beam current to 200 mA.

Although the MBAs enhance only the horizontal brilliance, this leads to electron bunches that are more nearly equal in their horizontal and vertical dimensions. This in turn permits better focusing optics and more nearly circular X-ray focal spots than in third generation sources. However, an inevitable consequence of the reduced emittance in the transverse, horizontal dimension is that the electron bunches in the buckets are increased in their longitudinal dimension, which increases the duration of the X-ray pulse from ~100 ps to ~200 ps. This will affect experiments requiring the highest time resolution.

A final characteristic of an X-ray beam (or indeed a beam at any wavelength) is *coherence*. The classical indicator of coherence is the ability to form interference fringes. More formally, coherence is present in a radiative field if the phase difference between any two points in the field remains constant in time. Coherence can be spatial when the phase difference between points on a plane normal to the beam propagation is considered. It depends on source size and collimation. Coherence can be temporal when points separated along the propagation direction are considered – points reached at different times by the beam. Temporal coherence requires a monochromatic beam with a very narrow bandwidth. High spatial coherence is required in SPI, where 2D diffraction snapshots from many particles are captured and pieced together to provide a full 3D diffraction from the particle (Spence et al. 2004; Schlichting and Miao 2012). Fourth generation SR sources provide significantly increased coherence compared with third generation sources. It remains to be explored how increased coherence can be most effectively deployed in structural biology.

Spatial and temporal coherence is a characteristic feature of XFELs, much higher than at even the most advanced fourth generation SR sources.

6.3 X-ray Free Electron Lasers

XFELs are based on novel insights (Pellegrini 2012) on the interaction between the electromagnetic field associated with X-rays and an electron beam as it traverses an undulator. The role of undulators in XFELs is quite different from that of the undulators used as stand-alone insertion devices in SRs. In an XFEL the undulator length is extended from the few m in SRs to perhaps 100 m. These very long undulators are traversed by low emittance electron beams diffraction-limited in divergence and size (Attwood and Sakdinawat 2017). Interference effects between the electron beam and X-rays produced at the upstream end of the undulator and collinear with the electron beam generate laser-like X-rays of high energy, exceptionally high brilliance and high coherence.

XFELs create electron pulses utilizing a linear geometry of accelerators rather than the polygonal geometry in an SR. When the first hard X-ray FEL was being planned, a large linear electron accelerator at the Stanford Linear Accelerator Center was available which was no longer used for high-energy physics. It could readily be adapted to provide small electron beam emittance, high electron beam energy with a small energy spread, high peak current and very short bunches in the few tens of fs range. This accelerator formed the foundation for the Linac Coherent Light Source (LCLS), the world's first hard X-ray laser. Despite considerable skepticism, the LCLS worked "straight out of the box." Highly novel X-ray experiments were promptly and successfully conducted at the LCLS in many disciplines including structural biology (Chapman et al. 2011; Seibert et al. 2011). These provided a model for new XFELs now in operation around the world, and upgrade of the original LCLS to LCLS-II. XFELs proved to be a truly disruptive innovation in technology (Bower and Christensen 1995; Moffat 2017). Their introduction has transformed dynamic experiments in structural biology.

XFELs are single-pass lasers: each electron bunch passes only once through the entire device and generates one pulse of X-rays. A large undulator ~100 m in length amplifies the natural jitter in the longitudinal positions of the electrons in that bunch. Interaction between the electron bunch and X-rays generated at the upstream end of the undulator eventually redistributes the electrons into a sequence of microbunches separated longitudinally by almost exactly one wavelength of the undulator radiation (Figure 6.3). Because all electrons in these microbunches

(a)

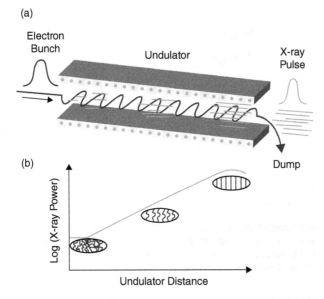

Figure 6.3 Formation of electron microbunches. An electron bunch enters the undulator at the left. Interaction between the emitted X-rays and the electrons leads to the segmentation of the bunch into microbunches that are separated by almost exactly one X-ray wavelength in the laboratory frame. Radiation from the microbunches adds coherently. Bostedt et al. 2016/ Reproduced with permission from American Physical Society.

emit X-rays in phase, the X-ray beam is coherent; its intensity is proportional to the *square* of the number of electrons in the full bunch. By contrast, the electrons in each bunch in an SR do not emit X-rays in phase as they traverse an undulator; they do not emit a highly coherent, laser-like beam. The peak brilliance of an undulator X-ray beam at an SR is proportional simply to the number of electrons in the bunch. Phase coherence is the origin of the staggering increase in peak brilliance shown in Figure 6.1, discussed further below.

The formation of microbunches from the initial electron bunch is based upon its noise structure; that is, on random fluctuations in the longitudinal position of the electrons within the full bunch as it enters the long undulator. If these positions were distributed uniformly (say, a completely smooth Gaussian), there would be no way to break its symmetry to form microbunches. Interactions between the electromagnetic field of the X-rays co-propagating along the axis of the undulator, collinear with the electron beam, cause microbunching to arise from these noisy, random fluctuations. Amplification of the microbunching by the self-amplified spontaneous emission process (SASE) forms the basis of lasing.

The longitudinal, noise-derived structure of the electrons in the bunch is imprinted on each X-ray pulse. The noise varies dramatically from bunch to bunch. XFEL radiation generated by the SASE process is therefore also extremely noisy in several characteristics. For example, X-ray pulses differ markedly in intensity. One pulse can be of nearly maximum intensity and its successor of zero intensity. Importantly for experimenters, each pulse itself has a spiky, random intensity profile with variable spectral distribution (Figure 6.4). There is no correlation in the noise structure of these characteristics from pulse to pulse. Substantial noise poses unusual experimental difficulties in making accurate measurements. To begin to

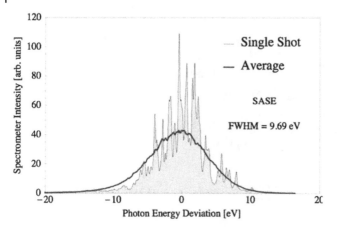

Figure 6.4   Energy spectrum of a SASE pulse. Energy spectrum for a single SASE pulse compared with the average over many pulses. These curves correspond to specific undulator configuration described in the original publication. The benefits for both dynamic and static experiments of making pulses more uniform are obvious. Emma et al. 2017/Reproduced with permission from AIP Publishing.

address these, SASE is being supplemented by external seeding that imposes a controllable, microbunch-like profile on the electron bunch at the upstream end of the undulator. This is intended to ensure that the SASE amplification process leading to lasing will be closely similar from pulse to pulse. Indeed, a compact XFEL promises sub-fs pulses (Chapter 9.4.2) that will show almost no pulse-to-pulse variation (Graves et al. 2017).

The extraordinary peak brilliance of fully focused fs XFEL pulses destroys all samples: only one shot can be obtained per sample. This has required the transformative new method of data collection called diffraction before destruction, introduced in Chapter 4.3.3 (Chapman et al., 2011). In a nutshell, a sufficiently short pulse will be fully diffracted by a sample before any secondary X-ray radiation damage has occurred, and before primary radiation damage becomes significant. The noted virus crystallographer Michael Rossmann coined a term well suited to this approach: "The American method of data collection: shoot first, ask questions later." It follows that diffraction from many individual samples must be measured and pieced together to assemble a full 3D pattern. A closely related advantage of these intense pulses is that tiny crystals of low scattering power with volumes as small as a few μm^3 give rise to readily observable diffraction. The necessity for rapid sample replacement has led to the development of serial femtosecond crystallography and novel methods for introducing samples (Chapter 4.2). Indeed, serial methods for sample introduction pioneered at XFEL sources have been usefully extended to SR sources (Pearson and Mehrabi 2020).

For structural dynamics, the most transformative characteristic of XFELs is the exceptionally short duration of the X-ray pulses. Durations in the 10–100 fs range are roughly one thousand times shorter than those at storage rings of ~100 ps. The pulse lengths at XFELs are governed by the length of the generating electron bunches and are adjustable over this range in the initial configuration of the LCLS (now being upgraded to LCLS-II). This short duration enables pump–probe, time-resolved structural dynamics on the fundamental, chemical time scale of <1 ps on which bond breaking, bond making, isomerization, and electron transfer occur (Chapter 5.1).

A limiting feature of LCLS operation is that pulses arrive at 120 Hz, which establishes the highest rate at which diffraction from individual samples can be measured. The initial success of the LCLS led briskly to the design, construction, and completion of other XFELs worldwide, with improved capabilities and different timing structure. For example, the European XFEL (EuXFEL) uses superconducting RF cavities to accelerate brilliant electron bunches to a maximum energy of 17.5 GeV. Its broad tunability allows a wide range of X-ray wavelengths from 0.05 to 4.7 nm. The EuXFEL also offers a much higher pulse rate, although with a complicated pulse time structure in which short stretches of 5 MHz pulse trains are interspersed with longer dark periods. Figure 6.5 shows the time structure of the pulse trains at EuXFEL. Each train comprises 2700 X-ray pulses and lasts 0.6 ms; pulse trains are separated by 99.4 ms dark (idle) time. The interval between pulses in a train is 220 ns, corresponding to a pulse frequency of ∼5 MHz.

The Single Particles, Clusters and Biomolecules and Serial Femtosecond Crystallography (SPB/SFX) instrument is the most relevant EuXFEL instrument for biological studies (Mancuso et al. 2019). It is primarily concerned with 3D structure determination of both crystalline and non-crystalline micrometer-size and smaller objects. Each X-ray pulse of about 100 fs duration contains about 10^{12} photons at 10 keV, corresponding to a total pulse energy of about 1.6 mJ (Kirkwood et al. 2022). Pulses are diffraction limited to about 100 nm, but in practice appear to be limited to about 200 nm diameter FWHM focus at 12 keV. A broad bandwidth ($\Delta E / E = 10^{-3}$) pink SASE mode is available for maximum flux. Bandwidth can be reduced to below 10^{-4} by adding a crystal monochromator.

The layout of SRs and XFELs is very different. SRs place insertion devices in each straight section of the ring polygon, 34 in the case of the APS, which allows many experimental end stations at SRs to be simultaneously usable. Their capabilities are specialized from station to station and are scientifically diverse. The experimental capacity and diversity of SR sources, measured for example by experiments and users per year, is therefore very high. In contrast, many XFELs have a single, very long undulator. The number of instruments that can receive an X-ray beam at (nearly) the same time, and hence the number of simultaneous user experiments, is severely limited. To enhance the experimental capacity of an XFEL, both temporal and spatial beam sharing are employed. At the LCLS, temporal sharing happens when the Matter under Extreme Conditions beamline has to wait

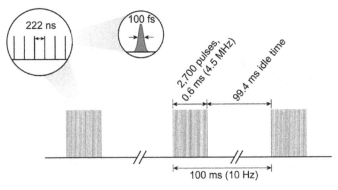

Figure 6.5 Timing structure of X-ray pulses at the European XFEL. Details of the timing structure are described in the text. Reproduced from the public Euprorean XFEL website.

10 minutes between visible laser pulses, during which time the Coherent X-ray Imaging (CXI) beamline can use the X-ray beam. The facility accomplishes temporal sharing by using two undulators and kicking the electrons into one branch or the other. The LCLS employs three methods for spatial sharing. First, thin crystals are used to diffract a single monochromatic slice out of the beam and delivers it to one of the monochromatic instruments. For example, the X-ray Pump–Probe (XPP) beamline can share with the CXI or Macromolecular Femtosecond Crystallography (MFX) beamlines in this mode. Second, an X-ray mirror is inserted halfway into the beam. This is a form of spatial sharing where CXI and MFX each receive around one-half of the beam. Although this destroys the wavefront, the beam is still useful for crystallography. Third, the spent beam is refocused. This is how CXI and the SSC chamber operate, as do the SPB and SFX instruments at EuXFEL.

In the operation of an XFEL, to a first approximation

$$\lambda = \frac{\lambda_u}{2\gamma^2} \tag{6.3}$$

where λ is the wavelength of the X-ray beam, λ_u is the period of the undulator (for example, 10 cm), and γ is the relativistic factor relating the energy of a relativistic particle to its energy at rest, introduced in Equation 6.2. The factor γ also relates length measured in a moving system to the same length measured at rest. Using this idea one can readily rationalize the factor of $1/\gamma^2$ in Equation 6.3. One factor of $1/\gamma$ describes transformation into the electron frame, where the undulator period is contracted. The second factor describes the blue shift of the emitted radiation in the laboratory frame. Adjusting the electron beam energy allows very significant changes in λ. For example, the operating range of the electron bunch in the original LCLS linac was 3.3–14.3 GeV – a fivefold range, which could generate an X-ray energy range from 0.48 to 9.6 keV, a factor of 25. This range is being greatly expanded in the upgrade to LCLS-II. The full version of Equation 6.3 contains a secondary, additional factor that depends on the magnetic field strength in the undulator. Variable gap undulators allow this field to be adjusted and thus provide additional fine tuning of the X-ray energy.

Nevertheless, the number of simultaneous experiments that can be conducted at an XFEL source is limited by their fundamental, linear geometry to a value inherently lower than at SR sources. The linear geometry arises because lasing requires a very high-quality electron beam with GeV energy, very small emittance and energy spread. For many technical reasons, a linear accelerator is better able to provide electron pulses with the required characteristics than a polygonal SR source. However, it has been suggested (Feng and Zhao 2017) that the continued improvement of electron optics at SRs may lead to electron beams of the higher quality necessary to permit lasing, and construction of an XFEL at such facilities.

The advent of XFELS has rejuvenated X-ray-based structural biology. Difficulty in growing large, well-ordered crystals is a common limitation, so the ability to obtain excellent diffraction patterns from crystals only a few μm across opens up crystallographic analysis of a whole new array of proteins. This has been of particular benefit for membrane proteins, for which good crystals are notoriously hard to obtain.

Structural dynamics at XFELs is thus superior to that at SRs in two highly significant ways. Ultrafast scattering experiments in the fs, chemical time range become accessible to direct structural study. The chemical time range is inaccessible to SR sources, limited by their pulse length to a much longer time resolution of ~100 ps. Even shorter pulses in the attosecond range will come on-line at LCLS-II and the EuXFEL. It remains to be seen whether, and if so how, attosecond pulses will be biologically relevant. And, reaction initiation by a laser pump pulse is more readily and more uniformly achieved in the tiny crystals studied at XFELs (Tenboer et al. 2014).

6.4 Detectors

6.4.1 Introduction

Position-sensitive, 2D X-ray detectors play a critical role in time-resolved experiments on both crystals and solutions. Dramatic developments in detectors over the last two decades have transformed experiments at SR and, more recently, XFEL facilities. Detectors measure intensity (X-ray photon flux on each pixel); position of the pixel in the detector array; time of arrival of the X-ray pulse; and sometimes, its energy. The differences in the scattered X-rays from SR and XFEL sources place very different demands on detectors. The peak brilliance of the X-ray beams from SRs is high, but sufficiently low to enable detectors at SRs to work well in single-photon counting mode for both crystallography and solution scattering experiments. The many orders of magnitude higher brilliance at XFELs means that photons arrive too fast at each pixel to be individually counted. The much higher dynamic range characteristic of charge-integrating detector modes is required. However, for SPI experiments at XFELs, very weak scattering of less than one photon per pixel per X-ray pulse is normal and the single-photon counting mode remains applicable.

Until the advent of the EuXFEL with its exceptionally high peak pulse rate, the readout speed of the detector was not a critical factor. At earlier facilities such as the LCLS operating at 120 Hz, individual scattering patterns could be acquired and processed prior to the arrival of the next pulse and pattern. The 4.5 MHz peak pulse rate at the EuXFEL placed huge new demands on the frame rate and readout speed of detectors. These capabilities are particularly important for time-resolved experiments that require the ability to collect data in a brief time window corresponding to the shutter speed of the experiment.

The number and size of the pixels is also important. Coherent XFEL beams can resolve very large spatial frequencies requiring high angular resolution at the detector, usually achieved by increasing the sample to detector distance. In contrast, the collection of crystal data at large Bragg angles requires a short sample to detector distance. The demand for small pixels in large arrays with rapid readout continues to pose technical challenges to detector designers.

The pixels in contemporary detectors utilize small, X-ray-sensitive, semiconductor tiles assembled in a large array. Each pixel is backed by an application-specific integrated circuit (ASIC) that processes its output. The Cornell-SLAC Pixel Array Detector (CSPAD) was a pioneering device of this type installed at the LCLS. The gain could be selected based on the experimental conditions: single-photon counting, or the high dynamic range enabled by charge integration (Blaj et al. 2015). The EIGER detector, developed at the Paul Scherer Institute (Casanas et al. 2016) and commercialized by Dectris, represents a high point of the ASIC technology. A hallmark of the EIGER ASIC is continuous readout that enables a kHz frame rate with a duty cycle greater than 99%, well matched to the ~0.1 kHz frame rate of the LCLS. Every pixel features a digital counter for noise-free photon counting. The CXI beamline at the LCLS uses a JUNGFRAU detector; an upgraded version of the EIGER described below.

6.4.2 The Adaptive Gain Integrating Pixel Detector: AGIPD

A key recent advance has been the development of adaptive gain individual pixel detectors (AGIPDs), in which individual pixels independently adjust their gain depending on the immediate X-ray flux (Graafsma et al. 2015). This enables great flexibility of operation, ranging from single-photon counting to saturation-free, integrated measurement of intense signals such as those from XFELs. The AGIPD and JUNGFRAU detectors are outstanding examples of adaptive gain detectors.

The AGIPD is a hybrid pixel detector developed for use at the EuXFEL (Allahgholi et al. 2019a). It has a large dynamic range, from single-photon to 10^4 photons per pixel at 12.5 keV. Each pixel is associated with a charge-sensitive amplifier with three adaptively selected gains. An in-pixel analog memory can store 352 images, recorded at up to 4.5 million frames per second during the complicated pulse trains characteristic of the EuXFEL (Figure 6.5). Transfer of images from storage and acquisition of data is carried out during the much longer, 99.4 ms dark period between the pulse trains.

The core component is the AGIPD ASIC which consists of 64×64 pixels, each 200×200 μm^2. These ASICS are grouped into a module of 128×256 pixels, and modules are tiled to produce a 1-megapixel array. The 1-megapixel detector has a noise level corresponding to a signal-to-noise ratio of >10 σ for a single 12.4 keV photon (Allahgholi et al. 2019b). Development of the AGIPD platform into even larger arrays and to allow use at higher X-ray energies is actively under way.

The AGIPD has performed well in several TR SFX experiments. For example, Schmidt and colleagues (Pandey et al. 2020) examined the photocycle of PYP at the EuXFEL over the time range from 3 ps to 100 ps. The photocycle was pumped by laser pulses of 240 fs at a wavelength of 420 nm. Elaborate controls were in place for all aspects of laser activation and X-ray reproducibility. The measured size of the X-ray focal spot in the interaction region was 2×3 μm^2, and each X-ray pulse was about 700 μJ. Data were recorded using an AGIPD detector 118 mm downstream of the sample, which enabled data at a resolution of 1.6 Å to be recorded at the corner of the detector.

6.4.3 The JUNGFRAU Detector

The JUNGFRAU detector, a further outstanding example of the AGIPD principle, was developed for use at the Swiss XFEL and is also used at the LCLS and several SR beamlines. Its capabilities are shown in Figure 6.6. The JUNGFRAU detector has the following design and performance characteristics (Mozzanica et al. 2018):

- Two-dimensional, modular/tilable, vacuum-compatible imaging detector
- 75 μm pixel size, comparable to that of photon counting detectors
- Single photon sensitivity
- High dynamic range with up to 10,000 12 keV photons/frame/pixel
- Low noise over the full dynamic range
- Good linearity of detection over the full dynamic range (nonlinearity < 1%)
- Frame rates: maximum 2.2 kHz; 1.13 kHz with current readout board
- Minimum energy for single photon detection: 1.5 keV
- Saturation: ~11,000 12 keV photons per frame per pixel

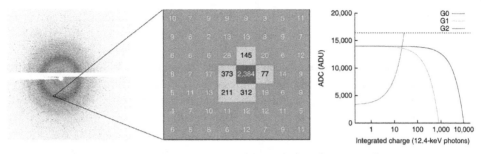

Figure 6.6   JUNGFRAU detector. Left: a diffraction image from a lysozyme crystal measured without beam attenuation at the X06SA beamline, Swiss Light Source. Middle: zoomed-in view of a Bragg peak showing the number of photons detected, where the central pixel was measured in low gain (red), tails of the peak were measured in medium gain (yellow), and the background was measured in high gain (blue). Right: the relationship between a charge integrated by the pixel and its analog-to-digital converter (ADC) count output for three gains (high (G0), medium (G1), and low (G2)) in analog-to-digital converter units (ADU). The dashed line represents the highest count permitted by a 14-bit ADC: $2^{14} - 1 = 16,383$. Leonarski et al. 2018/Reproduced with permission from Springer Nature.

The basic JUNGFRAU chip contains a 256×256 array of $75\ \mu m \times 75\ \mu m$ pixels. Arrays of 2×4 chips are tiled to form modules of about 500 kilopixels. Several modules can be combined to provide 1 megapixel coverage or larger.

A useful comparison of the EIGER and JUNGFRAU detectors is provided in (Chapman 2018).

6.5   Beamlines and Experimental Stations

6.5.1   Introduction

The raw X-ray beam generated within an SR or XFEL emerges from the high vacuum region through a window or a differentially pumped aperture. Downstream of the window, the beamline staff are responsible for processing the X-ray beam to make it suitable for specific classes of experiments. The beam is directed to an instrument hutch (also known as an end station) where users conduct their experiments with a variety of X-ray and other instruments and devices. In practice beamlines are developed, equipped, and maintained by beamline scientists associated with the facility, in close consultation with the community of users. In an ongoing process, the beamlines are constantly reconfigured and upgraded in response to users' requests.

The beamline components tailor the raw X-ray beam emitted by the source. Key tasks are tunable monochromatization and focusing, both of which are carried out by grazing incidence mirrors, specialized crystal X-ray optics, and other devices. The high power of the X-ray beam imposes a major heat load on the optical elements, which are designed to dissipate the heat without significant distortion that would degrade their optical properties. The end station contains the point where the X-ray beam is delivered to the sample. In TR experiments the end station also incorporates systems such as visible lasers and mixing devices that initiate the reaction (Chapter 3.3); injector systems (Chapter 3.4.2) that deliver the sample to the beam; and devices that measure and control the timing of the pump and probe pulses.

Beamlines at the LCLS are described at https://lcls.slac.stanford.edu/instruments/maps. Beamlines at the EuXFEL are described at https://www.xfel.eu/facility/beamlines/index_ eng.html. No two beamlines are identical. Rather than trying to cover the whole variety, we describe a representative LCLS instrument in some detail.

6.5.2 Beamline Example: The Coherent X-ray Imaging Facility

The CXI instrument at the LCLS contains both a beamline and an experimental end station. Its layout is shown in Figure 6.7, with more specific details of the optical elements in Figure 6.8. CXI enables many techniques in static SFX, TR-SFX, SAXS/WAXS and X-ray emission spectroscopy, and pioneering studies on coherent SPI. The diverse scientific successes of its users led us to select it as our exemplar.

CXI is particularly well suited to the diffraction-before-destruction regime. The LCLS source enables extremely short pulse durations below 10 fs, and employs two exchangeable

CXI Beamline Layout

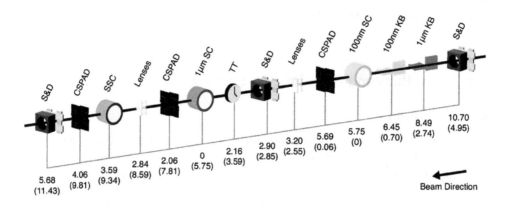

JOURNAL OF SYNCHROTRON RADIATION

Liang et al.

Volume 22 | Part 3 | May 2015 | Pages 514–519 | 10.11107/S160057751500449X

Figure 6.7 **Layout of the CXI instrument**. Distances are indicated in meters from the 1 mm sample chamber (SC) and in parentheses for the 100 nm sample chamber. Each chamber is colored to match its corresponding KB pair. One set of Be lenses can be used to further control the focus in the 1 mm sample chamber and another set to refocus the unscattered beam to the serial sample chamber at 3.59. There are slits and diagnostic (S&D) tools along the beamline and a timing tool (TT) for fine timing between optical laser and X-ray beams for pump–probe experiments. The sample is located approximately 440 m downstream of the undulators. Liang et al. 2015 / International Union of Crystallography / Licensed under CC BY 4.0.

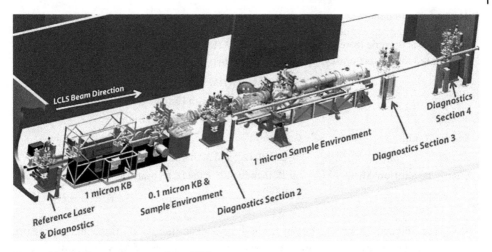

Figure 6.8 Main components of the CXI instrument.

pairs of Kirkpatrick-Baez (KB) mirrors (yes, the father of *Joan* Baez) that focus X-rays by total external reflection from their specially coated surfaces. One pair delivers a 1 μm focus and the other, an even tighter 100 nm focus. These deliver the full power of the incident beam to samples of various sizes. Focusing is also available through sets of compound-focusing, transmission lenses of beryllium which can deliver a 10 μm focal spot. Compound lenses make use of the fact that the real part of the X-ray refractive index for all materials is slightly less than unity, often written 1-δ. In contrast to visible light, a concave X-ray lens will be focusing. Because δ is numerically very small, the focusing power of a single lens is low. To achieve high focusing power and a tighter focus, a sequence of such lenses is required – a compound lens.

Key specifications of CXI are in Table 6.1.

Table 6.1 Operating characteristics of the CXI beam line at the Linac Coherent Light Source.

Photon Energy	5–10 keV for the 1st harmonic*
	Up to 25 keV for 2nd and 3rd harmonic
Repetition Rate	120, 60, 30, 10 Hz, single shot mode
Pulse Duration	40–300 fs (high pulse charge mode)
	<10 fs (low pulse charge mode)
Photons per Pulse	$\sim 1 \times 10^{12}$ (high pulse charge mode @ 8.3 keV)
	$\sim 1 \times 10^{11}$ (low pulse charge mode @ 8.3 keV)
Focusing Capability	KB1 mirrors (1.3 μm focus)
	KB01 mirrors (~100 nm focus)
	Beryllium Lenses in XRT (10 μm focus)
	Beryllium Lenses in Hutch 5 (~1 μm focus)

(Continued)

Table 6.1 (Continued)

Beam Size at Sample (8 keV)	$1.3 \times 1.3\ \mu m^2$ FWHM with 1 micron KB pair (KB1)
(calculated for perfect optics)	$90 \times 150\ nm^2$ FWHM (V $\times$ H) with 100 nm KB pair (KB01)
	$10 \times 10\ \mu m^2$ FWHM with XRT Be Lenses
	$\sim 1 \times 1\ \mu m^2$ FWHM with Hutch 5 Be Lenses
	$750 \times 750\ \mu m^2$ FWHM unfocused beam
Energy Range	2–10 keV (1st harmonic)
	10–25 keV (2nd and 3rd harmonic)
Energy Resolution $\Delta E/E$	$\sim 0.2\%$ (bandwidth of the LCLS beam)No monochromator currently

CXI is an in-vacuum system but in limited cases experiments at atmospheric pressure are possible. For scattering experiments such as SFX and fast SAXS, it mainly uses a forward-scattering geometry. There are two primary sample chambers (labeled SC in Figure 6.8), at the 1 µm focus and at 100 nm. In addition to maximize use of the beam, a serial sample chamber (labeled SSC) well-suited for SFX has been installed downstream of the first CSPAD detector (Chapter 6.4.1) for the 1 µm chamber, which uses beryllium lenses to refocus the unscattered beam from the 1 µm chamber. Three different mounting positions for the CSPAD detector are shown. A JUNGFRAU detector is also available.

6.5.2.1 Beam Monitoring and Use

The CXI beamline is equipped with devices to monitor X-ray beam characteristics. Some such as the pop-up intensity monitor are inserted into the beam and can only be used diagnostically, not during the actual experiment. The critical intensity-position monitor (termed Wave8 at the beamline) measures the total LCLS pulse energy on a pulse-by-pulse basis during experiments and estimates the beam position. The sample changer at the 1 µm focus supports key experimental arrangements to control the environment of the sample in the chamber, where the pressure is typically 10^{-4}–10^{-7} Torr. Samples mounted on thin supports can be positioned accurately into the LCLS beam using a long-range microscope which views the sample directly along the X-ray axis. This is advantageous since alignment of the tightly focused X-ray beam on a sample of a few µm dimensions is challenging, as is the related problem of aiming a pump laser at the sample and ensuring that the sample lies exactly at the focus of the probe X-ray beam. A particle injector allows gas phase or other samples to be delivered to the X-ray beam without the need for any substrate on which a sample might be mounted, to minimize background X-ray scattering. A mechanical system (originally developed by CXI users from Arizona State University) is compatible with multiple liquid jet delivery systems and can be used in either sample chamber. The 1 µm SSC is the experimental hub for CXI. Many focal spot sizes are possible: the 1 µm KB focus; the focus of either 10 µm or 0.1 µm from one of the set of X-ray focusing beryllium lenses; with an unfocused beam; or with the refocused, already used beam from the upstream 10 µm KB focus. For ultrafast TR studies, CXI also provides an X-ray shutter system that can transmit only a single XFEL pulse to the sample chambers.

The CXI beamline provides two optical pump lasers as part of several standard laser configurations. In the ns range, an Opolette HR 355 delivers <8 ns pump laser pulses with wide tunability from 410 to 2200 nm. This laser beam is fiber-coupled then propagated in free space into the sample chamber. For experiments with lower demands on time resolution, the laser beam can be orthogonal to the X-ray beam. For ultrafast experiments with the highest time resolution, the laser beam propagates collinearly with the X-ray beam. Collinearity ensures that the pump laser beam and the probe X-ray beam strike the same region of the sample, and minimizes temporal walk-off across the sample. For pump beams in the fs range, the fundamental (800 nm) or second harmonic (400 nm) of a Ti:Sapphire laser are available, in which ~50–150 fs pulses are delivered to the sample collinearly with the X-ray beam. The instrument is also equipped with a Spectra Physics TOPAS optical parametric amplifier, which can generate wavelengths from the deep UV through the IR (189–2000 nm) range.

Effective use of the fs pump laser requires accurate knowledge of the time interval between the arrivals of the pump laser and probe X-ray pulses. CXI provides a timing tool (TT) for measuring the actual interval for each pump–probe pulse pair. The principles and use of the TT are well-described on the data analysis portion of the LCLS website.

Pump optical pulses are linked to probe XFEL pulses by RF locking technology. Its intrinsic jitter of more than 100 fs arises in part from the very long distances (tens to hundreds of meters) between the optical and X-ray sources and the sample. System drift can increase the timing uncertainty to more than 1 ps over hours. A substantial hardware research effort has now successfully incorporated feedback from the TT to measure the actual interval between pump and probe pulses on a shot-by-shot basis. This fully compensates for the ps drifts and leaves only jitter as a source of timing uncertainty. A broadband optical probe pulse, called a super-continuum, is chirped so that different frequencies arrive at different times. This generates a profile of wavelength versus time. The chirped probe pulse is passed simultaneously with an X-ray pulse through an optically transparent material such as Si_3N_4. The X-ray pulse, shorter in time than the probe, generates a change in the complex index of refraction of the material, which in turn modulates the transmitted spectrum. The time point marked by this modulation marks the X-ray pulse arrival time. The spectral TT is now part of the standard beamline configuration. Nearly every time-resolved experiment performed at the LCLS now uses the TT to temporally sort data. Time stamps are automatically calculated and saved into the data stream as experiments are being performed. A useful discussion of timing in pump–probe experiments appears in (Glownia et al. 2019).

With direct measurement of the time difference between each pump and probe pulse, the diffraction patterns can be resorted from their assumed time order into an actual time order, then re-binned into time intervals as needed. An example is given by (Pande et al. 2016). In Chapter 7.3.4 we discuss powerful new computational algorithms that can dramatically improve the experimental temporal resolution even beyond that provided by the TT. These algorithms may remove the need for any experimental TT, since they derive accurate timing information from analysis of the raw X-ray data itself.

The versatility and capability of the CXI beamline and instruments are evidenced in the broad range of successful SFX (Chapman et al. 2011; Boutet et al. 2012; Kang et al. 2015) and TR studies (Barends et al. 2015; Pande et al. 2016) emerging from it.

References

Allahgholi, A., Becker, J., Delfs, A. et al. (2019a). The adaptive gain integrating pixel detector at the European XFEL. *Journal of Synchrotron Radiation* 26 (1): 74–82.

Allahgholi, A., Becker, J., Delfs, A. et al. (2019b). Megapixels @ Megahertz–the AGIPD high-speed cameras for the European XFEL. *Nuclear Instruments and Methods in Physics Research Section A: Accelerators, Spectrometers, Detectors and Associated Equipment* 942: 162324.

Als-Nielsen, J. and McMorrow, D. (2011). *Elements of Modern X-ray Physics*. John Wiley & Sons.

Attwood, D. and Sakdinawat, A. (2017). *Soft X-rays and Extreme Ultraviolet Radiation: Principles and Applications*. Cambridge university press.

Barends, T.R.M., Foucar, L., Ardevol, A. et al. (2015). Direct observation of ultrafast collective motions in CO myoglobin upon ligand dissociation. *Science* 350: 445.

Blaj, G., Caragiulo, P., Carini, G. et al. (2015). X-ray detectors at the linac coherent light source. *Journal of Synchrotron Radiation* 22 (3): 577–583.

Bostedt, C., Boutet, S., Fritz, D.M. et al. (2016). Linac coherent light source: the first five years. *Reviews of Modern Physics* 88: 015007.

Boutet, S., Lomb, L., Williams, G.J. et al. (2012). High-resolution protein structure determination by serial femtosecond crystallography. *Science* 337 (6092): 362–364.

Boutet, S. and Yabashi, M. (2018). X-ray free electron lasers and their applications. In: *X-ray Free Electron Lasers* (ed. S. Boutet, P. Fromme, and M.S. Hunter), 1–21. Springer.

Bowers, J. and Christensen C. (1995). Disruptive Technologies: Catching the Wave. *Harvard Business Review* (Jan-Feb): 43–53.

Casanas, A., Warshamanage, R., Finke, A.D. et al. (2016). EIGER detector: application in macromolecular crystallography. *Acta Crystallographica Section D: Structural Biology* 72 (9): 1036–1048.

Chapman, H.N., Fromme, P., Barty, A. et al. (2011). Femtosecond X-ray protein nanocrystallography. *Nature* 470 (7332): 73–77.

Chapman, H.N. (2018). A detector for the sources. *Nature Methods* 15 (10): 774–775.

Emma, C., Lutman, A., Guetg, M. et al. (2017). Experimental demonstration of fresh bunch self-seeding in an X-ray free electron laser. *Applied Physics Letters* 110 (15): 154101.

Feng, C. and Zhao, Z. (2017). A storage ring based free-electron laser for generating ultrashort coherent EUV and X-ray radiation. *Scientific Reports* 7 (1): 1–7.

Gembicky, M., Oss, D., Fuchs, R. et al. (2005). A fast mechanical shutter for submicrosecond time-resolved synchrotron experiments. *Journal of Synchrotron Radiation* 12 (5): 665–669.

Glownia, J.M., Gumerlock, K., Lemke, H.T. et al. (2019). Pump–probe experimental methodology at the linac coherent light source. *Journal of Synchrotron Radiation* 26 (3): 685–691.

Graafsma, H., Becker, J. and Gruner, S.M. (2015). Integrating hybrid area detectors for storage ring and free-electron laser applications. *Synchrotron Light Sources and Free-Electron Lasers*.

Graber, T., Anderson, S., Brewer, H. et al. (2011). BioCARS: a synchrotron resource for time-resolved X-ray science. *Journal of Synchrotron Radiation* 18: 658.

Graves, W., Chen, J., Fromme, P. et al. (2017). *ASU compact XFEL*. 38th Int. Free Electron Laser Conf.

Kang, Y.Y., Zhou, X.E., Gao, X. et al. (2015). Crystal structure of rhodopsin bound to arrestin by femtosecond X-ray laser. *Nature* 523 (7562): 561–567.

Kirkwood, H.J., de Wijn, R., Mills, G. et al. (2022). A multi-million image serial femtosecond crystallography dataset collected at the European XFEL. *Scientific Data* 9 (1): 1–7.

Lattman, E.E., Grant, T.D., and Snell, E.H. (2018). *Biological Small Angle Scattering: Theory and Practice*. Oxford University Press.

Leonarski, F., Redford, S., Mozzanica, A. et al. (2018). Fast and accurate data collection for macromolecular crystallography using the JUNGFRAU detector. *Nature Methods* 15 (10): 799–804.

Liang, M., Williams, G.J., Messerschmidt, M. et al. (2015). The coherent X-ray imaging instrument at the linac coherent light source. *Journal of Synchrotron Radiation* 22 (3): 514–519.

Mancuso, A.P., Aquila, A., Batchelor, L. et al. (2019). The single particles, clusters and biomolecules and serial femtosecond crystallography instrument of the European XFEL: initial installation. *Journal of Synchrotron Radiation* 26 (3): 660–676.

Martensson, N. and Eriksson, M. (2018). The saga of MAX IV, the first multi-bend achromat synchrotron light source. *Nuclear Instruments and Methods in Physics Research Section A: Accelerators, Spectrometers, Detectors and Associated Equipment* 907: 97–104.

Moffat, K. (2017). Small crystals, fast dynamics and noisy data are indeed beautiful. *The International Union of Crystallography of Journal* 4 (4): 303–305.

Mozzanica, A., Andrä, M., Barten, R. et al. (2018). The JUNGFRAU detector for applications at synchrotron light sources and XFELs. *Synchrotron Radiation News* 31 (6): 16–20.

Pande, K., Hutchison, C., DM Groenhof, G. et al. (2016). Femtosecond structural dynamics drives the trans/cis isomerization in photoactive yellow protein. *Science* 352 (6286): 725–729.

Pandey, S., Bean, R., Sato, T. et al. (2020). Time-resolved serial femtosecond crystallography at the European XFEL. *Nature Methods* 17 (1): 73–78.

Pearson, A.R. and Mehrabi, P. (2020). Serial synchrotron crystallography for time-resolved structural biology. *Current Opinion in Structural Biology* 65: 168–174.

Pellegrini, C. (2012). The history of X-ray free-electron lasers. *The European Physical Journal H* 37 (5): 659–708.

Schlichting, I. and Miao, J. (2012). Emerging opportunities in structural biology with X-ray free-electron lasers. *Current Opinion in Structural Biology* 22 (5): 613–626.

Seibert, M.M., Ekeberg, T., Maia, F.R. et al. (2011). Single mimivirus particles intercepted and imaged with an X-ray laser. *Nature* 470 (7332): 78–81.

Spence, J., Weierstall, U., and Howells, M. (2004). Coherence and sampling requirements for diffractive imaging. *Ultramicroscopy* 101 (2–4): 149–152.

Szebenyi, D.M., Bilderback, D.H., LeGrand, A. et al. (1992). Quantitative analysis of Laue diffraction patterns recorded with a 120 ps exposure from an X-ray undulator. *Journal of Applied Crystallography* 25 (3): 414–423.

Tenboer, J., Basu, S., Zatsepin, N. et al. (2014). Time-resolved serial crystallography captures high-resolution intermediates of photoactive yellow protein. *Science* 346 (6214): 1242–1246.

Watson, W. and Perlman, M. (eds.) (1973). *Research Applications of Synchrotron Radiation: BNL Document 50381*. Upton, NY: Brookhaven National Laboratory.

7

Data Analysis and Interpretation

7.1   Introduction

In Chapter 5 we discussed the acquisition of accurate structure factor amplitudes in crystallography and scattering profiles in solution scattering. How should these raw measurements in reciprocal, scattering space be quantitatively analyzed to understand the dynamical behavior in real space of the system of interest? We emphasize issues of principle in analysis and interpretation.

In time-resolved crystallography, innovations are being rapidly introduced into data analysis and interpretation of DED maps. The overall goal is to identify in real space the time-independent structure of each intermediate and hence the structural changes between reactants, intermediates, and products along the functional trajectory, the rate coefficients with which these changes occur, and the chemical kinetic mechanism which the structures populate. Likewise, in time-resolved solution scattering the overall goal is to identify the number of distinct intermediates that comprise the total scattering at all time points, to extract the time-independent scattering curves associated with each intermediate, and from these scattering curves, to model the structure of each intermediate. The time dependence of the formation and decay of each scattering curve gives the corresponding rates for each intermediate.

In both experimental approaches, the overall goal is to identify in structural terms what happens, how fast, and through what chemical kinetic mechanism. The final step of interpretation – touched on in Chapter 8 – is to explain by complementary chemical, theoretical, and computational approaches why the reaction proceeds via these particular intermediates, with the observed rates.

All analyses of kinetics experiments have a critical limitation (Chapter 1.5). No kinetics experiment can **prove** a mechanism; it can only **disprove** a mechanism. If a candidate mechanism with the observed number of states is shown to be incompatible with the time-dependent structural data and rates, that mechanism is disproved. If all such mechanisms can be disproved except one, that one is the best that can be achieved compatible with the data. The sole remaining mechanism is proved. If, however, several mechanisms cannot be disproved and remain as candidates, any preference among them must be based on non-kinetic experiments or even on subjective judgments such as Occam's Razor.

Dynamics and Kinetics in Structural Biology: Unravelling Function Through Time-Resolved Structural Analysis, First Edition. Keith Moffat and Eaton E. Lattman.
© 2024 John Wiley & Sons Ltd. Published 2024 by John Wiley & Sons Ltd.

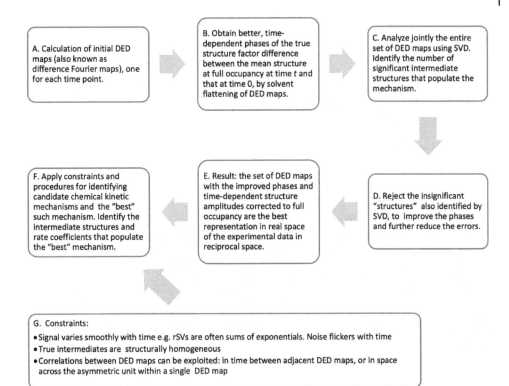

Figure 7.1 **Flow chart of steps in the improvement and analysis of DED maps.**

The steps in the overall analysis of time-resolved crystallographic data are summarized in the panels of Figure 7.1. After briefly discussing the analysis approach and general constraints that apply, we consider the individual steps.

The first step generates the set of time-dependent DED maps (panel A) from a sequence of structural changes as the reaction proceeds through an as yet undetermined number of intermediates via an unknown chemical kinetic mechanism. The maps are calculated using Equations 7.2 and 7.3 below. Ideally, each map displays at full occupancy the difference in electron density between the structure at that time point and of the reactant at time zero, prior to reaction initiation. However, initial DED maps contain significant errors since all are calculated using the time-independent X-ray phases of the static, reactant structure. That is, they use the difference Fourier approximation (Henderson and Moffat 1971). Errors are present both in those phases and in the experimental structure amplitudes at time zero and later time points after reaction initiation. Later time points also reflect the occupancy of intermediate structure(s) at that time point. Their occupancy may be substantially less than 100%, which introduces a systematic error into both the phase and the amplitude of the true difference structure amplitudes (Figure 7.2). Minimization of these errors by application of simple constraints to the DEDs involves cycling between reciprocal and real space.

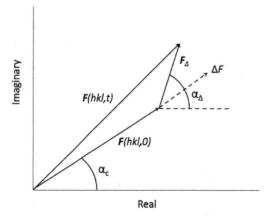

Figure 7.2 **Structure factor diagram for DED map coefficients** F. $F(hkl,0)$ is the structure factor for the zero-time or ground state structure, for which the phase α_c is normally known. $F(hkl,t)$ is the structure factor at time t. F_Δ is the structure factor for the true DED map between times zero and t, for which α_Δ is the true phase. Thus, $F(hkl,0) + F_\Delta = F(hkl,t)$. $\Delta F = |F(hkl,t)| - |F(hkl,0)|$ is the amplitude normally used in DED maps, along with the phase α_c. Importantly, the phase α_c is *uncorrelated* with the correct DED phase α_Δ When α_Δ Is near 90° ΔF will be small, much less than $|F_\Delta|$. Also, this diagram assumes full occupancy, that every molecule has been affected by the light pulse or has bound ligand. If there is partial occupancy, scaling is required to give the difference map peaks correct weight.

The second step (panel B) sketches the process of improving these phases. The third step (panel C) is to identify the number of significant intermediate structures. This number imposes a stringent constraint on candidate mechanisms and can be obtained by singular value decomposition (SVD), a mathematical technique from linear algebra presented in Chapter 7.4. SVD enables the number of significant structures present in the set to be identified, to within a specified error level. SVD also helps to identify density fluctuations – insignificant structures – that arise from errors in both the phases and the structure amplitudes (panel D). Panel E highlights these improved maps. Finally, panels F and G cover the use of constraints to elucidate candidate mechanisms.

7.2 General Constraints on Analysis and Interpretation

Several physical and chemical constraints apply to analysis of all dynamic, structural experiments and can greatly help to distinguish signal from noise.

1) True signal varies smoothly with time but almost all noise flickers, varying sharply with time. This constraint is hard to apply if only a few time points are available but is very useful when, for example, three or more time points have been collected per decade of time. A powerful corollary is that the more signals resemble each other, the more likely they are to be adjacent in time. This constraint can be used to identify timing errors and to sort in time raw signals such as individual diffraction frames (Chapter 7.3.4).

2) True intermediate structures are homogeneous; they are not a heterogeneous mixture of structures. True structures must be refinable as a single structure with standard stereochemical parameters. However, standard parameters are derived from structures at thermodynamic equilibrium and may not be directly applicable to structures far from equilibrium, such as structure(s) immediately after absorbing a photon. Short-lived structures in the fs time range often exhibit extreme strain such as a torsion angle about a double bond near 90° during an isomerization reaction (Pande et al. 2016). Since strain is of great chemical interest, its existence must be carefully confirmed. Structural heterogeneity or inability to refine longer-lived, unstrained structures shows that the chemical kinetic mechanism which gave rise to these structures is wrong. That mechanism is convincingly disproved.

3) In SVD analysis (Chapter 7.4), it should be possible to fit all right singular vectors (rSVs) representing the time dependence by a sum of exponentials, if the mechanism contains first-order rate coefficients and a series of simple steps. If a sum of exponentials is found, a chemical kinetic model holds. However, more complicated functions may be required if one or more rSVs have an odd, unexpected time dependence. For example, functions other than simple exponentials will be needed if a bimolecular reaction is known to be present, initiated by mixing of reactants. Another example is a stretched exponential of the form $\exp(kt^{\alpha})$ where α differs from 1.0. If this is present, a chemical kinetic mechanism with a small number of distinct states does not hold. A candidate state assumed to be distinct may in fact be a group of closely similar structures populated from the preceding, single state and structure by parallel decays over a range of first-order rate coefficients. Unless these rates are sufficiently different to be cleanly separated (a notoriously difficult analytical problem), the time course of parallel decays from the preceding state will be fitted well by a single, stretched exponential where $\alpha > 1.0$. Stretched exponentials may also occur if, for example, another parameter such as temperature that influences rate coefficients inadvertently varies across the time series. This may occur for short time points in the fs to few ps time range, since absorption of a photon by a chromophore deposits significant energy. Much of the energy appears promptly as vibrational energy which dissipates throughout the chromophore, its pocket, and the entire protein. The process corresponds to a very rapid temperature jump followed by a slower decay as the vibrational energy dissipates, ultimately into the solvent.

A further exception to sum-of-exponentials behavior arises from secondary radiation damage. This may appear in data acquired at an SR with long exposures, or when several diffraction patterns are acquired from one crystal. The expected smooth variation of primary radiation damage is linear with dose/time; that of secondary damage adds a quadratic term (Teng and Moffat 2002). Since the time dependence of true signal and total radiation damage are quite different, they should be readily distinguishable.

Here, we ignore any effects arising from coherent atomic motion, which are likely to be present in the fs time range up to 1 ps. Coherent motion is unlikely to decay via a sum-of-exponentials time course, may display oscillatory behavior, and is under active investigation (Hosseinizadeh et al. 2021).

In all the above cases, the true signal varies smoothly with time, though the mathematical form of this variation depends on the source of the signal: structural changes, or another physical or chemical effect such as diffusion or radiation damage.

4) In crystallography, lack of isomorphism or the presence of major, time-dependent crystal disorder would seriously confound analysis. As a rule of thumb, the larger the structural change, the higher the associated energy barrier and the more slowly it occurs. For example, changes in quaternary structure are generally much slower than changes in tertiary structure. Changes in unit cell dimensions that introduce non-isomorphism occur over length scales up to the dimensions of the crystals and are likely to be much slower than even quaternary changes. Any phase change involves a very large number of molecules and is much slower than structural changes in individual molecules. This suggests that time-dependent non-isomorphism in TRX and TR-SFX is unlikely to be a complicating factor.

5) Correlations in time between DED maps close together in time or in space across the asymmetric unit within one DED map can be exploited (Chapter 9.4.7).

7.3 Difference Electron Density Maps

DED maps have been widely used in conventional protein crystallography since its earliest days and are now principally applied to refinement. These DED maps display the lack of agreement between the current model of a structure with calculated phases and structure amplitudes and the model with observed amplitudes:

$$\Delta\rho(\boldsymbol{x}) = \frac{1}{V}\sum_{\boldsymbol{h}}\left[mF_o(\boldsymbol{h}) - DF_c(\boldsymbol{h})\right]exp(i\alpha_c)exp(-2\pi i\boldsymbol{hx}). \tag{7.1}$$

Here $\Delta\rho$ is the difference electron density in real space, $\boldsymbol{h}$ is the reciprocal lattice point hkl, F_o and F_c are the observed and calculated structure factor amplitudes, α_c is the phase calculated from the current model. The weight m is the figure of merit for the current phase, and is approximately equal to the cosine of the phase error (Blundell et al. 1976). The weight D, related to the parameter σ_A in small molecule crystallography, accounts for the completeness of the model (Pannu and Read 1996).

Analogous DED maps play a major role in time-resolved studies. The structural changes of molecules in the asymmetric unit during a TRX or TR-SFX experiment are largely based on analysis of DED maps calculated from data collected at time t_j and the pre-pump (dark or ground-state) point $t = 0$. They may also be based on adjacent time points t_j and t_{j+1}, or indeed on any two time points. The idealized difference map in this context is given by

$$\Delta\rho(\boldsymbol{x},t) = \frac{1}{V}\sum_{\boldsymbol{h}}\left[F(\boldsymbol{h},t) - F(\boldsymbol{h},0)\right]exp(-2\pi i\boldsymbol{hx}) \equiv \frac{1}{V}\sum_{\boldsymbol{h}}\Delta F(\boldsymbol{h},t)exp(-2\pi i\boldsymbol{hx}). \tag{7.2}$$

Equation 7.2 recognizes that the phases of $F(\boldsymbol{h},t)$ change with time and assumes that these phases are known. It also assumes that the structure amplitudes $|\,\boldsymbol{F}\,|$ are known without error, and that all are on the same scale. These assumptions are idealized; none is entirely valid.

The scaling of structure factor amplitudes is quite straightforward and is a feature of many software packages. See for example Chapter 8 in (Rupp 2009). The more serious errors arise from approximation in the phase. The phase α_c of $F(h, 0)$, the ground-state structure factor, has usually been well refined and is thus fairly well known. In the difference Fourier approximation, this phase is used in place of the correct but as yet unknown phase α_Δ of $\Delta F(h,t)$ in Equation 7.2; see Figure 7.2. This gives rise to the equation

$$\Delta\rho(x,t) = \frac{1}{V}\sum_h \big[|F(h,t)| - |F(h,0)|\big]exp(i\alpha_c)exp(-2\pi ihx)$$

$$\equiv \frac{1}{V}\sum_h \Delta F(h,t)exp(i\alpha_c)exp(-2\pi ihx), \qquad (7.3)$$

which presents a time-dependent isomorphous difference map. Equation 7.3 illustrates the initial situation in which we do not have any phase information about $\rho(x,t)$ beyond that calculated from the known structure at time zero. Indeed, the purpose of calculating the difference map is to allow us to model $\rho(x,t)$.

Figure 7.2 shows that the difference Fourier approximation uses $\big[|F(h,t)| - |F(h,0)|\big]$, including the sign, as the difference structure factor amplitude, with phase α_c. The (unsigned) structure amplitude is (to a close approximation) the projection of $F(h,t)$ on to $F(h,0)$.

7.3.1 Hidden Advantages

In Equation 7.3, use of $exp(i\alpha_c)$ for the unknown phase factor is a fraught approximation. The use of a phase calculated from a model introduces bias toward that model. As shown in Figure 7.2, the desired phase of ΔF is also uncorrelated with the phase of the ground state structure. The sign of the term $\big[|F(h,t)| - |F(h,0)|\big]$ contains all the real phase information that we have. The figure of merit m for the phase choice α_c is $2/\pi$ (Lattman 1972). The figure of merit is approximately given by $<\cos(\Delta\alpha)>$, where $\Delta\alpha$ is the phase error. Because the cosine has the same value for $\Delta\alpha$ and $-\Delta\alpha$, one can usefully identify a mean phase error $\Delta\alpha = $ arc $\cos(m)$. For m = $2/\pi$, $\Delta\alpha$ is 50°. Since α_c itself is in error, the true error is even larger.

A couple of hidden advantages in DED maps compensate for these errors. The difference amplitudes contain an intrinsic weight that reduces the impact of terms for which the phase error ($\alpha_\Delta - \alpha_c$) is large. Figure 7.2 shows that when the difference between phase α_Δ of the true difference coefficient and α_c is particularly large, the magnitude of ΔF in Equation 7.2 is much reduced compared with that of the true difference coefficient F_Δ. Thus, terms that contain the largest phase errors are automatically down-weighted.

In addition, certain errors present in a conventional electron density map cancel in the DED map. For example, a small error in the $t=0$ model appears in the phase angle α_c, is retained in all individual maps at later time points t that use the same phase angle and will not appear in the DED maps.

The relative signal in DED maps is often much larger than in conventional electron density maps. The mean-square value of electron density is

$$\rho^2 = \left(\frac{1}{V^2}\right)\sum F^2(h),$$

and the corresponding value for DED is

$$\Delta\rho^2 = \left(\frac{1}{V^2}\right)\sum\Delta F^2(\boldsymbol{h}).$$

Thus, the magnitude of a typical DED peak is much smaller than the peak in a conventional map. However, a real peak arises from the density of atoms or chemical groups and should be of comparable magnitude in both conventional and DED maps. Even though peaks in DED maps appear at half weight, such a peak represents a relatively much larger signal in the DED maps. The signal to noise ratio in a DED map is thus much superior to the ratio in a conventional electron density (ED) map (Henderson and Moffat 1971).

7.3.2 Phase Improvement

Despite these hidden advantages, better approximation to the true phase α_Δ of $\Delta\boldsymbol{F}(\boldsymbol{h},t)$ is essential. How can this be determined? In a completely error-free DED map, no authentic difference features would appear in the solvent channels between the molecules. Solvent structure distant from the molecular surface will not vary in time; distant solvent does not contain signal. Phase errors cause spurious features to appear in the solvent channels. Solvent flattening is therefore the first, simple but powerful step in phase improvement. Identify a mask for the intermolecular solvent region; set all DEDs within this mask to zero; recalculate phases; and combine them with the experimental values of $\Delta F(\boldsymbol{h}, t)$ to calculate a second DED map. Most features in the solvent will be reduced but some will remain; reset them to zero; and repeat the process to convergence, which requires only a few cycles.

Although the solvent masks are identical for all time points, solvent flattening proceeds independently for each map and thus may introduce small time-dependent errors into the set. In contrast, the original set of DED maps used the same (erroneous) phase α_c for each time point, which introduces a larger but time-independent error.

Some DED features within the molecule itself may also be spurious. In the earliest time points of a light-sensitive system, authentic difference features tend to cluster near the chromophore and its immediate environment. In an enzyme, they tend to cluster near its active site. Additional flattening of DED features within the molecule but distant from the chromophore or active site may therefore also be attempted, by applying a time-dependent, spatially localized mask. Flattening features within the molecule is risky because identification of regions of the molecule lacking true signal may be in error. Flattening true signal eliminates authentic features which may be difficult to recover subsequently.

SVD also identifies components of noise in the DED maps which can be removed via SVD-flattening, discussed below (Chapter 7.4).

7.3.3 Partial Occupancy

Reaction initiation may not occur at all sites in the crystal, resulting in partial occupancy of reactive molecules. For example, an optical pump pulse may not be absorbed by all chromophores; or if absorbed, not all molecules enter the functional trajectory since the quantum yield for entry is often substantially less than 1.0. For mix-and-inject experiments that depend on diffusion in the crystal, chemically identical sites that are not symmetry-related may have very different accessibilities to a diffusing ligand. Diffusion of a reactant into the crystal takes time; the concentration of reactant varies with location in the crystal and time. Molecules in its interior see a lower ligand concentration than those at the surface, at a later time. The location actually illuminated by X-rays may vary from crystal to crystal. For example, an X-ray pulse may skim the surface of one crystal, and the next pulse may strike another crystal squarely. If the crystals have a range of sizes, the degree of partial site occupancy may vary markedly from one data frame to the next. In practice, these issues have not proved major obstacles. Refinement programs for static structures already incorporate routines for refining occupancy. In addition, the basis functions obtained in SVD are independent of occupancy; only the amplitudes of the time-dependent weights depend on occupancy (Chapter 7.4). The form of the final atomic models for intermediate structures is not affected.

Issues of partial occupancy date from the earliest days of protein crystallography. The binding sites for the heavy atoms used for phasing by isomorphous replacement were rarely fully occupied. The issue remains important today when attempting to identify metal ions in a protein crystal structure. A peak from a half-occupancy site containing an atom with N electrons will look much like the peak from a site fully occupied by an atom with N/2 electrons. Indeed, misidentification of metal ions in conventional protein crystal structure determination is very common (Grime et al. 2019).

Deriving an atomic model from a DED map can be a challenge, particularly in the case of partially occupied sites. A conventional ED map puts the features in the DED into context. Fortunately, maps may be created in which these features appear at approximately full weight, superimposed on the reactant (ground state) structure albeit with some additional noise. Such a map is simply an approximation to $\rho(\mathbf{x},t)$. As mentioned above, fully occupied peaks in a DED appear at roughly half weight. In this case, adding a 2X-weighted DED to the map for the zero-time state creates an ED map of the structure at time t in which all peaks appear roughly at full weight (Equation 7.4). Thus,

$$\rho(\mathbf{x},t) = \rho(\mathbf{x},0) + 2\Delta\rho$$

$$\rho(\mathbf{x},t) = \frac{1}{V}\sum_{h}\left\{|\,F(\mathbf{h},0)\,| + 2\big[|\,F(\mathbf{h},t)\,| - |\,F(\mathbf{h},0)\,|\big]\right\}exp(i\alpha_c)exp(-2\pi i\mathbf{hx}). \qquad (7.4)$$

$$\rho(\mathbf{x},t) = \frac{1}{V}\sum_{h}\big[2|\,F(\mathbf{h},t)\,| - |\,F(\mathbf{h},0)\,|\big]exp(i\alpha_c)exp(-2\pi i\mathbf{hx}).$$

Here $\Delta\rho(\mathbf{x},t)$ is the DED between the states at times 0 and t (Equation 7.3). The factor of 2 that weights the DED above does not include consideration of partial occupancy or the effect of phase errors. Crystallographers rapidly realized that operationally improved maps could be obtained by using extrapolated maps of the form $\rho(\mathbf{x},t) = \rho(\mathbf{x},0) + \beta\Delta\rho(\mathbf{x},t)$, where the weight β is chosen by a mix of empirical considerations including, for example, spectroscopic data. In practice it is more convenient to extrapolate structure factor amplitudes to represent the full occupancy structures. Much consideration has been given to the best way to do this (Genick 2007; Gorel et al. 2021). The Supplementary Material in (Pandey et al. 2020) offers a practical method applicable even when the extent of reaction initiation is low.

7.3.4 Timing Uncertainty

In all pump–probe approaches, at first glance the overall time resolution is limited by the convolution of the time profile of the pump pulse with that of the probe X-ray pulse. Additional uncertainty arises from such processes as the slow diffusion of a chemical probe within the crystal lattice, and ultrafast jitter and slower drift in the interval between pump and probe pulses. That is, errors are present in both the time and spatial domains of the DED maps. The effects of timing uncertainty are mitigated by the assumption, backed by many observations, that molecules spend most of their time near energy minima along their functional trajectory. Each minimum is associated with a reaction intermediate. There are only a limited number of intermediate states to be described, each with a reasonably well-defined lifetime.

Tight control and experimental measurement of the time delay between pump pulses from optical lasers and probe XFEL pulses have long been thought important for precise reconstruction of dynamical trajectories. The TT (Chapter 6.5.2.1) was developed to assist this measurement. Recently, highly sophisticated mathematical techniques have been applied to improve dramatically the re-sorting of time series beyond the apparent time resolution of the data, even after experimental correction of the nominal timing measurements by the TT (Ourmazd 2019b; Hosseinizadeh et al. 2021). These techniques are applied to the raw experimental data to computationally estimate time delays and are underlain by the mathematical structure known as a manifold. A manifold is a generalization of a 3D curved surface. It is an N-dimensional space locally similar in all neighborhoods to Euclidean N-dimensional space but curved over larger distances. Manifolds are particularly useful in dealing with data that span all N dimensions, but whose intrinsic dimensionality is lower. An example arises in the treatment of diffraction data obtained from a series of single frames, as in SFX. If the detector has P pixels, then each diffraction frame can be regarded as a P-dimensional vector whose components are the counts recorded in each pixel for that frame. A typical experiment will generate very large numbers, many millions in some cases, of such vectors. Superficially it may seem that the set of vectors could span the whole manifold. However, the diffraction data are periodic functions of the angles that specify the orientation of the crystal. The diffraction data fundamentally lie on a subspace in 3D, albeit a highly curved one.

Individual diffraction frames in an SFX experiment may show only a small number of (often weak) Bragg reflections. The issues of identifying and indexing such peaks have been the subject of intense exploration and development (Lyubimov et al. 2016; Hadian-Jazi et al. 2021).

The development of artificial intelligence (AI) and machine learning has greatly enhanced the utility of the manifold formalism. From the many frames obtained experimentally, AI algorithms "learn" the curved manifold that best spans the data. Tools such as nonlinear SVD or nonlinear Laplacian analysis (which operate in a differential geometry mode) are then applied to the manifold to sort and analyze the data that arises from signal and of equal importance, to identify and reject outliers that arise from noise. The scope of this analytical tool has been developed during the last decade. A landmark was the analysis of spectra on molecular nitrogen from studies carried out at the LCLS in an atomic physics experiment. The experiment followed an optical pump with an X-ray probe (Fung et al. 2016) in the face of extreme timing uncertainty of around 150 fs, largely set by the duration of the probe X-ray pulse and its timing jitter. Despite this uncertainty, the analysis provided a remarkably accurate dynamical description of the system. The final time uncertainty was reduced to around 2 fs, roughly 100-fold less than the experimental timing uncertainty.

This mathematical approach to the reduction of timing uncertainty – time sharpening – is completely general. Of direct relevance to structural dynamics, (Hosseinizadeh et al. 2021) extended the analysis to single crystal X-ray diffraction data. Though the entire data set of many patterns or frames offers high redundancy, each diffraction pattern provides only fragmentary, 2D sampling of the complete 3D scattering data. The original fs time-resolved experiment at the LCLS explored the dynamical behavior of single crystals of PYP immediately after entry into its photocycle (Pande et al. 2016). Entry was driven by absorption of a fs pump pulse from an optical laser. Diffraction data spanned the early stages of the photocycle from a few tens of fs to 3 ps. Careful application of the experimental timing tool to the nominal timing data and conventional pump–probe analysis achieved a time resolution of around 140 fs. This resolution was adequate to confirm that *trans* to *cis* isomerization of the chromophore was centered at a time delay around 550 fs but could not reveal details of the functional trajectory of isomerization.

Reanalysis of the identical data from this experiment by the manifold formalism sought to reduce the timing uncertainty and identify these chemically essential details. The potential energy surface for the earliest stages of the photocycle from fs to 1 ps is described by coherent motion toward a so-called conical intersection, a low-dimensional trajectory which connects the potential energy surface of the initial, electronically excited, photo-activated state to that of the first intermediate state. The familiar Born-Oppenheimer approximation assumes the independence of electron and nuclear motion and fails at conical intersections, where structural behavior is characterized by their cooperative motion. The data reanalysis (Hosseinizadeh et al. 2021) was spectacularly successful. The timing uncertainty was reduced to around 2 fs and a plausible trajectory through the conical intersection was identified. This provided an exceptionally detailed structural view of the traversal of a conical intersection, a first in any molecule.

These advances suggest that as algorithms and software mature in the near future, time resolution in dynamic experiments may no longer be determined solely by the experimental duration of the pump and probe pulses, nor by drift or jitter in the time delay between them. That is, the overall time resolution may be reduced to a few fs at XFELs. This would open up new experiments on the few fs time scale in biophysics, chemistry, and materials science, even though these experiments use much longer, intense, highly noisy X-ray pulses (Chapter 6.3).

The remarkable generality of these mathematical techniques for analysis of noisy time series is further illustrated by their application to a totally different problem: analysis of ultrasound images of human fetal development recorded over weeks and months during pregnancy. In effect, by estimating time zero – conception – with unprecedented precision, the approach accurately defines the date of full term and hence the extent of prematurity, data of great clinical importance (Fung et al. 2020).

The idea that the limitations of inaccurate, fuzzy timing, which appear to be inherent in the data, can be largely overcome by sophisticated mathematics is far from intuitive for most of us. Indeed, it can seem like magic. Thus, some simple – if not simplistic – thoughts on what makes this all possible may be useful. Those of us trained in crystallography, or in many other disciplines in which imaging or time series are important, are accustomed to the digital Fourier paradigm in which functions are evenly sampled on a lattice, and the minimum number of sample points required to represent the function can be determined. This approach is incredibly useful when data acquisition is expensive, and when straight-forward mathematical syntheses are welcome. However, it does not prepare us for cases in which functions are sampled irregularly in time, and both noisily and partially in space. Two key ideas underlie the mathematical techniques for time sharpening.

First, the methods apply over times during which neither the system nor the signals that characterize it change very much. Thus, the displacements of the system from its initial time state can typically be described by a few major eigenvectors or eigenfunctions. For example, the nonlinear analysis of the conical intersection in PYP revealed only five modes above noise. We do not have to locate all the few thousand atoms in PYP, just the few eigen-functions and eigenvalues that describe their motion. Likewise, the intensities of the Laue spots in the diffraction patterns – the raw signals – do not change much with time as the reaction proceeds. The changes present are strongly coupled, not independent. About two centuries ago Laplace pointed out that, given all initial positions and velocities, the state of a system could be predicted far into the future using Newton's Laws. If N numbers specify the system at time zero, it takes only N additional numbers to specify it for later times. Dynamics is less complex than it seems. Of course, in practice even tiny errors in these initial values propagate so that prediction fails after longer times. The N numbers must be accurately specified – but how accurately, for a given system and time scale of interest? For sufficiently short times, far fewer than N additional numbers may satisfactorily describe evolution of the system.

Second, large numbers of frames are collected. Jitter samples time very finely, although irregularly. In SFX only a modest number of diffraction spots appear on each frame, but there are many frames near each time point. Thus, even data that are seriously incomplete in the number of diffraction spots and noisy in the intensity of each spot may be sufficient to determine the far lower number of independent variables in the system. Sophisticated pattern recognition methods like AI/manifold embedding provide the relevant pathway for the discovery of these variables.

7.4 Singular Value Decomposition (SVD)

In systems that are evolving in time during an experiment, the scattering or diffraction data will necessarily contain contributions from all states present in the volume of the sample exposed to the X-ray beam. This situation is equally true when variables other than time such as temperature, pH, or concentration of a perturbant, differ between frames. How are the structures of the individual states present in the sample to be extracted from experimental frames arising from the structures of all states, under variable conditions? SVD, an application of linear algebra, comes to the rescue (Henry and Hofrichter 1992).

We present a general description of SVD in the time-resolved crystallographic context, introduced by Schmidt and colleagues (Schmidt et al. 2003; Rajagopal et al. 2004). Box 7.1 (Doniach 2001) bases a more formal description on matrix algebra, and applies SVD to SAXS data to identify conformations in solution. Chapter 5.2.2 provides a discussion of the delicate data processing required to generate the $I(q)$ curves needed for SAXS applications.

SVD is a generalization of a familiar concept: a vector V in n dimensions can be fully represented by its projections on a set of n orthogonal basis vectors. For example, a vector in 3D can be represented using its three components or basis vectors, namely x, y, and z. SVD evaluates a set of M such vectors in a high-dimensional space that fully describes the experimental data. SVD simultaneously uncovers the basis vectors on which the M vectors are projected, and the weights and coefficients that allow the reconstruction of the M vectors from their components on the basis vectors. Basis vectors are both orthogonal and normalized; thus, SVD is described as an orthonormal resolution. It is important to realize that the basis functions themselves do not necessarily represent actual changes in the system. Each basis function arises from a linear combination of the actual changes. Usually a physical model is needed to specify the weights associated with each basis function and thus reconstruct the changes themselves. We discuss this further below.

SVD requires the vector in M dimensions to be a scalar summation of the components along each of the basis vectors. For crystallography, this fact holds in real space. At all positions in the asymmetric unit, the overall DED is the scalar sum of the DEDs of each of the structural components present, weighted by the fraction of molecules in that component. Thus, Equation 7.4 can be regarded as a linear (scalar) sum of $\Delta\rho(x,t)$ maps (DED maps), imaging structural changes associated with each intermediate or component present at time t. (Recall that the Fourier transform of a sum of functions is the sum of the Fourier transforms of each function.) The linear relation required by SVD however does not hold in diffraction space. A total structure factor is the phased, vector (not scalar) summation of the partial structure factors, one from each component. The observed intensity of each diffraction spot is the square of the resultant vector summation.

SVD remains fully applicable when the data matrix A (introduced below) is complex. Thus, technically, SVD can be used with phased (complex) structure factors F as data. However, we know of no case where this has been done. The reason for preferring DED values as data is that the various constraints that we apply to improve phases are much more readily applied in direct space, solvent flattening for example. In addition, the interpretation of DED maps is much more visual than the interpretation of differences in complex structure factors.

In SAXS experiments however, the SVD input data comprise the set of set of scattering curves $I(q,t)$ rather than DED maps. Equation 5.1 reminds us that $I(q)$ is expressible as a scalar sum of terms like $f_i f_j sin(qr_{ij})/qr_{ij}$, where f_i is the scattering from the i-th scatterer, and r_{ij} is the distance between scatterers i and j. If scatterers i and j are located within the component k present with weight w_k, then a further (scalar) summation over k must be taken to obtain the total $I(q)$ curve. Thus, the $I(q,t)$ curves in time-resolved SAXS are also appropriate input for SVD.

SVD analysis in time-resolved crystallography begins by creating a data matrix A in which the columns of A contain all DEDs in the crystallographic asymmetric unit for a given time point, presented as a 1D string. That is, the 3D DED map of an asymmetric unit is presented in 1D; each column contains the DED map for a separate time point. The locations at which the difference densities are sampled is that typical of conventional ED maps, roughly 1/3 of the crystallographic resolution. In an SAXS/WAXS solution scattering experiment, each column contains the spectrum $I(q, t)$ as a function of q (or scattering angle), at a given value of the variable parameter such as time t, pH, temperature, or concentration of a reactant. In time-resolved crystallography, each row of A thus contains the time dependence of the DED at that single location in the asymmetric unit in real space; each column of A contains the variation of the DED in real space across the asymmetric unit. In time-resolved solution scattering, each row of the data matrix A contains the time dependence of the scattered intensity at that value of q; the data in each column varies in q. In other scientific contexts, the column might contain a gene sequence, or any data in which there are believed to be common elements.

The data matrix A is not square and can be large. A contains N columns, one for each value of the time, and M rows, one for each sample point in the DED map. M is very large since there can easily be several thousand spatial locations in the asymmetric unit. Data may be acquired at tens of time points spanning the complete functional trajectory, thus N is quite large but much smaller than M.

It is often useful to apply SVD to subsets of the data. Examples that reduce the size of A may include only regions of the asymmetric unit near the chromophore or active site where the signal is expected to be larger and the signal to noise ratio superior (reduced M); or only a specific time period such as the ultrafast time region with time points less than 1 ns (reduced N). Alternatively, the overall size of A may be retained but SVD is applied to a truncated subset of matrix elements, such as only those locations in the asymmetric unit where the magnitude of the DED exceeds a specified value such as 3σ in one or more time points (reduced M; σ is the rms value of the DED across the asymmetric unit). Values at all other elements of A are set to zero.

The mathematical process of SVD decomposes the data matrix A into a product of three matrices U, V and W, each with very useful characteristics. The decomposition itself is carried out efficiently in Mathematica, MatLab, or many other mathematics packages, not elaborated on here. Thus

$$A = UWV^T.$$

Matrix U contains the orthogonal basis vectors on which the DEDs are projected, known as the left singular vectors (lSVs). The matrix W is square and diagonal. Its elements are known as the singular values (SVs), appear in decreasing order, and give the weight or significance of the corresponding basis vector. The matrix V^T contains the coefficients for reconstructing the original time dependence, known as the right singular vectors (rSVs). The decomposition is shown schematically in Figure 7.3.

There are N SVs in all, corresponding to the N time points in the data matrix. The power of SVD lies in the fact that not all SVs are significant. Significant SVs represent the time dependence of authentic structural changes; insignificant SVs represent noise. The

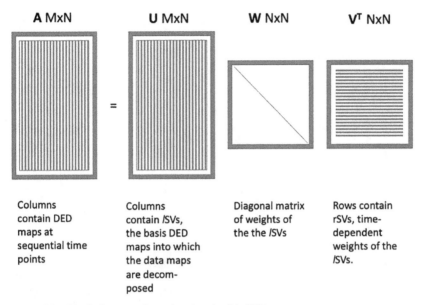

A MxN	**U** MxN	**W** NxN	**V**^{T} NxN
Columns contain DED maps at sequential time points	Columns contain *l*SVs, the basis DED maps into which the data maps are decomposed	Diagonal matrix of weights of the the *l*SVs	Rows contain rSVs, time-dependent weights of the *l*SVs.

Figure 7.3   Block diagram of matrices involved in SVD.

number of significant SVs is frequently small, very much less than N. There are perhaps three to five significant SVs in a crystallographic experiment and rarely more than three in a SAXS experiment. The small number may be because the system in fact has little variability and the signal is described well by only a few components; or because the data are inherently noisy. In general, most of the SVs are small in magnitude, judged to be insignificant and representing noise. It remains to be seen whether more SVs will be judged significant when phase improvement and manifold embedding techniques are applied to the raw data.

The distinction between SVs deemed significant (signal) and insignificant (noise) is not clear-cut. Their values fall on a declining continuum (Figure 7.4) rather than having a sharp breakpoint. The initial distinction is made based on the value of the SVs but a useful, finer distinction is added by examining the corresponding rSVs and lSVs. Three constraints presented above stand out (Chapter 7.2). First, signal varies smoothly in time (or with another parameter being varied such as pH or concentration), but noise varies sharply and unphysically. Thus in a time-resolved experiment, rSVs arising from signal vary smoothly, but rSVs arising from noise tend to vary sharply with time; they flicker. Second, signal (in crystallography) is spatially confined to the molecules, but noise is distributed across both the molecules and the intermolecular solvent. lSVs arising from signal have features largely confined to the molecules, but lSVs arising from noise have features in both the molecules and the solvent. (Recall that in SVD, the 3D grid of DED values across the asymmetric unit is stretched out into a single 1D data vector when forming the data matrix A. The resultant lSVs can be unpacked on to the 3D grid to regenerate a DED map for each lSV.) Third, rSVs associated with signal have a specific mathematical form. For example, if the system follows chemical kinetics, the signal corresponds to relaxation times that vary as a sum of exponentials (Chapter 1). Jitter in the time domain of the raw data ultimately appears as

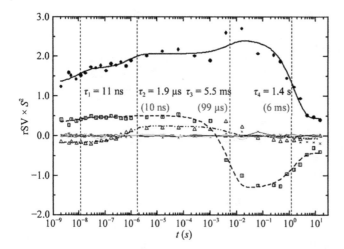

Figure 7.4 TRX study of the photocycle of PYP as a function of temperature. Right singular vectors (rSVs) at 233 K. The rSVs are weighted by the square of their respective singular value S. Four kinetic processes $\tau_1 \ldots \tau_4$ are globally observed (dashed vertical lines). Solid spheres, open triangles, and squares: first, second, and third significant rSVs. Colored thin lines around zero: remaining less significant rSVs. Solid black line, dashed line, and dashed double dotted line: global fit of the significant rSVs by four exponential functions with the same set of relaxation times but different amplitudes. Relaxation times obtained at room temperature (298 K, red) are shown in parentheses for comparison. Schmidt et al. 2013/International Union of Crystallography/Licensed under CC BY 2.0.

jitter in the rSVs. A process of rotation seeks to maximize the temporal correlation within a set of rSVs; that is, to maximize the smoothness of the rSVs and minimize jitter.

Combining the value of the SVs with these constraints on the nature of the corresponding lSVs and rSVs enables SVs to be distinguished with reasonable confidence between signal and noise. This distinction is very important: the number of significant SVs is the number of orthonormal basis vectors, which is in turn the number of independent states represented in the data.

SVD played a critical role in a study of PYP in which both time and temperature were varied systematically (Schmidt et al. 2013). Using the van t'Hoff–Arrhenius equation, a well-validated initial model for the PYP photocycle and other constraints, the authors were able to reconstruct the energy landscape for PYP, including the structures of intermediates. Figure 7.4 shows a sample set of rSVs from the SVD analysis, weighted by the square of corresponding element of **W**, the matrix of singular values. There are only 4 significant vectors in this set across the entire time range, and the remaining 11 vectors are distributed randomly about zero and judged to be insignificant. If for example the time range were restricted to time points >100 μs, only two SVs would be revealed. The top four rSVs are very smooth, a credit to the high quality of the experimental data. All four rSVs emerging from the analysis can be well fit by a sum of exponentials using a common set of relaxation times, represented in Figure 7.4 by the four lines (one solid and three dashed) through the data points that span the complete time range. The four vertical dashed lines represent the four relaxation times. The common set of exponentials convincingly establishes that the data arise from a chemical kinetic mechanism. The temperature dependence of the relaxation times and of the microscopic rate constants in the favored chemical kinetic mechanism followed linear Arrhenius behavior (Chapter 2; data not shown here) and offered further support for that specific mechanism (Schmidt et al. 2013).

In practice, those SVs deemed to be insignificant and arising from noise are rejected by setting their values to zero in the singular value matrix $\boldsymbol{W}$ and renaming it $\boldsymbol{W'}$. Noise removal is substantial and constitutes SVD-flattening. The full data matrix $\boldsymbol{A}$ can then be recalculated and renamed matrix $\boldsymbol{A'}$ in which $\boldsymbol{W'}$ replaces $\boldsymbol{W}$:

$$\boldsymbol{A'} = \boldsymbol{U'W'V'^T}$$

Each individual DED map in a column of $\boldsymbol{A'}$ has had its difference density modified by SVD-flattening. Back transformation results in a new set of phases which when combined with the experimental differences in structure amplitudes, produces new, individual DED maps and reconstitution of a third data matrix $\boldsymbol{A''}$. The entire process of SVD flattening and phase modification can be repeated to convergence.

We reemphasize that the basis vectors, the significant lSVs, are not themselves the structural intermediates. They do not correspond directly to the individual states. Rather, they are a linear combination of the structures of the intermediates, whose (as yet unknown) coefficients in that linear combination contain the information from which these desired intermediate structures may be expressed. The relationship between the significant lSVs and the desired intermediate structures – the coefficients – depends on a physical model. In time-resolved crystallography, a typical model to be tested would be a candidate chemical kinetic mechanism with the appropriate number of states. For a model chemical kinetic mechanism with S significant SVs and S states (Chapter 1.5), the time-dependent concentration of each state can be formally calculated from the rate coefficients of each elementary step and the conservation of concentration. The goal is to relate these model-dependent state concentrations and rate coefficients to the observed lSVs and rSVs. This requires estimation of the coefficients to be applied to the lSVs. A specific example of this process from SVD analysis of SAXS data is illustrated in Equations 7.6 through 7.9 of Box 7.1 (Doniach 2001). Only some models for the chemical kinetic mechanism – perhaps only one – will show satisfactory agreement with the SVD data. These models are retained; others are rejected.

In time-resolved crystallography, SVD has been successfully applied to different sets of wild type and mutant PYP data. The photocycle of PYP has several intermediates whose lifetimes span the time range from ns to a few s, in a relatively complicated mechanism with parallel paths. (See for example (Ihee et al. 2005; Jung et al. 2013; Schmidt et al. 2013).) These papers are based on data from SR sources such as BioCARS, which laid the foundation for complete analysis of the PYP photocycle including the ultrafast fs to 100 ps domain at the LCLS XFEL (Pande et al. 2016). A much simpler example of SVD analysis on the related PAS domain protein FixLH confirmed that there was no detectable intermediate between the reactant and product states (Key and Moffat 2005).

Dynamics over short time periods can be usually represented by a few basis functions (see also the example in Box 7.1, below). As a specific illustration, the (inverse) vibrational frequencies of α-helices are in the ps regime. The velocity of sound in water is

1500 m s^{-1}, 15 Å ps^{-1}. These two values make clear that structural changes in times below the ps range remain spatially localized and propagate in simple ways. In binding or catalysis where collective motions of significant portions of a protein are involved, the functional trajectory will remain simple, even though an ensemble of structurally distinct microstates are all adjacent to and consistent with the trajectory.

Box 7.1 Example: Application of SVD to SAXS

A review (Doniach 2001) gives a more formal, algebraic treatment of the application of SVD. In one experiment, SAXS was used to probe the unfolding of cytochrome *c* by measuring the scattering spectra as a function of the concentration of the denaturant, guanidine hydrochloride (Segel et al. 1998).

In Equation 7.5, *A* (*s*, *m*) denotes the full data matrix **A** in which *s* indexes the data points in one SAXS spectrum *I*(*s*) (columns), and *m* indexes the spectra as a function of the concentration of the denaturant, over *P* values (rows). Formally speaking, the index $k \leqslant P$ runs through the diagonal elements w_k of the singular value matrix **W**. The vector $u_k(s)$ is the *k*-th column of the lSV matrix **U**, and represents one of the eigenfunctions needed to reconstruct the data matrix **A**. The v_k^m are the elements of the rSV matrix V^T. The rows of V^T give the coefficients needed to reconstruct the data from the eigenfunctions.

$$A(s,m) = \sum_k w_k v_k^m u_k(s).$$ (7.5)

The number of eigenfunctions needed to reconstruct the cytochrome *c* data was determined to be three. The effective dimensionality of the system may become evident in several ways. There may be only a limited number of eigenfunctions (columns of **U**) that are significantly different from zero. Or, there may be only a small number of significant elements in the diagonal element in **W**.

Segel and colleagues (Segel et al. 1998) therefore created a thermodynamic model incorporating three species. The idea is that at each concentration of denaturant c_m, the profile of the total scattering $I(s,c_m)$ can be expressed in terms of the (as yet undetermined) spectra of the pure states, where the proportions of the three states are related by a chemical equilibrium equation.

$$I(s,c_m) = \sum_a f_a(c_m) I_a(s).$$ (7.6)

Here *a* runs through the set of three singular values, and I_a is spectrum of the *a-th* component of the mixture. The weights $f_a(c_m)$ come from the chemical equilibrium model relating the various species, outlined below. Each of I_a can in turn be expressed in terms of the basis vectors from the matrix **U**.

$$I_a(s) = \sum_k w_k b_k^a u_k(s).$$ (7.7)

Box 7.1 (Continued)

Here the b_k^a are numerical coefficients to be determined. If we substitute Equation 7.7 into Equation 7.6. we find that

$$I(s,c_m) = \sum_a f_a(c_m) \sum_k w_k b_k^a u_k(s) = \sum_k w_k u_k(s) \sum_a f_a(c_m) b_a^k. \qquad (7.8)$$

By comparison with Equation 7.4 we see that the experimentally determined v_k^m should be equal to $\sum_a f_a(c_m) b_k^a$ from the model. This allows a least squares refinement to be carried out, which minimizes the difference between the modeled and observed elements of the matrix $\boldsymbol{V}^T$ through the residual

$$\chi^2 = \sum_k^K \sum_m^M \frac{w_k^2 \left(v_k^m - \sum_a f_a(c_m) b_k^a\right)^2}{\sigma_k^2(m)}. \qquad (7.9)$$

χ^2 is minimized with respect to the b_k^a. The parameters that determine the form of $f_a(c_m)$ come from a simple model in which the free energy differences between folding states are a linear function of the concentration of denaturant. We omit the details of this model but it involves only four parameters, two for each of non-native states, that determine the free energy difference between the non-native and native states as a function of concentration. These differences are given by $\Delta G_\alpha = \Delta G_\alpha^0 - m_\alpha[GdnCl]$, and the parameters m_α and ΔG_α^0 are included in the refinement. A key link to experimental reality in this work was to calculate Kratky plots (Chapter 5.2.2.3) for the SVD basis functions, to confirm that they provided plausible contributions to the overall SAXS pattern.

7.5 Features Commonly Found in DED Maps

The most obvious feature of experimental, time-dependent DED maps is also the most important. The structural changes from time t to time zero prior to reaction initiation are small; the DED maps are largely flat with a few modest features. In general, the magnitude of these features forms a Gaussian distribution centered at zero, supplemented by a few large positive and negative outliers that clearly arise from signal. If structural changes include ligand binding which adds well-ordered atoms and their electrons, the Gaussian may be centered at a slightly positive value; if the changes increase disorder, at a slightly negative value. If the Gaussian is centered far from zero, inaccurate scaling of the difference coefficients should be suspected.

The rms value of the DED across the asymmetric unit (sometimes modified by omitting these large outliers) is denoted σ. Thus, most DED maps display only those features whose magnitude exceeds 3σ to 4σ. On statistical grounds alone, these features are clearly signal. However, the individual features in the DED maps are not all independent; strong correlations between them are present (Chapters 7.2 and 9.4.7). Correlations are distinctive; every group that moves has a leading edge and a trailing edge. This results in a pair of peaks of opposite signs in the DED map flanking the group that has moved, positive at the leading edge and negative at the trailing edge. Neither peak might attain a magnitude of 3σ and statistically, both peaks might well arise from noise. However, the pairing of positive and negative peaks surrounding a known chemical group provides a dramatic increase in confidence in the interpretation of the peaks as arising from a small shift. This logic has been expanded into the program Ringer (Lang et al. 2010) that systematically samples the DED in the vicinity of amino acid side chains in search of patterns of peaks characteristic of various rotamers. Many unreported rotamers have been detected in reexamined proteins. The approach could be developed to detect multiple conformations of regions existing along a functional trajectory in a time series of DED maps. The method has also been applied to cryo-EM structures (Barad et al. 2015).

The program qFIT (Keedy et al. 2015) seeks to explain local DED features by exploring a vast set combinations of sidechain conformers and backbone fragments. The program can be used to trace the presence of conformations through time, and to explore previously unnoticed conformational excursions such as a peptide flip. Features in DED maps may also be correlated over longer distances. For example, motion of an α-helix as a rigid body is accompanied by positive and negative DED features along its length flanking the moving groups. Correlations may also exist over time. For example, if structural changes extend over all atoms in a compact domain in a correlated motion, all DED features in that domain will exhibit the same time course and identify a "motional domain" (Kostov and Moffat 2011). Spatial and temporal correlations in DED maps deserve further exploration as powerful aids to analysis (Chapter 9.4.7).

Radiation damage is another common feature, which may not be easy to separate from time-dependent difference features associated with mechanism. Chemistry helps if features appear near moieties that are known to be subject to radiation damage such as S–S bonds, acidic side chains, and metal centers. Indeed, metal ions were misidentified in more than half of a sample of 30 protein crystal structures (Grime et al. 2019). Many of these difficulties also apply to DED maps where a metal can change electronic state during a functional process, because of radiation damage, or both. Since time-dependent radiation damage is dose-dependent, it is expected to follow a different time course than functional processes. Probes such as X-ray spectroscopies that explore metal centers (Chapter 8.4) in parallel with the structural experiments help to tease apart the functional processes and damage.

7.6 Refinement of Intermediate Structures

Knowledge of a previously refined structure of the reactant (or dark or ground state) molecule is the basis for refinement of candidate intermediate structures, which often differ little from the structure of the reactant. It is important to distinguish the crystal structures which might be refined. One common case is the average structure in the crystal at some general time t, which may well be a time-dependent mixture of two component structures such as the first and second intermediate states. At later times the occupancy of the first intermediate may have dropped to zero. Such refinements are carried out against the structure amplitudes $F(hkl,t)$. The capacity to simultaneously include two or more conformers is essential, and the result in principle is a refined structure of one or more intermediates.

If instead a time-independent state is not a pure intermediate but rather, a mixture of two components in a quasi-equilibrium, stationary state, the structure and occupancy of the two components may be refined simultaneously. This requires that the two structures are sufficiently different, the structure amplitudes of the mixture are of high quality, and the minor component has sufficient occupancy.

A different case arises when a pure, time-independent intermediate state is present in the crystal over an extended time range. Refinement of such a time-independent structure may be carried out exactly as if it were a conventional, static structure. Refinement may be based either on difference Fourier methods in which atoms are shifted in accordance with the DED maps (Afonine et al. 2018), or on Fourier methods applied to conventional derivative maps (Rupp 2009).

The most interesting – but still understudied – situation arises when the mechanism and rate coefficients are such that no time period exists when only a single intermediate structure is present. Figure 7.4 is an example of the time course of such a reaction. A mixture of structures is present at all time points in this period, and the occupancy of each varies with time. Nevertheless, at each time point useful information is present on the structures of all components whose occupancy is sufficient. The set of adjacent time points also contains information on the rate coefficients for formation and decay of each component. Initial estimates of these rate coefficients are available from the rSVs in SVD analysis, for the candidate mechanism being considered. It should therefore be possible to refine the structures of all components, their occupancies, and hence the rate coefficients simultaneously, against data spanning the time period when each component has a significant occupancy. This challenge remains to be attacked.

7.7 Example: The Photosynthetic Reaction Center

A recent study of the photosynthetic reaction center (Dods et al. 2021) shows the power of using a multipronged approach in TR-SFX. Photosynthetic reaction centers and photosystems are complex, integral membrane proteins that use the energy of an absorbe photon of visible light to create a charge-separated state. Charge separation initiates a multistep electron transport process that leads ultimately to the synthesis of ATP. The authors studied the photosynthetic reaction center of the purple non-sulfur bacterium *B. viridis* (*Bv*RC) in which electron-transfer reactions originate at a special pair (SP) of strongly interacting bacteriochlorophylls with an absorption maximum at 960 nm. Ultimately, two protons are transported across an energy-transducing membrane for every photon absorbed, which contributes to the energy gradient that drives ATP synthase. The electron transport chain contains many components, but we concentrate on the elucidation of certain of the very rapid reaction steps. Figure 7.5 (Dods et al. 2021) is a simplified diagram of the electron transfer process. Note the arrows show which regions of the complex are perturbed at various times.

The electron liberated at the SP moves within a few ps to a nearby bacteriochlorophyll BPh_L, and onward in about 1 ns to an adjacent quinone Q_A. The motion of electrons in a molecule is not governed by the same rules of mechanics that apply to (relatively) massive atoms connected to other atoms by spring-like bonds. In the 1950s, Marcus theory was developed to explain the rates of electron transfer reactions that are not coupled to the breaking of a chemical bond (Di Giacomo 2020). It proposes that in the case of outer sphere electrons, the solvent plays a critical role in controlling the rate, rather than the nature of

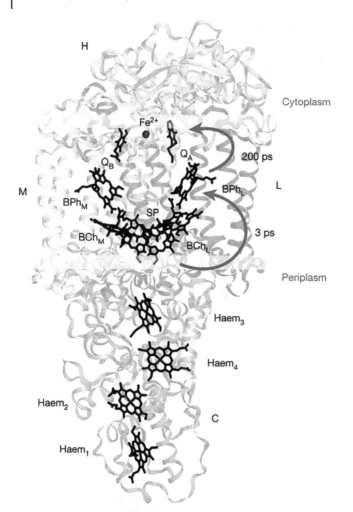

Figure 7.5   Electron-transfer steps of the photosynthetic reaction center of *B. viridis*. Cartoon representation of the H, L, M, and C subunits. Cofactors are shown in black including the special pair (SP) of strongly interacting bacteriochlorophylls where electron-transfer reactions originate. Also shown are two monomeric BCh molecules, two BPh molecules, a tightly bound Q_A molecule, a mobile Q_B molecule, a non-heme iron (Fe^{2+}), and four hemes. The approximate boundaries of the membrane are suggested in blue. The electron-transfer pathway: SP $\rightarrow$ BPh$_L$ $\rightarrow$ Q_A is referred to as the A-branch. Approximate timescales for the first two electron-transfer events, from SP to BPh$_L$ and from BPh$_L$ to Q_A, are shown. Dods et al. 2021/Reproduced with permission from Springer Nature.

the donor or receptor as in transition state theory. Marcus theory in many cases also leads to exponential kinetics. When quantum effects such as electron hopping come into play, rates can be faster than simple Marcus theory might predict. Indeed, spectroscopic studies have shown that a two-step hopping mechanism accounts unambiguously for the primary electron-transfer step from the SP to BPh$_L$. This 2.8 ± 0.2 ps process spans about 1 nm and is more rapid than conventional Marcus theory. These and other observations motivate studies at timescales significantly shorter than photo-induced reactions in which the initial event requires bond breaking or isomerization.

Working at the LCLS, the group used 960 nm pump pulses of 150 fs duration to excite the special pair, and 40 fs X-ray probe pulses generated single diffraction snapshots from tens of thousands of microcrystals. Post-initiation time points at 1 ps, 5 ps, 20 ps, 300 ps, and 8 μs were explored, and significant information emerged at even the shortest times. The authors' summary states that "Time point $\Delta t = 1$ ps populates the photo-excited charge transfer state of the SP in which charge rearrangements have occurred within the bacteriochlorophyll dimer but are before the primary electron-transfer step; $\Delta t = 5$ ps and 20 ps are after the initial charge-transfer step and SP is oxidized and BPhL is reduced; $\Delta t = 300$ ps is longer than the time constant for electron transfer to Q_A and menaquinone is reduced; and $\Delta t = 8$ μs corresponds to a metastable, charge-separated state." Figure 7.6 illustrates

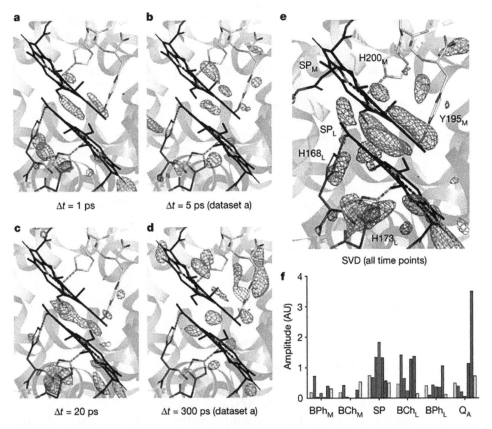

Figure 7.6   Light-induced electron density changes in *BvRC* at the site of photo-oxidation. Panel a, Experimental F_o(light) – F_o(dark) difference Fourier electron density map for $\Delta t = 1$ ps. **b**, Difference Fourier electron density map for $\Delta t = 5$ ps (dataset a). **c**, Difference Fourier electron density map for $\Delta t = 20$ ps. **d**, Difference Fourier electron density map for $\Delta t = 300$ ps (dataset a). **e**, Principal component from SVD analysis of all seven experimental difference Fourier electron density maps. All maps are contoured at ±3.2σ. Blue, positive difference electron density; gold, negative difference electron density. **f**, Relative amplitudes of difference electron density features integrated within a 4.5 Å sphere20 centred on the BvRC cofactors. The color bars represent (from left to right): cyan, $\Delta t = 1$ ps; blue, $\Delta t = 5$ ps, datasets a and b (in that order); purple, $\Delta t = 20$ ps; red, $\Delta t = 300$ ps, datasets b and a (in that order); mustard, $\Delta t = 8$ μs. AU, arbitrary units. Dods et al. 2021/Reproduced with permission from Springer Nature.

both the DED maps at the shortest time points, and the principal SVD lSV components extracted from the four time points spanning 1–300 ps.

The maps zoom in on the SP of bacteriochlorophyll molecules that form the site of initial photo-oxidation, and that respond rapidly with subtle changes in ring planarity. For reference the SP appears just above the center of Figure 7.5. Note the increasing spatial range of the nonzero regions of the DED maps with increasing time. Panel 7.6e shows a map including all significant lSVs arising from SVD analysis of the data. It is a useful reality check that the regions of significance in panel 7.6e coincide well with those in the DED maps in panels 7.6a through 7.6e. Panel 7.6f summarizes the temporal sequence of the appearance and disappearance of transient features of the maps and charts the flow of structural and electronic changes. The analysis indicates that the distortion of the SP precedes the charge-transfer event and could promote its efficiency.

To develop structures for the intermediate states, the authors used the refinement package Phenix (Adams et al. 2010) starting with a model of the dark, reactant structure placed into the unit cell using rigid body refinement. A critical step was the introduction of potential second conformers for portions of the structure involved in the reaction. The occupancy of these conformers was fixed at 30%.

The mechanism suggested by the structural analysis was probed for consistency using a number of other techniques, in which QM/MM computer simulations played a critical role. Such simulations treat the factors involved with charge separation and electron transfer using quantum mechanics, while the remainder of the molecule is treated classically using molecular dynamics. The authors also searched for possible artifacts by creating DEDs using a variety of subsets of the data and found that the results were robust. Critically, all DEDs were generated using data measured during the same experiment.

The authors eloquently summarize their work: "These findings demonstrate that *Bv*RC is not a passive scaffold but rather that low-amplitude protein motions engage in a choreographed dance with electron movements taking the lead and protein conformational changes following. Conversely, as the structure of the protein adjusts to stabilize these charge rearrangements, the energetic barriers that hinder the reverse electron-transfer reaction increase, thereby extending the lifetime of the charge-separated species and enhancing the overall efficiency of photosynthesis."

7.8 Making a Molecular Movie

In Chapter 1.4 we introduced the connection between kinetics and dynamics, and molecular movies. How can molecular movies be created using time-resolved structural data? The movies we view in the everyday world are composed of still images of a single object such as an actor, collected at equally spaced time points. Each still is usually crisp, although rapidly moving objects such as golf balls may leave contrails. Our brains fuse the rapidly presented series of stills into a smoothly evolving image, a pure functional trajectory of a dynamic process. Animations form another class of everyday movies. Here each image (known as a cel) is drawn by an artist or computer, and big changes may occur between frames, such as Wile E. Coyote being flattened by an anvil in a single step. Some molecular movies echo these classes of everyday movies. They show the motion of a representative single molecule or particle as it moves along its functional trajectory. The preparation of movies of this type requires the discovery, modeling, and refinement of time-independent intermediate structures, such as those directly revealed by crystallographic studies in chemical kinetics. Typically, the portions of these movies that depict the passage over an energy barrier cannot be based upon experimental structural data from chemical kinetics. No such data exist. Rather, passage over a barrier is based on computational extrapolation from the two intermediate states that flank the barrier. A more straightforward class of movie is based directly on chemical kinetics and shows how the ensemble of molecules in the crystal evolves. In the latter class of movie, many frames will contain mixtures of structures of intermediates, whose populations rise and fall as the structures themselves remain constant. Both classes of movies are valuable. It is thrilling to watch a single molecule go through its paces, but it is easy to be hypnotized into thinking that the typical trajectory is all there is.

The modalities for molecular movies are quite varied. Model-free efforts lie at one extreme in which $\rho(x,t)$ and/or the corresponding difference map $\Delta\rho(x,t)$ are viewed directly, perhaps with a stick model of the ground state structure superimposed. More sophisticated approaches incorporate refined models for the structures of intermediates, with time-dependent weights that describe their appearance and disappearance. This class of movie typically requires information from many sources: spectroscopy and related methods to establish the kinetics of the system, SVD and DED map analysis to identify the structure of intermediates, and refinement of the intermediate structures. We reemphasize that knowledge of the structures of intermediates (which occur at local free-energy minima) provides no information about the structures along the functional trajectory that connects adjacent intermediates. Molecular dynamics and QM/MM simulations represent more rigorous ways of looking at the functional trajectory between states. A powerful application of these methods to the study of the light-driven sodium pump is given by (Skopintsev et al. 2020). A particular strength of the nascent field of single-particle imaging by X-rays is that both the structures of intermediate and of the higher energy states along the functional trajectory between them are present in the diffraction snapshots, and therefore appear in the reconstructions. These issues have been eloquently addressed (Ourmazd 2019a).

7.9 Does the Mechanism in the Crystal Represent the Mechanism in Solution?

Successful demonstration of the existence of activity in the molecules in the crystal may nevertheless reveal quantitative differences between the time course of activity in the crystal and in dilute solution (Parkhurst and Gibson 1967; Yeremenko et al. 2006). Does this invalidate studies of mechanism in such crystals? The quantitative differences may be quite large, perhaps a difference of one or two orders of magnitude in certain rate coefficients, with consequences for the overall mechanism. For example, the rate coefficient for breakdown of an intermediate may be increased by two orders of magnitude in the crystal compared with solution. The peak population of that intermediate is greatly decreased in the crystal, perhaps to the point of undetectability. Short-lived intermediates that follow longer-lived intermediates may not attain a peak population high enough to be detectable. From a purely crystallographic perspective, the transient structure does not exist. This is naturally interpreted as a clear qualitative difference between crystals and solution. There may however be sound chemical, spectroscopic, theoretical, or computational evidence for the existence of this intermediate, a circumstance which is not uncommon. Such evidence does not invalidate time-resolved structural studies but does suggest caution in interpreting their results.

As we reemphasized in the introduction of this chapter, kinetic studies cannot prove a mechanism, only disprove mechanisms. Selection of a particular mechanism from several candidates may require Occam's Razor (Chapter 1), but crystallographers, spectroscopists, and theoreticians often shave in different ways. Interpretation of spectroscopic data may be implicitly based on a linear mechanism throughout. This may be appropriate for a reaction which is strongly downhill in energy in its earliest stages such as those characteristic of photoreceptors, but inappropriate in later stages. A combination of these circumstances may account for notable disagreements over the structural interpretation of DED maps associated with short-lived intermediates in the photocycle of PYP (Schotte et al. 2012; Jung et al. 2013). See also an interesting dialogue in which a critique (Kaila et al. 2014) is followed by a reply (Jung et al. 2014).

In a more recent example, also from PYP, the rate coefficients differ quantitatively in the crystal and solution throughout its photocycle, but the intermediates themselves appear to be substantially identical (Konold et al. 2020). An additional intermediate is inferred in the crystal in the ns time range, and the longest-lived solution intermediate in the s time range is absent in the crystal. The ns intermediate is of particular interest. Rather than reflecting an artifactual difference between the mechanisms in crystal and solution, it may be associated with a tertiary structural change detectable by crystallography (a global structural probe; Chapter 1.3) but lacking electronic changes in the chromophore detectable by absorption spectroscopy (a local probe). Alternatively, it may represent a side path off the main photocycle, favored by crystal contacts but absent in solution. The absence of the longest-lived intermediate in the crystal is very likely to be a real difference in mechanism. Constraints of the crystal lattice may not permit it to be formed. Rather, the penultimate intermediate decays directly to the dark, ground state in a short-circuit of the photocycle. Crystals such as those of PYP are almost always obtained in the ground state, which provides a larger thermodynamic free energy directed toward that state and favors a short-circuit of the photocycle.

References

Adams, P.D., Afonine, P.V., Bunkóczi, G. et al. (2010). PHENIX: a comprehensive Python-based system for macromolecular structure solution. *Acta Crystallographica Section D: Biological Crystallography* 66 (2): 213–221.

Afonine, P.V., Poon, B.K., Read, R.J. et al. (2018). Real-space refinement in PHENIX for cryo-EM and crystallography. *Acta Crystallographica Section D: Structural Biology* 74 (6): 531–544.

Barad, B.A., Echols, N., Wang, R.Y.-R. et al. (2015). EMRinger: side chain–directed model and map validation for 3D cryo-electron microscopy. *Nature Methods* 12 (10): 943–946.

Blundell, T.L., Johnson, L., and Johnson, L.N. (1976). *Protein Crystallography*. Academic Press.

Di Giacomo, F. (2020). *Introduction to Marcus Theory of Electron Transfer Reactions*. World Scientific.

Dods, R., Båth, P., Morozov, D. et al. (2021). Ultrafast structural changes within a photosynthetic reaction centre. *Nature* 589 (7841): 310–314.

Doniach, S. (2001). Changes in biomolecular conformation seen by small angle X-ray scattering. *Chemical Reviews* 101 (6): 1763–1778.

Fung, R., Hanna, A.M., Vendrell, O. et al. (2016). Dynamics from noisy data with extreme timing uncertainty. *Nature* 532 (7600): 471–475.

Fung, R., Villar, J., Dashti, A. et al. (2020). Achieving accurate estimates of fetal gestational age and personalised predictions of fetal growth based on data from an international prospective cohort study: a population-based machine learning study. *The Lancet Digital Health* 2 (7): e368–e375.

Genick, U.K. (2007). Structure-factor extrapolation using the scalar approximation: theory, applications and limitations. *Acta Crystallographica Section D: Biological Crystallography* 63 (10): 1029–1041.

Gorel, A., Schlichting, I., and Barends, T.R. (2021). Discerning best practices in XFEL-based biological crystallography–standards for nonstandard experiments. *IUCrJ* 8 (4): 532–543.

Grime, G.W., Zeldin, O.B., Snell, M.E. et al. (2019). High-throughput PIXE as an essential quantitative assay for accurate metalloprotein structural analysis: development and application. *Journal of the American Chemical Society* 142 (1): 185–197.

Hadian-Jazi, M., Sadri, A., Barty, A. et al. (2021). Data reduction for serial crystallography using a robust peak finder. *Journal of Applied Crystallography* 54 (5).

Henderson, R. and Moffat, J. (1971). The difference Fourier technique in protein crystallography: errors and their treatment. *Acta Crystallographica Section B: Structural Crystallography and Crystal Chemistry* 27 (7): 1414–1420.

Henry, E. and Hofrichter, J. (1992). Singular value decomposition: application to analysis of experimental data. *Methods Enzymol* 210: 129–192.

Hosseinizadeh, A., Breckwoldt, N., Fung, R. et al. (2021). Few-fs resolution of a photoactive protein traversing a conical intersection. *Nature* 599: 697–701.

Ihee, H., Rajagopal, S., Srajer, V. et al. (2005). Visualizing reaction pathways in photoactive yellow protein from nanoseconds to seconds. *Proceedings of the National Academy of Sciences of the United States of America* 102 (20): 7145–7150.

Jung, Y.O., Lee, J.H., Kim, J. et al. (2013). Volume-conserving trans-cis isomerization pathways in photoactive yellow protein visualized by picosecond X-ray crystallography. *Nature Chemistry* 5 (3): 212–220.

Jung, Y.O., Lee, J.H., Kim, J. et al. (2014). Reply to 'Contradictions in X-ray structures of intermediates in the photocycle of photoactive yellow protein'. *Nature Chemistry* 6 (4): 259–260.

Kaila, V.R., Schotte, F., Cho, H.S. et al. (2014). Contradictions in X-ray structures of intermediates in the photocycle of photoactive yellow protein. *Nature Chemistry* 6 (4): 258–259.

Keedy, D.A., Fraser, J.S., and van den Bedem, H. (2015). Exposing hidden alternative backbone conformations in X-ray crystallography using qFit. *Plos Computational Biology* 11 (10): e1004507.

Key, J. and Moffat, K. (2005). Crystal structures of deoxy and CO-bound bj FixLH reveal details of ligand recognition and signaling. *Biochemistry* 44 (12): 4627–4635.

Konold, P.E., Arik, E., Weißenborn, J. et al. (2020). Confinement in crystal lattice alters entire photocycle pathway of the photoactive yellow protein. *Nature Communications* 11 (1): 1–12.

Kostov, K. and Moffat, K. (2011). Cluster analysis of time-dependent crystallographic data: direct identification of time-independent structural intermediates. *Biophysical Journal* 100: 440–449.

Lang, P.T., Ng, H.-L., Fraser, J.S. et al. (2010). Automated electron-density sampling reveals widespread conformational polymorphism in proteins. *Protein Science* 19 (7): 1420–1431.

Lattman, E. (1972). The mean figure-of-merit for a difference Fourier synthesis. *Acta Crystallographica Section A: Crystal Physics, Diffraction, Theoretical and General Crystallography* 28 (2): 220–221.

Lyubimov, A.Y., Uervirojnangkoorn, M., Zeldin, O.B. et al. (2016). Advances in X-ray free electron laser (XFEL) diffraction data processing applied to the crystal structure of the synaptotagmin-1/SNARE complex. *Elife* 5: e18740.

Ourmazd, A. (2019a). Cryo-EM, XFELs and the structure conundrum in structural biology. *Nature Methods* 16 (10): 941–944.

Ourmazd, A. (2019b). Dynamics from data with extreme timing uncertainty. *APS* 2019 (P03): 003.

Pande, K., Hutchison, C.D.M., Groenhof, G. et al. (2016). Femtosecond structural dynamics drives the trans/cis isomerization in photoactive yellow protein. *Science* 352 (6286): 725–729.

Pandey, S., Bean, R., Sato, T. et al. (2020). Time-resolved serial femtosecond crystallography at the European XFEL. *Nature Methods* 17 (1): 73–78.

Pannu, N.S. and Read, R.J. (1996). Improved structure refinement through maximum likelihood. *Acta Crystallographica Section A: Foundations of Crystallography* 52 (5): 659–668.

Parkhurst, L.J. and Gibson, Q.H. (1967). The reaction of carbon monoxide with horse hemoglobin in solution, in erythrocytes, and in crystals. *Journal of Biological Chemistry* 242 (24): 5762–5770.

Rajagopal, S., Schmidt, M., Anderson, S. et al. (2004). Analysis of experimental time-resolved crystallographic data by singular value decomposition. *Acta Crystallographica Section D: Biological Crystallography* 60 (5): 860–871.

Rupp, B. (2009). *Biomolecular Crystallography: Principles, Practice, and Application to Structural Biology*. Garland Science.

Schmidt, M., Rajagopal, S., Ren, Z. et al. (2003). Application of singular value decomposition to the analysis of time-resolved macromolecular X-ray data. *Biophysical Journal* 84 (3): 2112–2129.

Schmidt, M., Srajer, V., Henning, R. et al. (2013). Protein energy landscapes determined by five-dimensional crystallography. *Acta Crystallographica Section D-Structural Biology* 69 (12): 2534–2542.

Schotte, F., Cho, H.S., Kaila, V.R.I. et al. (2012). Watching a signaling protein function in real time via 100-ps time-resolved Laue crystallography. *Proceedings of the National Academy of Sciences of the United States of America* 109 (47): 19256–19261.

Segel, D.J., Fink, A.L., Hodgson, K.O. et al. (1998). Protein denaturation: a small-angle X-ray scattering study of the ensemble of unfolded states of cytochrome c. *Biochemistry* 37 (36): 12443–12451.

Skopintsev, P., Ehrenberg, D., Weinert, T. et al. (2020). Femtosecond-to-millisecond structural changes in a light-driven sodium pump. *Nature* 583 (7815): 314–318.

Teng, T.-Y. and Moffat, K. (2002). Radiation damage of protein crystals at cryogenic temperatures between 40 K and 150 K. *Journal of Synchrotron Radiation* 9 (4): 198–201.

Yeremenko, S., van Stokkum, I.H.M., Moffat, K. et al. (2006). Influence of the crystalline state on photoinduced dynamics of photoactive yellow protein studied by ultraviolet-visible transient absorption spectroscopy. *Biophysical Journal* 90 (11): 4224–4235.

8

Other Structural Biology Techniques

Joseph Sachleben

University of Chicago, authored the NMR section (8.5) of this chapter

8.1 Introduction

Several experimental methods complement X-ray-based techniques of time-resolved crystallography and solution scattering. We emphasize those important for time-resolved experiments, particularly those that provide atomic level information. These include single-particle cryo-electron microscopy (cryo-EM), X-ray absorption (XAS) and emission (XES) spectroscopies, nuclear magnetic resonance (NMR), and hydrogen–deuterium exchange (HDX). Each method has unique strengths that distinguish it from crystallography and solution scattering, but with some important limitations. We also consider energy landscape analysis (ELA), a powerful method of analyzing and visualizing the results of all dynamic structural experiments.

8.2 Single-Particle Cryo-Electron Microscopy

8.2.1 Dizzying Progress

Within the last decade rapid progress in single-particle cryo-EM has transformed structural biology, enabling determination at near-atomic resolution of the structures of large protein and protein–nucleic acid complexes that are difficult if not impossible to crystallize. Several critical technical advances enabled the leap from low resolution structures to the near-atomic level (Cheng 2018). Innovations in detectors offered direct detection of scattered electrons; sophisticated algorithms were developed for classification and analysis of raw, low-dose images; and these algorithms were incorporated into widely available, user-friendly software. Improvements continue at a dizzying pace, with reports of structure determination at 0.125 nm resolution (Yip et al. 2020). Further, cryo-EM methods are now capable of imaging molecules well below the historical size limitation of 100 kDa (Fan et al. 2019; Herzik et al. 2019).

Applications of cryo-EM to dynamic processes come in two forms. In the first, an ensemble originally at equilibrium in which several structural states are present – for example,

Dynamics and Kinetics in Structural Biology: Unravelling Function Through Time-Resolved Structural Analysis, First Edition. Keith Moffat and Eaton E. Lattman.
© 2024 John Wiley & Sons Ltd. Published 2024 by John Wiley & Sons Ltd.

various conformational states of the ribosome – is frozen and imaged. The limitation placed by the Boltzmann distribution on the observation of such states is discussed in Section 8.3.3. In the second, a triggering event (a pump) displaces the equilibrium; structures are captured by rapid freezing at appropriate time-points after triggering and then imaged (a probe). Pump–probe experiments on single particles reveal dynamic structures distributed over the entire energy landscape. Structures are not confined to the energy minima characteristic of kinetic studies of large populations of molecules (Chapter 2). Each particle on the cryo-EM image provides a snapshot of a single object such as an individual molecule or larger complex as it proceeds along its functional trajectory, independent of all other molecules in the image. That is, cryo-EM images directly visualize dynamics by exploring all points along the reaction coordinate, in contrast to kinetic measurements. Population averaging during the measurement itself is an intrinsic feature of kinetics but is absent in the raw image data from a single object, an intrinsic feature of dynamics. Averaging over single molecules is applied later in the analysis of cryo-EM images, across the molecules in one image or many images.

We give a brief introduction to cryo-EM to clarify these two forms of imaging dynamic processes and their limitations. Structures with lifetimes shorter than tens of ms cannot be captured and imaged by cryo-EM at present. However, progress is very rapid and major developments in cryo-EM dynamics can be anticipated (Chapter 9.4).

8.2.2 Outline of the Experiment

As with any form of microscopy, proper sample preparation is critical. Excellent review articles describe various techniques and their pitfalls (Noble et al. 2018; Lyumkis 2019; Li et al. 2020). Importantly, rapid freezing does not automatically preserve the active biological structure at ambient temperature and damaged particles are imaged along with native particles. We assume that a highly purified and homogeneous sample is available while recognizing that in the real experimental world, images could contain false positives such as impurities or damaged particles. These must be identified and removed in data analysis. A droplet of this sample is applied to an EM grid and excess liquid removed, classically by blotting the EM grid. A complication is that particles may migrate rapidly to the air–water interface where their structures change prior to freezing and, in the worst case, denature (Glaeser 2018). The grid containing the sample is then rapidly plunged into a liquid cryogen, frequently ethane, leading to the formation of a vitreous ice film ~100 nm thick on the surface of the grid. The formation of vitreous rather than crystalline ice helps to preserve native-like biological structure. Low temperature reduces radiation damage by limiting the mobility of highly reactive free radicals. Radiation damage is substantially reduced in an electron beam by comparison with a hard X-ray beam (Henderson 1995). Particles on the grid are then imaged using a low electron dose regime. In some cases, the stage may be tilted before imaging to circumvent the common issue of the particles lying on the grid in preferred orientations, and to acquire the family of images that spans all particle orientations. Sample preparation, freezing and collection of images are largely automated in major cryo-EM facilities. Though the techniques obviously differ, the organization of these facilities and access by outside users was modeled on and now resembles that of longer-established SR or XFEL beamlines.

8.2.3 Reconstruction of Structure from Single Particles

The next step is to analyze the images to arrive at structural reconstructions of the sample in 3D. Each image of a particle represents a 2D projection along the EM beam axis. The reconstruction process requires many such projections, viewed from many directions. The FT of a projection of a 3D object provides a central section through the 3D FT of the whole particle. If the orientation of the particle can be determined, these sections can be pieced together to give the complete diffraction volume of the object – its FT in 3D. Unlike structure determination by crystallography, cryo-EM has no phase problem. Since the primary data are images in real space, the computed FT has an associated phase. The true phase is altered by the microscope contrast transfer function (CTF), which is dominated by defocus values that vary from image to image. These values are determined and modulated in the reconstruction process. This FT is then inverted to produce a 3D volume in real space.

Building up the 3D diffraction pattern requires that the orientation of each particle be accurately determined. Solving this problem is no mean feat when dealing with large numbers of noisy, low-dose images, where not all images are of authentic, undamaged particles (Frank and Ourmazd 2016). The images must also be aligned to a common origin. The orientation problem has been actively studied for at least 50 years. Until recently, image reconstruction required many independent, manual operations. Current EM software automates most operations, reduces human intervention and bias (at least for more straightforward samples), and provides effective solutions. Packages aimed at single-particle cryo-EM carry out steps including selection of suitable particles (picking), particle alignment, 3D reconstruction and eventually 3D classifications (Scheres 2010; Stabrin et al. 2020). The open-source package RELION (Scheres 2012; Zivanov et al. 2018) has been widely adopted and is a prime mover in the development of cryo-EM.

8.2.4 Time-resolved Studies

Cryo-EM studies that examine time-resolved, dynamic processes or the reconstruction of functional trajectories are much less common than studies of static structures. Most dynamic studies to date involve the ribosome as a model system (Dashti et al. 2018; Frazier et al. 2021). The ribosome is a robust test object of great biological relevance with many local energy minima, each associated with a distinct intermediate structure, and with complicated transitions between these structures along the functional trajectory.

The first powerful method for analyzing functional trajectories of molecules or molecular complexes through cryo-EM involves ELA (Dashti et al. 2014, 2020). ELA provides an effective route for analyzing dynamic SPI data by both X-ray and cryo-EM approaches (Chapter 8.3). This contrasts with the routes for analyzing kinetic, many particle data from crystallography, solution scattering, and (for example) spectroscopies (Chapter 7). Briefly, ELA relies on the Boltzmann distribution of structures of differing free energy (Chapter 2.5.2). The method asserts that under the conditions of thermal equilibrium prior to freezing, all structures along a functional trajectory are present, even in the absence of any explicit reaction initiation.

There used to be a popular puzzle in which frames from a cartoon strip were cut out and placed in a random order, and the solver was asked to reassemble them in logical sequence.

In this spirit, ELA requires not only the determination of many distinct structures present at the same (clock) time on the cryo-EM grid, but their placement into a functional order corresponding to reaction time: their position on the functional trajectory. The assumption is that the more similar structures are, the closer together they lie on the functional trajectory. Although there is no explicit time dependence in the experiment, similarity in structure is inferred to reveal similarity in time, along a functional trajectory (and ultimately a reaction coordinate) that initially is poorly determined.

The second method is more in line with traditional time-resolved studies based on reaction initiation. Introduction of the mix-and-spray technology for reaction initiation by Frank and many colleagues (Frank 2017) has allowed cryo-EM to determine time-resolved structures with lifetimes in the ms range and longer. A microsprayer chip mixes sample and reactant components at a controllable time. The mixer deposits a spray plume containing sample molecules on a cryo-EM grid held by sharp-tipped tweezers. The grid is held in a motorized plunging device that quickly submerges it in a cryogen such as liquid ethane maintained at ~90 K to vitrify the sample and its surrounding solvent. The minimum plunge time is ~2.5 ms, and the sample and solvent cool from 290 K to 90 K in a few ms. The overall time resolution lies in the 10 ms range (Bock and Grubmüller 2022).

In a key proof-of-principle experiment, Frank and colleagues examined the association of the small (30S) and large (50S) ribosomal subunits of *E. coli* at 60 ms and 140 ms after initiation of the reaction (Frank 2017). The association of the two subunits is known to involve several intermediates including the formation of inter-subunit bridges that break and re-form as the subunits reorient. The experiments captured the subunit association reaction in two stages before equilibrium was reached and detected three distinct conformations of the complete 70S ribosome formed by association. The imaging reconstructions demonstrate that all bridges are formed within 60 ms in the non-rotated form of the ribosome. The authors suggest a lower limit of 10 ms for the time resolution of transient ribosomal structures. More recently, cryo-EM has been used to visualize later steps in the initiation of bacterial translation by adding initiation factors to preparation of the sample. These generate native-like initiation complexes of the 30S and 50S subunits, and of the complete 70S ribosome (Jobe et al. 2019).

Recently a device called a Spotiton has been introduced to enhance TR cryo-EM (Dandey et al. 2020). The Spotiton has two independent streams of droplets. Using a piezoelectric syringe, a first ~50 pl drop is placed on a nanowire cryo-EM grid and within 10 ms, a second drop is placed on the first to initiate a reaction by mixing. The reaction is terminated ~100 ms later by plunging the grid and mixed drop into liquid ethane. Although the Spotiton is convenient, its use is restricted to slow reactions in the 100 ms time range. To enhance the time resolution to the μs range, a quite different approach of rapid melting and revitrification has been suggested (Voss et al. 2021).

8.3 Energy Landscape Analysis

ELA is based on an approach to dynamics that operates at equilibrium. It uses images of many individual particles, derived from cryo-EM or X-ray scattering, to populate the entire energy landscape with experimentally derived 3D structures. This distinguishes ELA from other current applications where structures can typically be determined only for a few intermediates along the reaction coordinate. The latest version of ELA presents a truly new framework for analyzing protein structure and dynamics. We judge this framework to be promising, emphasize ELA, and discuss some unfamiliar aspects of data analysis.

8.3.1 Landscapes

Earlier examples of ELA-like analysis appear in studies of enzyme mechanism, introduced in Chapter 2. A sophisticated 2D landscape for a catalytic network is shown in Figure 8.1 (Benkovic et al. 2008). In particular, the funnel energy landscape that depicts protein folding (Dill and MacCallum 2012) has had a powerful impact.

The free energy of an enzyme molecule is displayed on the z-axis as a function of its reaction coordinate along the x-axis. A reaction coordinate displays distinct states of the enzyme and their functional interconversions during its catalytic cycle, such as substrate binding, transition state(s), conversion to product in chemical steps, and product release. The z-axis represents the free energy change of the enzyme with any associated ligands along the reaction coordinate. The third, y-axis represents conformational variation in the enzyme

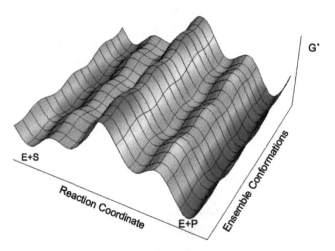

Figure 8.1 Schematic representation of the standard free-energy landscape for the catalytic network of an enzyme reaction. Conformational changes occur along both axes. The conformational changes occurring along the reaction coordinate axis correspond to the environmental reorganization that facilitates the chemical reaction. In contrast, the conformational changes occurring along the ensemble conformations axis represent the ensembles of configurations existing at all stages along the reaction coordinate, leading to many parallel catalytic pathways. If a plane parallel to the axis labeled "ensemble conformations" bisects this catalytic mountain range by cutting along the red mountain top, the substrates (E + S) are on one side of the plane and the products (E + P) are on the other. This figure illustrates the multiple populations of conformations, intermediates, and transition states. Strong coupling can occur between the reaction coordinates and the conformational ensembles; i.e. the reaction paths can slide along and between both coordinates. For real enzymes, the number of maxima and minima along the coordinates is expected to be greater than shown. The dominant catalytic pathways will be altered by external conditions and protein mutations. The figure was created by Sarah Jane Edwards. Benkovic et al. 2008/Reproduced with permission from American Chemical Society.

not directly related to its mechanism. For example, the exact conformation of a disordered loop may have no effect on ligand binding, although such variability contributes to the overall free energy through an entropic term. Our understanding of this traditional ELA-like presentation is greatly enriched by knowledge of structures, often from crystallography, of the enzyme at transiently stable intermediate states along the pathway, such as complexes between the enzyme and a substrate or substrate analog. Information about the transition state may also be available, either from the structures of transition state analogs or from kinetic data. However, a serious limitation of this form of ELA is that most structures along the continuous reaction coordinate are not experimentally accessible, and so functional trajectories are not revealed. Similarly, the distinction between configurational changes relevant to mechanism (along the x-axis) and orthogonal to it (along the y-axis) remains conceptual rather than experimental. Time is not an explicit variable in ELA but comes into play in a qualitative sense. Molecules can move rapidly across flat regions of nearly constant energy but take longer times to surmount an energy barrier. Many abortive attempts may occur before the final attempt to cross a higher barrier succeeds.

Experiments in chemical kinetics give direct quantitative information about the rate at which populations of molecules fall and rise as transitions occur between one state and the next along the reaction coordinate, and the free energies of activation associated with these transitions (Chapter 2). Such experiments provide the basis for determining the heights of the valley bottoms and peaks in a free energy diagram. However, the functional trajectory between these bottoms and peaks – the states and transition states – is completely invisible to chemical kinetics. The power of ELA is that it exploits the new experimental ability to determine the structures of individual particles (clearly, a statistically small number!) with true dynamics, to reveal the structural bases of functional trajectories that lie between the states. ELA can reveal continuously varying conformations, whereas the mix-and-spray or light-initiated approaches to kinetics probe only a relatively small number of discrete states.

A functional trajectory is an authentic movie, even if uncertainties in the time dimension mean that frame speed of the movie camera is both variable and uncertain. How can this authentic movie be acquired?

8.3.2 Energy Landscapes Derived from Structural Fluctuations

ELA has been profoundly expanded and enriched by the work of Ourmazd and his colleagues and collaborators (Dashti et al. 2014, 2020; Frank and Ourmazd 2016; Copperman et al. 2017, 2018). This approach makes use of experimental developments in single-particle imaging by both cryo-EM and X-ray approaches that probe the structures of hundreds of thousands, even millions, of individual particles, trapped as they undergo constant fluctuations in energy and structure at thermal equilibrium. These structural fluctuations are driven by thermal fluctuations in the solvent that surrounds the particles. The solvent remains liquid in X-ray scattering, and is liquid prior to freezing and vitrification in cryo-EM.

We begin with a big-picture look. Consider a thought experiment that follows a single molecule in an equilibrium thermal bath. The most obvious behavior is that, bombarded by constant collisions with solvent, the molecule fluctuates wildly in structure and energy.

Cooper has provided a discussion of the role of thermodynamic fluctuations in biological systems (Cooper 1976, 1984). His approach invokes the methods of statistical thermodynamics to evaluate such fluctuations, an approach pioneered by Einstein among others, well presented in his biography (Pais 1982). The mean value of the internal energy $<E>$ of a system is given by

$$\langle E \rangle = \frac{\int_0^\infty E\exp(-\beta E)\omega(E)dE}{\int_0^\infty \exp(-\beta E)\omega(E)dE}, \tag{8.1}$$

where E is the instantaneous energy, $\beta = 1/kT$, and $\omega(E)$ is the density of states. The Boltzmann factor $\exp(-\beta E)$ gives the relative probability of the presence of a state with energy E. By differentiating Equation 8.1 with respect to β, one can show that

$$\varepsilon^2 \equiv \langle E - \langle E \rangle \rangle^2 = \langle E^2 \rangle - \langle E \rangle = -\frac{\partial \langle E \rangle}{\partial \beta} = k_B T^2 \frac{\partial \langle E \rangle}{\partial T}. \tag{8.2}$$

The quantity ε^2 is thus a measure of the mean-square fluctuation in internal energy of the system. Invoking the definition of C_V, the heat capacity at constant volume, we have

$$\varepsilon^2 = k_B T^2 m C_V, \tag{8.3}$$

where m is the mass of the system. Cooper presented as an example a protein molecule with mass of 25 kDa. Using a plausible value for C_V, he found that $\varepsilon_{rms} = 6.4 \times 10^{-20}$ calories per molecule. This is equivalent to 38 kcal mol^{-1} (159 kJ mol^{-1}) on a molar basis, if all molecules fluctuate simultaneously. Of course, simultaneous fluctuations are effectively impossible, but the molar basis puts the energy on a more familiar scale.

In practice most experimental studies of proteins are done at constant pressure where C_p, the heat capacity at constant pressure, is strictly speaking the more appropriate constant in Equation 8.3. However, as pointed out in a very useful review (Prabhu and Sharp 2005), under physiological conditions the amount of PV work that can be done on a protein molecule is very small, less than $k_B T$. Thus C_p and C_v are essentially identical, and the internal energy E and the enthalpy H can be used interchangeably.

This value of ε_{rms} is (perhaps surprisingly) large. It suggests that the fluctuations in internal energy at physiological temperature are larger in magnitude than the free-energy barrier to unfolding of the protein. For many proteins the energy barrier to unfolding is largely enthalpic, so that comparing an internal energy fluctuation to a free-energy barrier has justification. And of course, energy fluctuations also occur in the unfolded state. Indeed, MD simulations reveal that the energy stored in bonds and bond angles is similar in the native state ensemble and the denatured state ensemble. The significant point is that the ground state of a typical protein is marginally stable, and intermediate states of somewhat higher energy may be readily accessible by thermal fluctuations at equilibrium. It follows that even at physiological temperature, the system can sample a large region of its energy landscape (Chapter 2.5.2). These issues are less dramatic than they seem at first. A molecule visits many conformations during its energy fluctuations, and the probability is very low that any specified conformation is visited. Thus, even though the internal energy fluctuations of a molecule may be comparable to the free energy barrier for unfolding, most of the fluctuations represent conformations that do not lie on the functional trajectory for unfolding. The same is true for other events that present a high-energy barrier, such as the 180 degrees flip of a tryptophan side chain buried within a protein.

One of the critical roles of the free energy G (as distinct from internal energy E) is to account properly for irrelevant conformations. Entropy fulfills this role. We discuss this further below.

An intuitive way to visualize the energy fluctuation process is to recall that a macromolecule has a large number N of independent degrees of freedom. Normal mode analysis presents one way of modeling these processes. The energy of each of these degrees of freedom fluctuates on the scale of k_BT. The energy fluctuation is therefore a random walk among these N states, and thus the mean square fluctuation in energy is proportional to N (Berg 2018), a value closely related to the heat capacity of a single molecule.

The set of conformations visited by these fluctuations can be organized into an energy landscape in the following way. First, the conformations are binned; that is, molecules with very similar 3D structures are placed in a mental group. Second, using advanced (and, to many, unfamiliar) mathematics, the set of bins is analyzed collectively to generate an abstract coordinate system of eigenvectors that maximally separates the structural bins. This coordinate system provides a metric for the "distance" between two structures, and thus for the volume of a bin. In a simple example, if the molecule comprises two subunits that can rotate with respect to one another, one axis of this coordinate system would reflect this rotation. Conveniently, in the limited set of structures thus far examined, two eigenvectors are sufficient to account for much of the structural variation experimentally observed. However, there is no guarantee that conformations separated by a high barrier in the 2D landscape are not connected by a shortcut along another coordinate in a landscape of higher, more realistic dimensions (Ourmazd 2019; Maji et al. 2020).

We can now represent an energy landscape in 3D as follows. The horizontal x-y plane plots structure and the vertical z axis plots energy, as in the simpler version introduced above. The critical difference here is that each observed structural bin is mapped experimentally to an x-y point in the landscape. All points are in principle experimentally accessible. The energy associated with that structural bin is obtained by recalling that high-energy states are less common that low-energy ones, so that bins with low occupancy contain high-energy structures. The quantitative relationship is given by the classical Boltzmann factor, which says that the probability of a molecule having energy E is proportional to $\exp(-E/k_BT)$. We know the relative probability of each bin from its occupancy, i.e. the number of separate structures within that bin, and the occupancy therefore determines the energy of a typical particle within the bin.

The definition of bins and the correct application of the Boltzmann factor involve some subtleties and uncertainties. Immediately above, we rather arbitrarily used the internal energy E in the Boltzmann equation. It is not always clear whether the observed states represent samples based on internal energy, or Gibbs free energy G, or perhaps the energy of another thermodynamic ensemble. The distinction between energy fluctuations and free-energy fluctuations in ELA is a delicate matter that needs further exploration, although the practical impact is likely to be small. The classic Boltzmann factor applies to the internal energy E. If A and B are two states with energies $E_A < E_B$, then the relative probability of a molecule occupying these two states is given by the classical Boltzmann factor $\exp(-\Delta E / k_B T)$, where $\Delta E = E_B - E_A$. Use of the Boltzmann factor plays a key role in the derivation of Equation 8.3 via Equation 8.1. A similar relation pertains in the case of free energy (Chapter 2). If two states A and B of a molecule are in equilibrium, then $\Delta G = -RT \ln(K)$, where K is the equilibrium constant for the reaction and ΔG is the difference in free energy between the two states. The equilibrium constant in this case is simply the ratio of the occupancies of the two states. Taking the exponential of both sides of Equation 2.4, we see that the relative probability of molecule occupying these two states is given by the factor $\exp(-\Delta G / k_B T)$. This is closely analogous to the Boltzmann factor, where we have replaced R by k_B to put free energy on a per molecule basis. This topic has been analyzed with great clarity (Dashti et al. 2014, 2020).

MD simulations illuminate the distinction between these two relations. The potential energy function for MD is purely energetic and lacks any entropic terms (Chapter 5.3.1). Thus, we may expect that molecules in the simulation have a distribution of internal energies corresponding to the analysis (Cooper 1984) cited above. Free energy information that explicitly incorporates entropic factors can be gleaned from MD simulations only by analysis of ensembles of structures taken from the simulation trajectory (Van Gunsteren et al. 2002).

The meaning of bin size is also important. If for example we divide a bin containing one million particle structures into 10 smaller bins, does that mean the particle energy in these smaller bins is higher? Indeed, it does not. The energy landscape is based on the fraction of structures that lie in each small volume of conformational space. If we divide bins, we reduce both the conformational volume and the number of particles equally, so that fraction of structures per unit volume stays the same. Thus, for systems having discrete energy levels, the Boltzmann equation says that the relative probability of the system having energy E_j is proportional to $\exp(-E_j/k_B T)$. What we can count experimentally, however, is the number of particles in each conformational bin rather than the number in an energy window. However, the very design of the conformational analysis obviates this issue. In the simple example introduced above, one conformational coordinate maps the rotation of a subunit with respect to another around an axis between them. Equal conformational steps correspond to equal rotational steps in radians. But some of these angular steps will have very few observed structures in regions of high energy and will therefore give rise to high calculated energies through the inverse Boltzmann factor. The metric underlying the conformational binning allows construction of the landscape.

Only time will tell whether 2D landscapes will always be sufficient to derive good energies, or whether three or more conformational variables (dimensions) may be needed in some cases. As discussed in a thoughtful opinion piece (Ourmazd 2019), this approach implicitly reconceptualizes a structure as a point on a continuum, rather than as a unique entity.

The conformational sampling used in current ELA analyses is not as fine-grained as the single frames seen in MD. Although individual cryo-EM images or single-particle X-ray diffraction patterns do arise from a single molecular conformation, the data are fragmentary and are insufficient to represent the conformation exactly. As we shall see below, ELA analysis develops limited sets of basis functions to reconstruct the set of structures along the functional trajectory. (It is always possible to go back to the original raw, single particle images to reconstruct the 3D structures along the trajectory.) Use of this limited set effectively integrates over sets of closely similar structures. We do not see a ready route to delineate the ensemble created by this process, but it seems very likely that the occupancies will obey an exponential rule essentially identical to those mentioned above.

There are important assumptions in this outline. A clear limitation is that in contrast to a *gedanken* experimental approach, the analytical approach does not allow the structural changes in one molecule to be followed through time. Rather, the analysis relies on the ergodic hypothesis (Hill 1960) to stitch together the behavior of one molecule from observations of many. The assumptions are that similar structures are also similar in energy, closely grouped along the functional trajectory, and likely to occur close together in time. The primary data in cryo-EM are images of 2D projections of a very large numbers of individual particles in random orientations. From these 2D images, 3D structures are reconstructed using the methods discussed above (Chapter 8.2.3). In practice as elaborated below, the initial binning is done on the set of 2D images themselves. The images are grouped into bins, each corresponding to a view of the particle from a projection direction (PD) with a very narrow angular width of perhaps 1°.

The resultant energy landscape resembles a topographical map of a mountain range. Molecules, like hikers, mostly move through valleys and passes, rather than over the highest peaks or along the steepest paths. Thus, the functional trajectory of a molecule represents a hike along a low-energy path in the landscape, from one base camp to the next. A full biological cycle would be represented by a round trip in the landscape, a path from base camp across the landscape and back to the starting base camp.

This introduction to ELA leaves many open questions. In thermal equilibrium, why should a molecule move in any specific sense around the energy diagram? How can effectors, ligands, substrates, or absorption of a photon be incorporated? Although the diagram contains a sense of the arrow of time from the direction along which a molecule moves, no time scales are attached to these motions and even the order in time is not specified. Structure A may be demonstrably similar to structure B, but does structure A precede or follow structure B? The issue of transit times – the distribution of times taken by individual molecules to move from one state to the next – remains an active point of discussion. This is particularly acute for systems driven far from equilibrium, achieved for example when a photoreceptor absorbs a photon. Nonequilibrium thermodynamics comes into play in the subsequent, strongly downhill process.

To make clear the full impact of the ELA approach, we discuss the first major application of ELA to biological particles. This application yielded a 2D energy landscape of the ribosome (Dashti et al. 2014) that shows a low-energy trajectory corresponding to the ribosomal functional cycle. The landscape and trajectory are underpinned by structures representing a broad range of conformations, derived from large numbers of cryo-EM images, each of which is a 2D projection of a unique particle. Figure 8.2 shows the free-energy landscape and Figure 8.3 a flow chart for the full process of landscape generation. We emphasize that each rectangle in Figure 8.3 represents a significant creative, developmental, and/or experimental achievement, summarized very briefly here.

Results are encapsulated in the free-energy diagram supported by the identification of important ribosomal states and structural transitions with points and trajectories in that diagram. In the terminology of the cited authors, the system is a Brownian machine that undergoes energy fluctuations purely through interactions with the thermal bath in which

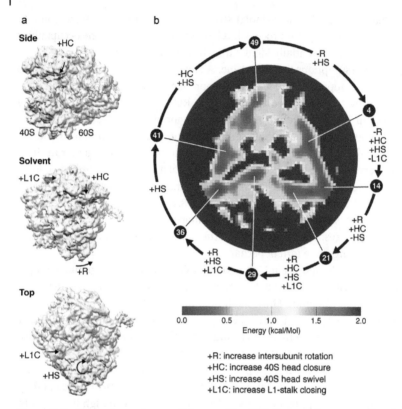

Figure 8.2 **Three views of a cryo-EM map of the 80S ribosome from yeast.** **Panel a** Arrows indicate four key conformational changes associated with the elongation work cycle of the ribosome. **Panel b** The energy landscape traversed by the ribosome. The color bar shows the energy scale. The energy range has been truncated at 2 kcal/mol to show details of the triangular trough. The error in energy determination along the closed triangle is 0.05 kcal/mol. The roughly triangular minimum free-energy trajectory is divided into 50 states. The arrows indicate the structural changes between 7 selected states, each identified by its place in the sequence of 50 states. Dashti et al. 2014 / Reproduced with permission from Proceedings of the National Academy of Science.

it is embedded (Dashti et al. 2014, 2020; Frank 2018). Strikingly, it emerges that the system visits conformations that represent important, intermediate, bound states of the ribosome as it interacts with other important components in the elongation cycle, described in the legend to Figure 8.2.

This provides a hint of how to deal with systems that cycle through changes in the chemical potential, driven for example by ligand binding or photon absorption. The energy landscapes for an activated state and for the ground state will "kiss" each other, thus allowing the functional trajectory to span the two landscapes through interactions with the thermal bath that surrounds it. We elaborate on this below. The energy landscape in Figure 8.2 spans a range of 3.8 ± 0.65 kcal mol^{-1} (15.9 ± 2.7 kJ mol^{-1}, corresponding to 6.5 ± 1.0 $k_B T$ at room temperature), a value sufficient to surmount significant energy barriers. This range could be expanded by increasing the number of experimental images so that conformations with even lower probability (corresponding to higher energies) are sampled. The landscape

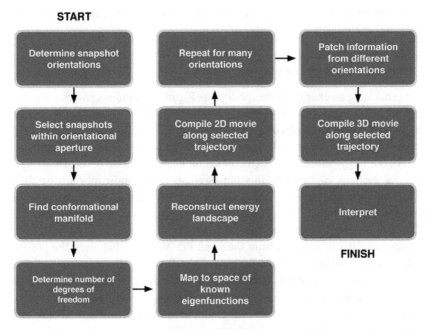

Figure 8.3   Flowchart representation of the overall computational approach. The approach used to determine the free-energy landscape, work-cycle trajectory, and associated continuous conformational changes from experimental snapshots of nanomachines in unknown orientational and conformational states. This figure reminds us of the scope of the undertaking. Dashti et al. 2014/Reproduced with permission from Proceedings of the National Academy of Science.

shows several connected low-energy pathways, but the roughly triangular pathway at the center of panel B of Figure 8.2 stands out. It contains about one-third of the conformations observed. The Ourmazd team generated 3D movies to follow the conformational changes occurring as a single ribosome moves around the closed, minimal path. These changes were compared with the differences between static crystal structures of the ribosome in chemically distinct states along the biochemically accepted pathway summarized in the reaction coordinate. Using rigid body fitting, they identified combinations of four motions previously assigned to points in the translation/elongation cycle such as a ratchet-like inter-subunit rotation, corresponding to a counterclockwise rotation of the 30 S subunit about an axis normal to the subunit interface (panel A of Figure 8.2, solvent view). Importantly, structural changes identified as the ribosome moves continuously along the trajectory correspond in each case to steps in the cycle. To put it another way, the structural changes along the trajectory follow biochemical time. Notably, the ribosomal structural states were determined by crystallography or cryo-EM using samples carefully prepared to isolate and image chosen, static, intermediate states. The particle reconstructions derived from ELA correspond closely in both structure and function to the static structures of pure intermediate states. This constitutes a critical confirmation of ELA. Further, the details of the functional trajectory gleaned from ELA are different from the comparable trajectory created by using MD to morph from one known static structure to the next. Morphing contains no new information beyond that in the static structures themselves. The ELA trajectory is much more information-rich and hence more realistic; it reveals the Brownian machine in action performing its work cycle.

8.3.3 Generation of the Energy Landscape: Details

The process of generating an energy landscape such as Figure 8.2 contains two major components. First, many landscapes are developed independently. These landscapes are based on identifying the orientation of individual particles from their images, and on grouping together images that represent essentially the same PD to within a small specified angular cone in a bin. A 2D energy landscape is then created for each PD, based on the small, systematic structural motions captured in this bin. Second, these landscapes are patched together, involving mathematical approaches which properly normalize the variation in structure across all the different PDs. These enable the structural frequencies and the Boltzmann factor to be accurately estimated.

Consider images of a particle viewed from the same PD. Methods that bring the projections into optimal superposition have been developed recently. Such superpositions allows pixels in the images of separate particles to be meaningfully associated with one another in a multidimensional manifold (Copperman et al. 2017), discussed in Chapter 7.3.4.

This approach provides the basis for the first component of the process: analyzing the structural fluctuations occurring along the chosen PD. Successful execution of the first component ensures that the variations in structure associated with the angular cone of uncertainty – the noise in the 2D projection – are small compared with the authentic variations in structure that occur in the functional trajectory – the signal. This does not mean that all sightings from the same PD are identical. Part of the power of the underlying mathematics is that the views of the particles from one PD can be placed in order representing the path of the particle's motion during the fluctuation. An example of dissecting a particle's motion within a 2D bin is given in Figure 8.4 (Frank and Ourmazd 2016).

In the second component of the process, the multidimensional manifolds derived from each different PD are related to one another, to generate a final energy landscape that describes the continuously varying conformations of the 3D structure and the occupancy of each conformation. By the Shannon-Nyquist sampling theorem (Sayre 1952), a sufficiently dense sampling of the structural changes in the particle is sufficient to generate a continuous description of these changes. An authentic movie can be generated that follows the system along a functional trajectory in the landscape (Figure 8.2). Each frame in this movie has an experimental basis, containing a structural representative of the bin from which it was drawn. Of course, no two particles will follow exactly the same functional trajectory through the landscape as they react. An infinite number of very closely related movies would depict the typical trajectory equally well (Frank and Ourmazd 2016).

It is a lot of fun (at least for a structural biologist) to create and watch a molecular movie but in the end, how does it help us advance science? The details of the answer are specific to the system under study but in general terms, the movie is a hypothesis-generating engine. It lets us observe hitherto invisible states and encourages us to design experiments that probe these states further. If for example drug designers are interested in inhibiting a protein, every point along the functional trajectory represents a different structure and hence a different druggable target. The design of a transition-state analog is a special case of this idea.

3D Projection view   2D Projection view

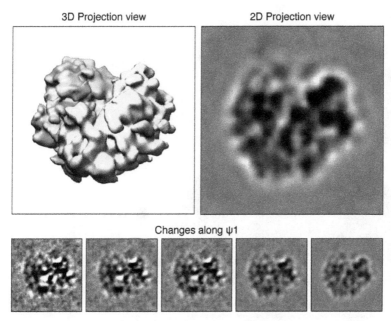

Changes along ψ1

Figure 8.4 Analysis of data falling into a single projection direction by manifold embedding. This projection direction shows a view between front and side view of the 80S ribosome (surface representation in upper left). The changes are encapsulated in five eigenvectors. The first eigenvector, ψ_1, can be recognized as the result of inter-subunit rotation (see lower 5 panels which, from right to left, show an increasing redistribution of mass consistent with the increasing rotation of the small subunit with respect to the large subunit). That is, through manifold embedding the projection data were sorted according to changes in angle between the two subunits. The other eigenvectors, ψ_2 through ψ_5, depict more subtle changes in conformation. Frank and Ourmazd 2016/With permission of ELSEVIER.

We end with an example. An important advance in ELA invokes single-particle images taken with and without the presence of ligands (Dashti et al. 2020). These images link high-energy forms of the apo particle poised in a ligand-binding conformation to the corresponding lower energy form of the holo-particle/ligand complex. Using cryo-EM data, the authors analyze the allosteric behavior of the ryanodine receptor type 1 (RyR1) in the absence and presence of key ligands. RyR1 is a calcium channel that plays a central role in excitation-contraction in striated muscle. RyR1 mutations are involved in a number of genetic disorders. Ligands such as ATP, caffeine, and Ca^{++} are well characterized activators of RyR1, which undergoes complex internal motions as it opens a calcium ion channel. However, the details of the activation process remain largely unknown. The work allowed comparison of ligand-binding pathways observed in the landscape derived from the ligand-bound structures with those generated by computer interpolation between discrete structures obtained by clustering techniques (Dashti et al. 2020).

Micrographs were obtained in the presence of ligands (caffeine, ATP, Ca^{++}) and their absence. The micrographs with and without ligand were pooled for manifold analysis, which enabled the two pools to be encompassed by the same set of conformational coordinates. In the event, four conformational coordinates emerged, of which only two were sufficiently robust to be analyzed further. Separate landscapes without ligand (Figure 8.5,

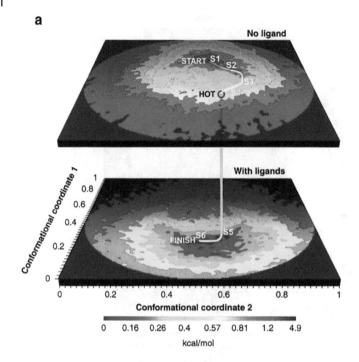

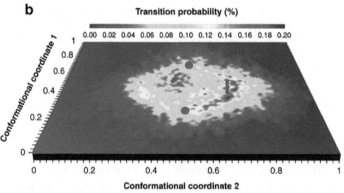

Figure 8.5 **Allosteric behavior of the ryanodine receptor type 1 (RyR1) in the absence and presence of key ligands**. Energy landscapes with and without ligands, and transition probability map. A. Energy landscapes without ligands (upper surface), and with ligands (Ca2+, ATP, and caffeine) (lower surface). The landscapes are described in terms of the most important two of a common set of orthogonal conformational coordinates. The curved path represents a high-probability route to the binding of ligands. This path starts at the minimum-energy conformation of RyR1 without ligands ("START"), follows the conduit of lowest energy to a point with a high probability of transition to the with-ligands energy landscape ("HOT"), and terminates at the minimum-energy conformation with ligands ("FINISH"). Six representative conformational states along the path, in white, are selected for validation with molecular dynamics simulations (S1–S6). B. Probability map for transitions from the energy landscape without ligands to the energy landscape with ligands. The axes are the same as in Figure 1a, with magenta discs indicating the positions of the minima of the energy landscapes. The magenta arrow indicates the position of the transition point. Dashti et al. 2020/Springer Nature/Licensed under CC BY 4.0.

panel A upper) and with ligand (panel A lower) are plotted using the same conformational coordinates. Careful analysis provided a plausible trajectory for ligand binding, which is marked by the sequence of states S1 through S6, which pass from the unliganded to the liganded landscape. Figure 8.5 panel B shows the probability of transition between corresponding points on the unliganded and liganded landscapes. As justified by detailed analysis, this probability is obtained at each point by taking the product of the density of occupied states on the upper, unliganded landscape and the probability of an unoccupied state on the lower, liganded landscape.

3D movies made along several representative functional trajectories visualized the motions associated with ligand binding, at very high resolution in some regions. The insights gained from these movies were deepened by a variety of MD simulations of specific points along the trajectory. Simulations confirmed the overall energetics from the landscape analysis and provided many intimate details.

Finally, the authors compared their results with those from a more conventional analysis, in which the images were clustered using the maximum likelihood algorithm embodied in the program RELION (Scheres 2012). MD simulations were used to morph between the clusters, and thus to provide an alternate vision of the ligand binding trajectory. There are a couple of key take-home lessons. The authors state that "...the extent to which maximum likelihood clustering is based on functionally relevant conformational coordinates is unknown...." In particular, the uncertain connection between the space in which the clustering takes place and energy landscape space spanned by the two principal conformation coordinates means that the positions of key points such as START and FINISH (Figure 8.5) may differ dramatically. Indeed, they found that the conformational trajectory emerging from the MD interpolation was dramatically different in some places from that derived from the dual landscape approach in the paper. This is consistent with the widely accepted view that morphing between static structures along a potential functional trajectory has little predictive power.

We anticipate that the ELA method will provide even more powerful insights into structural dynamics as instrumentation and algorithms develop.

8.4 X-ray Spectroscopy

8.4.1 Introduction: Spectroscopy of Transition Metals

Perhaps one-third of proteins of known structure contain metal ions, either bound directly or as part of organic cofactors such as heme. Some of the most important functional pathways in biology, including those involved in photosynthesis and vision, are mediated by redox reactions in transition metal ions such as iron, nickel, cobalt, and manganese. Iron-containing heme proteins fulfill multiple roles from oxygen transport or storage (hemoglobin and myoglobin) through redox intermediates (cytochrome c) through monooxygenase (cytochrome P450). Calcium ions play a critical role in the regulation of cellular life through proteins such as calmodulin that interact with and regulate the activity of many other cellular proteins depending on the occupancy of the four calcium-binding sites in calmodulin. Calcium ions also play a structural role in binding reactions, where they for example help to neutralize charge in protein–nucleic acid complexes. A proper understanding of the mechanisms of reactions involving these ions requires detailed knowledge of their coordination, spatial environment and oxidation state, and how these vary throughout the functional trajectory. Unfortunately, the ability of crystallography and cryo-EM to correctly identify metal ions in crystal structures is rather poor, and identification of the ionization state from the structure alone is genuinely problematic. Indeed, in a sample of the structures of 30 metalloproteins deposited in the PDB, over half contained misidentified ions. A recent survey work (Grime et al. 2019) stated: "Using ion beam analysis through particle-induced X-ray emission (PIXE), we have quantitatively identified the metal atoms in 30 previously structurally characterized proteins using minimal sample volume and a high-throughput approach. Over half of these metals had been misidentified in the deposited structural models. Some of the PIXE detected metals not seen in the models were explainable as artifacts from promiscuous crystallization reagents. For others, using the correct metal improved the structural models. For multinuclear sites, anomalous diffraction signals enabled the positioning of the correct metals to reveal previously obscured biological information."

Thus crystallography, and indeed every other form of 3D structure determination, require parallel use of other observational methods to provide a more complete and more accurate story of ions and their environment. Various forms of optical spectroscopy, including UV-visible, fluorescence, and CD, have a long and important history in monitoring biochemical reactions, including in enzymology and kinetics. Targets of spectroscopies include metal ions and chemical groups such as tryptophan and NAD, and secondary structural elements such as α-helices. The development in kinetics of the stopped-flow method by Chance, and its elaboration by Gibson and others (Chance et al. 2013), led to the unraveling of enormously complex biochemical pathways such as oxidative phosphorylation. However, identification of the elemental nature and oxidation states of transition metals present in proteins without corresponding knowledge of their location and chemical environment is also insufficient to provide a complete picture, especially one with predictive utility. Happily, recent developments in X-ray spectroscopy enabled by XFELs provide a pathway to TR studies that monitor 3D structure and transition metal chemistry and environment, sometimes simultaneously.

8.4.2 X-ray Absorption Spectroscopy

By far the most common spectroscopic method in use at SRs and XFELs today is X-ray absorption spectroscopy (XAS) in its various forms. As in all spectroscopies, XAS requires a tunable source. It was originally developed and remains widely used at SR sources. XAS is sensitive to elemental identity, electronic structure, and local coordination, and is thus ideally suited to dissecting complex metal centers and monitoring functional changes. Depending on the element under study, XAS is applicable in either the soft or hard X-ray regions, where it probes photo-induced electronic transitions from a core state to low-energy orbitals just above the highest energy occupied state. Figure 8.6 summarizes the various responses to the interaction of the sample with an X-ray photon. Figure 8.7 illustrates X-ray absorption near edge structure (XANES) and extended X-ray absorption fine structure (EXAFS) spectra, whose energy and oscillations near the absorption edge contain information about the identity, valence state, and nearest-neighbor interactions of the ion. We discuss X-ray emission spectroscopy (XES) in Chapter 8.4.5.

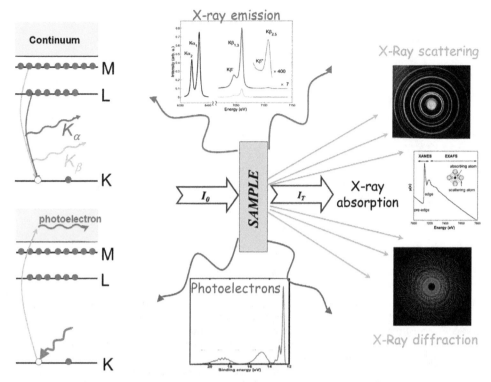

Figure 8.6   Processes triggered when an X-ray beam is incident on a sample. Beam is of intensity I_0. Different types of core-level spectroscopies are possible: X-ray absorption spectroscopy (XAS), and photon-in/photon-out spectroscopies, which are broadly classified as X-ray emission (XES) and photoelectron spectroscopy (photon on/electron-out). X-ray scattering is used to probe structures of molecules in solution and X-ray diffraction for crystalline systems. Chergui and Collet 2017/ Reproduced with permission from American Chemical Society.

8.4.3   XANES and EXAFS Spectroscopies

An XAS spectrum is dominated by a sharp absorption edge E_0 that identifies the promotion of an electron from inner K- or L-shell core orbitals of the absorbing atom to the continuum – that is, to ionization. This occurs when the energy of the incoming X-ray photon matches the binding energy of the K- or L-shell core electron. Since binding energies are element-specific, the position of the absorption edge identifies the absorbing element. The sharp edge is modulated by two additional features. Near-edge XANES resonances occur from just below the edge E_0 to a region 10–50 eV above it. XANES maps promotion of the core electron to empty or partially occupied valence states, which contain information about orbitals involved in chemical interactions and dynamic processes. XANES spectra are also sensitive to oxidation state, the chemical nature of bound ligands, the coordination geometry of these ligands, and the effective charge on the absorbing atom. At higher energies up to a few hundred eV above the edge, additional EXAFS resonances appear in the absorption spectrum. These resonances arise from a quantum phenomenon in which the wavelength of the outgoing photoelectron is smaller than the interatomic distance between the absorbing atom and its nearest neighbors. These photoelectrons are backscattered and interfere

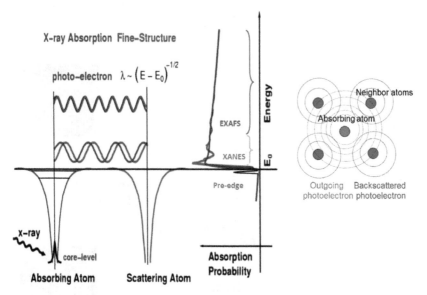

Figure 8.7 X-ray absorption spectroscopy. Excitation of a core orbital to the orbitals below the ionization potential (E_0) gives rise to pre-edge transitions, which probe the unoccupied density of states. Above E_0, there is a jump in the absorption cross section giving rise to an edge and a photoelectron is generated whose de Broglie wavelength depends on its excess energy ($E - E_0$). Low-energy electrons have high scattering cross sections, so if the absorbing atom is embedded in an assembly of atoms, multiple scattering events occur (right panel), giving rise to modulations of the spectrum in the above ionization region, called the X-ray absorption near edge structure (XANES). As the kinetic energy of the electrons increases, the cross section decreases and weaker modulations appear, which form the extended X-ray absorption fine structure (EXAFS). Chergui and Collet 2017/Reproduced with permission from American Chemical Society.

with the outgoing wave to produce the oscillations in the spectrum known as EXAFS. Analysis of the oscillations allows the determination of the coordination number of the absorbing atom, the bond distances to adjacent atoms, and their elemental types. These spectroscopies are comprehensively reviewed in (Chergui and Collet 2017). In the context of a protein whose overall structure is known, XAS thus provides a detailed structural and chemical map of the local environment of absorbing atoms such as transition metals.

8.4.4 Experimental Approaches to XAS

Applications of XAS to biological systems face experimental difficulties, even without the additional requirements of TR studies. These applications use XFEL or SR X-ray pulses and are constrained by the wavelength and spectral width of these pulses. Soft X-rays are strongly absorbed, which limits the sample thickness to a few μm. In the hard X-ray regime, samples in solution even at mM concentration provide only a weak absorption signal. However, there are proxies for the signal measured by direct absorption that greatly enhance the signal-to-noise in the spectrum. In these proxies the absorbed energy is largely redirected into various X-ray fluorescence modes, so that measurement of total fluorescent yield can provide a valid estimate of the absorption signal. Typically, the incident beam energy is scanned through an absorption edge while the detector is tuned to the lower energy of the fluorescence. By discriminating against the higher energy of the incident beam and elastically scattered X-ray photons, the background in the fluorescence spectra is minimized.

Fluorescence is measured by a 2D detector viewing the sample at right angles to the incident beam. Extrapolation to 4π solid angle gives the total fluorescent yield. A spectral width $\Delta E/E \sim 10^{-4}$ is needed to measure spectra with 1 eV resolution. Since the spectral width of a pulse at an XFEL is $\sim 10^{-3}$, a scanning monochromator is necessary in the incident beam to provide the narrow bandwidth needed. In practice, the monochromator may transmit as little as 1% of the incoming radiation. The difficulties are compounded by the pronounced spiky energy spectrum and pulse-to-pulse variation in intensity at XFEL sources (Chapter 6.3). The use of seeding (Geloni 2020) to generate more uniform and monochromatic pulses at XFELs is likely to assist XAS experiments and may eliminate the need for the incident beam monochromator (Samoylova et al. 2019).

Another approach to XAS uses a pair of dispersive spectrometers placed before and after the sample (Zhu et al. 2013). These spectrometers are thin, bent, perfect silicon crystals that transmit about 80% of the incoming radiation and reflect the remainder from their (111) planes. Comparing the spectra from a single X-ray pulse acquired before and after sample illumination provides a difference spectrum that internally controls for the varying profiles of SASE pulses. Although these devices are conceptually appealing, they have not yet come into widespread use.

8.4.4.1 Simultaneous Observation of XAS and Diffraction

Simultaneous XAS and diffraction in TR studies are challenging since XAS requires the wavelength to be scanned over a few 100 eV. TR studies at XFELs require measurement of XAS and diffraction signals from different samples, or correlation of XAS measurements with structural information from other sources.

An XAS study on the dynamics of the doming of heme in myoglobin (Levantino et al. 2015) illustrates both the power and the limitations of the method. The authors studied the photolysis of CO-myoglobin at the XPP instrument of the LCLS. They obtained roughly 100 fs time resolution in solution using a $\sim$70 fs pump laser pulse at 538 nm, followed by $\sim$30 fs probe X-ray pulses. X-ray fluorescence was measured as a function of time at three discrete X-ray wavelengths near the Fe K-edge. The XPP instrument was not suitable for rapid changes in wavelength, so TR XANES spectra could not be obtained. The time-resolved fluorescence data could be fit by a sum of two exponentials with time constants of 73 fs and 400 fs. Comparison with other studies suggests these two time constants represent a two-step process as the heme Fe converts from six-coordinate low spin to five-coordinate high spin upon CO dissociation and the heme then slightly domes.

The program nanoBragg (Sauter et al. 2020) has been developed to simulate SFX diffraction patterns with detail sufficient to model with great accuracy the XFEL spectrum, the mosaicity of the crystal, and other factors. Extending the approach to XAS, the authors used simulated data for the iron-containing protein ferredoxin taken with the full spectrum of a SASE pulse with no monochromator. This situation has been regarded as wildly unfavorable for reliable XAS measurements. For each atom in the model, application of a Bayesian algorithm allowed reconstruction of the XAS spectra that agreed with those entered into the simulation. This result suggests more straightforward XAS experiments are possible at XFEL sources.

8.4.5 X-ray Emission Spectroscopy (XES)

X-ray emission spectroscopy, also called non-resonant fluorescence spectroscopy, provides information that is in some senses complementary to XAS but is much easier to implement in a TR mode. In the hard X-ray regime, an X-ray energy of excitation well above the absorption edge generates photoelectrons that decay through pathways that together provide information about K-edge emission lines (Figure 8.8). These lines are sensitive to the oxidation state and the energies of ligand-binding orbitals. Emission is inelastic and occurs at energies lower than that of the absorbed photon.

Emission shifts to higher energy for higher oxidation of the absorbing atom but the shifts are small: for example, +2 eV for an increase of +1 in the oxidation state of iron. The emitted energy is scanned over the width of a fluorescence line. Various technical issues have limited the use of XES at SR sources (Alonso-Mori and Yano 2018), but application is more favorable at XFELs. The narrow energy bandwidth of XFEL pulses (Chapter 6.3) has no effect on the XES spectrum, and the use of 2D dispersive, von Hamos spectrometers (Kern et al. 2014) allows a complete (but very noisy) spectrum to be collected from each pulse, which simplifies normalization. At present limited by the weak fluorescence emission from dilute samples, the most stable results are obtained by using just the first moment of the emission spectra, rather than a more detailed analysis of the spectral shape (Jensen et al. 2017; Fransson et al. 2018). The first moment corresponds to the center of mass of the spectrum and estimates the oxidation state, primarily by detecting peak shifts arising from different effective numbers of metal 3d electrons.

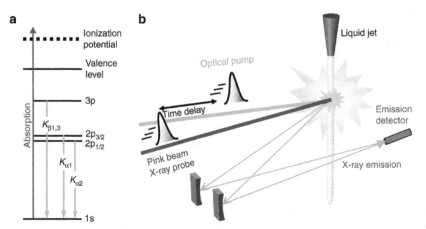

Figure 8.8 **Experimental setup for femtosecond X-ray emission spectroscopy. Panel a** Energy level diagram showing the origin of the K_α and K_β fluorescence after creation of a hole in the 1s (K) shell. The $K_{\alpha1}$ and $K_{\alpha2}$ lines originate from the splitting of the 2p orbital ($2p_{1/2}$ and $2p_{3/2}$), whereas for K_β these lines are degenerate, resulting in the line labeled $K_{\beta1,3}$. **Panel b** Experimental setup for the time-resolved X-ray emission spectroscopy measurements at the XFEL. The green line represents the optical (533 nm) pump pulse used to excite the sample and the red line represents the X-ray (centered at 8.168 keV) probe pulse at varying time delays (black arrow). The pulses are overlapped and intercept the sample at the green interaction region. A von Hamos geometry was used for these measurements. Kinschel et al. 2020/Springer Nature/Licensed under CC BY 4.0.

8.4.5.1 Photosystem II: An Example

One of the most important biological applications of XES has been to photosystem II (PS-II), the integral membrane protein complex that carries out the light-activated oxidation of water in plants and some bacteria. A recent review focusing on PS-II illustrates the state of the art (Fransson et al. 2018). Figure 8.9 shows a typical experimental layout. Collectively, the studies show the important contributions that XES can make to understanding the evolution of metal-cluster oxidation states. Equally, they show that these contributions are possible only in the context of atomic structures at higher resolution in a well-controlled environment. Analysis requires sophisticated modeling algorithms and data-processing software.

PS-II enables water oxidation by creating highly energetic charge separation across the insulating thylakoid membrane stacks in the chloroplast organelle. This energy is provided by sequential absorption of four photons of visible light (discussed in more detail below), coupled through the full photosynthetic complex to fix carbon dioxide into carbohydrates and other molecules, and simultaneously to liberate molecular oxygen. The ability to control the number and separation in time of light pulses allows specific intermediates in the photocycle to be generated. The overall process is known as the Kok cycle (Kok et al. 1970). The PS-II system has been the object of intensive study, partially motivated by potential applications to climate change, and has been recently reviewed (Vinyard and Brudvig 2017; Cox et al. 2020). PS-II is a dimeric complex in which each monomer contains 17 or 18 proteins subunits depending on the source, plus a plethora of cofactors including chlorophyll and all-trans β-carotene. Central to the electron transfer chain is an unusual Mn_4CaO_5 cluster at the heart of the oxygen-evolving complex, and a cluster of conjugated redox moieties such as plastoquinone. The high

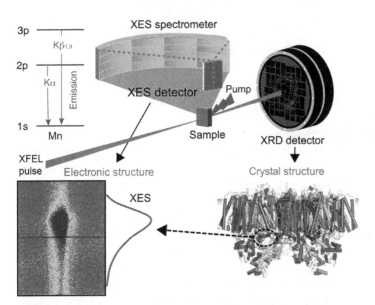

Figure 8.9 Schematic illustration of simultaneous XRD and XES measurements. Measurements use either a jet injector (Sierra et al. 2018) or the Drop-on-Tape (DOT) method (Fuller et al. 2017), giving both diffraction images and XES spectrum. This setup utilizes a von Hamos spectrometer for XES data collection, which enables the detection of the entire $K\beta_{1,3}$ spectra in an energy dispersive mode on a position-sensitive 2D detector on a shot-by-shot basis. Upper left: energy diagram of $K\alpha$ and $K\beta_{1,3}$ transitions. Fransson et al. 2018/Reproduced with permission from American Chemical Society.

resolution crystal structure of PS-II from a thermophilic cyanobacterium has been determined at 0.19 nm resolution, revealing the architecture of the molecule including the Mn-containing metal cluster (Suga et al. 2015). High-resolution structures have also been determined of the putative intermediate states formed after various numbers of individual light pulses are absorbed. The metal cluster undergoes a series of changes in oxidation state (+2 to +4) and in structure as it delivers a series of four electrons to the P680 complex. For a complete understanding, knowledge of the protein structures in all intermediate states, the cofactor structures and the oxidation state of the manganese is required for each step of the photocycle. These measurements have many challenges. In particular, X-ray radiation damage artifactually alters the Mn oxidation state and its structural coordination in the cluster.

8.4.5.2 XES Studies on PS-II

Much recent progress has been made on cofactor structure, Mn coordination and oxidation state, and how these vary from intermediate to intermediate. Important results on biomimetic model Mn systems (Jensen et al. 2017) set the stage for the study of truly biological clusters. For example, this work demonstrated the capacity of XES spectra to distinguish the two coordination compounds $[Mn^{IV}(OH)_2(Me_2EBC)]^{2+}$ and $[Mn^{IV}(O)\,(OH)\,(Me_2EBC)]^+$, where EBC denotes the ethylene-bridged dialkylcyclam ligand. These compounds have the same formal oxidation state, Mn^{IV}, but the latter contains the MnIV=O moiety, believed to be mechanistically important.

A large group working at the both the SACLA and LCLS XFELs (Alonso-Mori et al. 2016) has harnessed fs pulses to minimize secondary radiation damage (Chapter 4.3.2), and to measure simultaneously X-ray emission spectra and Bragg reflections from crystalline P-II at ambient temperature. The oxygen-evolving complex passes through 5 intermediate states during the Kok photocycle: the resting state S_0 and 4 intermediate states S_1 through S_4 that represent the number of oxidizing equivalents created by absorption of four successive photons of visible light, which drives the cycle. Intermediate states have been trapped by rapid freezing and their structures studied using SR. However, studies on unfrozen, active samples capable of undergoing the full photocycle are clearly desirable. Paying careful attention to issues of signal-to-noise and spurious XES from buffer ions, the group collected simultaneous XES and Bragg data on a two-pulse state of PS-II crystals, in which the XES spectra were obtained using a dispersive spectrometer. The XES spectrum from an individual crystal is extremely weak, and a useful final spectrum could only be obtained by averaging spectra over the many crystals used for collection of diffraction data.

This work confirmed that results at ambient temperature at XFEL sources were consistent with those obtained earlier on frozen samples at SR sources. A recent more XES-focused study shows the continued importance of SR sources when longer-lived intermediates exist (Davis et al. 2018). Working at BioCARS at the APS, the group used over 2 million X-ray pulses to measure Mn Kβ XES spectral emission lines from crystals prepared in the S_0 state through the S_1 to S_4 substrates. These spectra provide information about the number of unpaired *3d* electrons. The signal averaging enabled by the large number of individual spectra enabled the authors to provide the first analysis of primary PS-II XES data, in which the only manipulation was extraction of spectra via energy calibration of the detector. The authors note carefully that the first-moment estimation of noisy XES spectra is very sensitive to the details of background subtraction. Their spectra allowed them to develop and support a kinetic model for the step in which the covalent bond between two oxygen atoms is formed in the O_2 molecule ultimately released. The kinetics of this step and the related discussion of the nature of the S_4 substrate remain a key issue.

Further detail has been provided in a technically challenging XFEL study (Ibrahim et al. 2020). Combining room temperature crystallography and spectroscopy, the authors assembled frames of a molecular movie of the structural and oxidation state change steps underlying the transition between the S_2 and S_3 states. This work builds on the observation of a new oxygen ligand between Mn1 and Ca in the last metastable intermediate before the formation and release of molecular oxygen. We cannot resist reproducing their results in Figure 8.10. This impressive diagram illustrates the workflow from raw emission spectra,

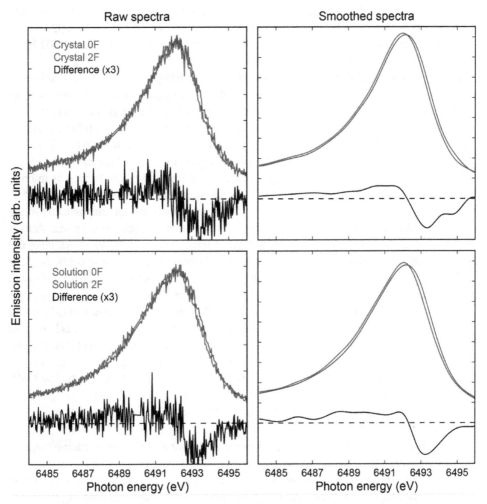

Figure 8.10 **Mn Kβ X-ray emission spectra of PS-II.** Spectra were collected at LCLS of zero-flash (0F) and two-flash (2F) dark-adapted PSII samples for one run of microcrystals and solution, as well as corresponding difference spectra (2F–0F). Raw spectra are shown to the left and smoothed spectra to the right, where the latter are constructed by binning to a grid of 0.75 eV, followed by a cubic spline to a resolution of 0.01 eV. Spectra are area-normalized within the interval 6485–6497 eV. Ibrahim et al. 2020/Reproduced with permission from Proceedings of the National Academy of Science.

collected separately from crystalline and solution samples, through to smoothed and fitted output. The very low signal to noise evident in the difference spectra requires exceptionally careful data acquisition and analysis, enabling fruitful conclusions to be drawn.

8.5 Nuclear Magnetic Resonance
Joseph Sachleben (University of Chicago)

8.5.1 Principles

Biological NMR is a powerful method for determining the 3D structure of macromolecules of small to intermediate size, and for monitoring local dynamics in biomolecules. There are many excellent introductions to this field such as the comprehensive tome from (Cavanagh et al. 2010), and the excellent heuristic guide by (Doucleff et al. 2011). Over 13,000 NMR structures have been deposited in the PDB.

There are dramatic differences between NMR and crystallography. NMR is carried out in solution, where the structures are not constrained by lattice contacts and the solvent can be readily adjusted. The range of time constants accessible in kinetic experiments is extraordinary, spanning up to 12 decades in time. However, NMR is less powerful than crystallography in its ability to follow detailed 3D structural changes as a function of time. It is also unable to visualize motions correlated across distant locations of the molecule.

The physical basis of NMR is entirely different from imaging methods such as microscopy or X-ray scattering. NMR is a spectroscopic technique that provides site-specific information in biomolecules due to the variation of electron density around magnetically active nuclei such as ^{1}H caused by the local chemical environment, called the chemical shift. A nucleus is magnetically active if it has a nonzero value of the nuclear spin, which gives the nucleus a magnetic dipolar moment causing it to interact with magnetic fields. Nuclear spin is quantized and thus has only discrete values, the most useful of which for biological studies are spin 1/2 nuclei such as ^{1}H, ^{13}C, ^{15}N, and ^{31}P. As shown in Figure 8.11, placing a spin 1/2 nucleus in a large external magnetic field, $\boldsymbol{B}_0$ (conventionally assumed to be in the +z direction) splits the energy of the ±1/2 states by $\Delta E = \gamma \hbar B_0$ The constant γ is known as the gyromagnetic ratio and is specific to each type of nucleus. To drive transitions between the ±1/2 states, an oscillating magnetic field whose frequency is given by $\Delta E = hf$ must be applied to the system. This defines the Larmor frequency f as $f = \dfrac{\gamma}{2\pi} B_0$. For typical NMR magnetic field strengths, f lies in the radio frequency (RF) spectrum. NMR spectrometers are often nicknamed for f. For example in a 900 MHz spectrometer, f for a ^{1}H nucleus is about 900 MHz.

In the absence of an external field $\boldsymbol{B}_0$, the number of spins pointing up and down is equal. In the presence of $\boldsymbol{B}_0$, however, the energy of the magnetic dipole E_D depends on the magnetic field and the spin quantum number, ±1/2, by $E_D = -m\gamma\hbar B_0$. Thus, spins pointing parallel to the magnetic field (+z) have slightly lower energy than those pointing antiparallel to the field (−z). The sign of γ is positive for ^{1}H so the (+1/2) state has a lower energy than the (−1/2) state. Thus, more nuclei align antiparallel to $\boldsymbol{B}_0$ than parallel to it. However, the fraction of spins pointing up and down differs only slightly from 0.5. This makes NMR an inherently insensitive technique compared to optical spectroscopy, but allows us to

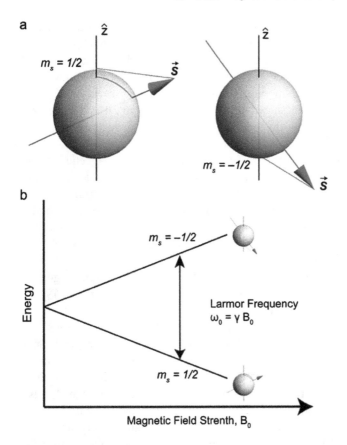

Figure 8.11   Nuclear spin. Panel a Every nucleus has an intrinsic spin angular momentum that is specified by the spin quantum number, which can have integer or half odd integer, $\frac{1}{2}, \frac{3}{2}, \frac{5}{2}, \frac{7}{2} \cdots$, values. The value of the spin quantum number specifies the number of states the spin possesses, $2s + 1$. Only nuclei with nonzero spin quantum numbers are NMR active. BioNMR concentrates on spin ½ nuclei which have two possible orientations. **Panel b** A nucleus with a nonzero value of the spin acts as a magnetic dipole and interacts with an externally applied magnetic field. In the spin ½ case, there are two possible orientations of the spin with the magnetic field, ±1/2. The difference in energy between these two states is $\hbar\omega_0$, where $\omega_0 = \gamma B_0$ is the Larmor frequency which is proportional to the magnetic field strength. γ is the gyromagnetic ration, the proportionality constant that is unique for each NMR active nucleus. Figure by Joseph Sachleben.

describe the effect of a macroscopic sample of spin 1/2 nuclei as a bulk magnetization, M, aligned with the large external magnetic field.

If we apply an on-resonance RF pulse at the Larmor frequency f to the NMR sample, the magnetic portion of the rf field, $\mathbf{B}_1$, can be decomposed into two circularly polarized components; one which rotates in the same sense as the spin, the other in the opposite direction. We imagine going into a reference frame rotating with the rf field in the sense of the precession of the spin. In this rotating frame, the rf field appears to the bulk magnetization as a static magnetic field perpendicular to a large external magnetic field which disappears due to a fictitious field created by transforming into the rotating frame. This fictitious field is analogous to the centrifugal force necessary to describe mechanical motion in a rotating reference system. According to the laws of electromagnetism, a magnetic field perpendicular to a magnetization results in a torque τ exerted on μ at right angles to both B_1 and M. That is, $\tau = M \times B_1$. Much as the gravitational force of the earth causes a spinning top to precess about the vertical direction, this torque causes the magnetization, M, to rotate around B_1 in a process called nutation, shown in Figure 8.11. In the typical 1D NMR experiment, the RF pulse is timed such that the magnetization is tipped by 90°, resulting in the magnetization lying in the xy plane perpendicular to the large external magnetic field. When the rf pulse ends, the magnetization feels the effect of the large external magnetic field and rotates about it at the Larmor frequency. This rotating magnetization can be detected by an antenna and digitized. Fourier transformation of this signal produces a spectrum.

In this presentation of NMR, the spins never come back to equilibrium. To reach equilibrium, two processes must occur: the bulk magnetization must stop rotating about the large external magnetic field and the magnetization must end up pointing along this field. The time constant for the return of magnetization along the large magnetic field is called the Longitudinal Relaxation Time, or T_1. The time constant for rotation of the bulk magnetization about the large magnetic field to die out is the Transverse Relaxation Time, or T_2. This process is illustrated in Figure 8.12. T_2 is inversely related to the half-width of NMR lines. A large value of T_2 means that spin vectors are tightly bunched in the x-y plane, and thus there is less spread in f. The relaxation pathways for T_1 and T_2 arise from varying magnetic fields caused by molecular motion. If molecules are tumbling rapidly by rotational diffusion, these fields are averaged out, thus increasing T_1 and T_2 and narrowing the line width. T_1 and T_2 depend on the correlation time for rotational diffusion of the molecule. T_1 has a minimum value when the correlation time is such that the molecular motions can drive transitions between the spin states. Thus T_1 is long in both very small and very large molecules. T_2 is a monotonically decreasing function of the correlation time. Large molecules that tumble very slowly generate greatly broadened spectra in which peaks tend to overlap and have a reduced peak height. Overlapping peaks make certain important NMR measurements such as NOESYs (discussed below) difficult or impossible. The applicability of certain NMR measurements is subject to size limitations on the molecules in the sample.

Each atom in the molecule has its own value of T_1 and T_2, which depend on both the overall tumbling of the molecule and the local dynamics of that atom. Importantly, this allows site-specific measurement of dynamics by NMR. See Table 8.1 for additional discussion of T_1 and other dynamics terms.

This outline is a simplification in many ways. For example, it remains unclear what useful information the NMR signal provides. The most important factor omitted is that the Larmor frequency f slightly differs for each atomic position in the molecule. The external field B_0 creates local electrical currents in the molecule, for example in the conjugated rings of phenylalanine or tryptophan. These currents in turn create a small local magnetic field σB_0 that opposes B_0. The local magnetic field B may then be written as

$$B = B_0 - \sigma B_0 = (1 - \sigma)B_0.$$

The local value of the magnetic field is determined by the immediate, structural environment of the nucleus. This affects each Larmor frequency individually and changes the

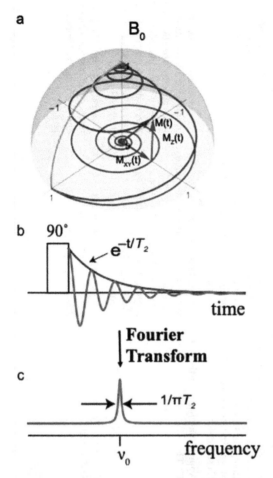

a

B_0

M(t)

M_z(t)

M_{XY}(t)

b **90°**

e^{-t/T_2}

time

Fourier Transform

c

$1/\pi T_2$

ν_0 frequency

Figure 8.12 The pulsed NMR experiment. Panel a The magnetization path for the single pulse NMR experiment in a frame rotating close to the Larmor frequency. The green curve shows the path followed by the magnetization as it is rotated from along the external magnetic field, B_0, into the xy plane by the radio frequency pulse. The red curve shows the path the magnetization vector follows as it precesses about the large external magnetic field and relaxes by T_1 and T_2 processes back to equilibrium. M(t) is the magnetization vector at time, t. M_{XY}(t) and M_z(t) are the XY and Z components of M(t). M_{XY}(t) decays exponentially with a time constant T_2 to zero while M_z(t) grows as $1 - e^{-t/T_1}$. T_2 is thus called the transverse relaxation time while T_1 is the longitudinal relaxation time. **Panel b** The pulse diagram of the single pulse experiment. The rectangle represents the radio frequency pulse whose length is chosen to rotate the equilibrium magnetization by 90° into the xy plane. The subsequent oscillating magnetization produces an oscillating voltage in a receiving coil called the free induction decay (FID). The envelope of the decay, blue curve, of the FID is exponential with a time constant T_2. c) Fourier Transformation of the FID produces the NMR spectrum. The width of the NMR resonance is $1 / \pi T_2$. Joseph Sachleben (author).

frequency by a few parts per million (ppm). The change (in units of ppm) is called the chemical shift of the nucleus in question. A 1D NMR spectrum plots the intensity of RF re-emission (monitored by RF detector coils) as a function of chemical shift. If we imagine a protein in which partial denaturation is induced by the addition of urea, then the protons

Table 8.1 NMR time constants.

T1 relaxation, a.k.a. spin lattice relaxation or longitudinal relaxation	Time needed for magnetic moment to return to the z-direction after perturbation by an rf pulse. Measured in 100s of ms to s for proteins.	Measured by the growth of magnetization along the magnetic field direction (z-direction) after inversion or saturation.	T_1 is a U-shaped function of molecular weight.	The larger T_2 the smaller the range of frequencies in the magnetization bundle for that nucleus, and the narrower the NMR peak.
T2 relaxation, a.k.a. transverse relaxation, or spin–spin relaxation	Time needed for the NMR signal to decay when signals from a given nucleus that are initially in phase drift apart in frequency and become incoherent. T_2 is inversely related to the line width of the NMR peak.	Measured by a Hahn echo (90°-τ-180°-τ) or CPMG sequence. Homonuclear J-coupling interferes with the measurement of T_2 so it is typically measured on ^{15}N in biomolecules.	T_2 is a monotonically decreasing function of molecular weight. T_2 always less than T_1. T_2 determines width of NMR peaks and limits the sensitivity in 2D and 3D NMR experiments of large molecules.	In HN NOE, rigid sections of the protein give a NOE ratio of about 0.8, while motional sections can give much lower values including negative values.
Heteronuclear NOE	The change in steady-state magnetization of one nucleus when another is saturated. Provides a measure of internal motions.	Typically measured in $^1H^{15}N$ or $^1H^{13}C$ systems by applying a long saturating rf pulse to 1H and measuring the ratio of the change in 15N or 13C magnetization to that without 1H saturation.	The effect depends upon the relative signs of the gyromagnetic ratios. 15N has a negative sign while 13C is positive. Thus in 15N, this effect leads to a decrease in magnetization while in 13C, to an increase.	Small values of τ_c efficiently average the internal magnetic fields affecting the NMR nucleus under study. This reduces them to zero over timescale of the Larmor interaction, and narrows the NMR peaks.
τ_c global correlation time. Large for large molecules and small for small ones	Time constant for the autocorrelation of the orientation of a molecule. Loosely, the time needed for a molecule in one orientation to lose memory of a previous orientation.	Approximately proportional to T2/T1.	In a rigid protein that has no internal motion, all residues have the same τ_c that is due to its rotational diffusion. Additional degrees of freedom are needed to describe proteins with internal motions.	S^2 measured from ^{15}N relaxation has been associated with internal entropy of the backbone of the protein.
S^2 order parameter	Tells fraction of motion that comes from global tumbling. Thus $1-S^2$ gives amplitude of motion from local fluctuations.	Calculated by the Lipari and Szabo model.	$S^2 < 0.7$ means significant additional motions are needed to describe.	
τ_e local correlation time for fast motions	Describes additional rapid motion of atom not captured by above.			

in some residues will alter their environment as they move from the interior of the molecule to full solvent exposure. The changes in environment will be revealed by changes in their chemical shifts.

1D NMR spectra are inherently limited in their resolution. In large molecules such as proteins, resonances from the many different environments of a common nucleus such as a proton, ^{1}H, overlap in the spectrum. Only a few resonances at the fringes of the spectra originate in environments sufficiently unusual to distinguish these resonances from the bulk. The introduction of 2D ^{1}H–^{1}H NMR spectra was transformative for NMR. 2D spectra separated the resonances from individual protons and ultimately enabled the determination of 3D protein structures. The form of 2D NMR spectroscopy known as Nuclear Overhauser Enhancement Spectroscopy (NOESY) displays peaks corresponding to protons that are in close spatial proximity, even if they are widely separated in the sequence. If the magnetization of a particular proton is excited, magnetization can be transferred through space to nearby protons by dipole–dipole interactions, which fall off as $1/r^6$ where r is the distance between the protons. Conceptually, a NOESY spectrum comprises a 1D proton spectrum for each value of the chemical shift, with the great simplifying benefit that only protons that have experienced magnetization transfer appear. The pulse sequences and other technology behind NOESY are discussed in detail (Post 2003; Marion 2013). The diagonal of the NOESY plot is the 1D proton spectrum, while off-diagonal points arise from pairs of protons separated by ~0.5 nm or less.

Protein structures are determined by estimating the distances between large numbers of pairs of protons in the protein from the NOESY spectrum. Although amide protons are very important, C–H, methyl, and aromatic protons are particularly useful. For a protein whose sequence is known, an algorithm based on distance geometry uses these data to determine plausible 3D structures of the protein, consistent with the set of NOESY distance measurements.

However, NOESY distances are sensitive to dynamic fluctuations in the protein structure and also have large intrinsic uncertainties. To reflect this variation, NMR structures are often reported as a set of conformations and coordinates, which more realistically represent the larger ensemble of solutions consistent with the NMR data. This contrasts with the single set of (time- and space-average) coordinates characteristic of X-ray structures.

NOESY spectra are only one of the extensive set of tools used to provide more accurate and comprehensive 3D structures. *J*-couplings (described below) are used to obtain backbone torsion angle restraints. Residual dipolar couplings (RDCs) represent another powerful tool (Prestegard et al. 2004; Chen and Tjandra 2011). Since proteins typically have a weak anisotropic magnetic susceptibility, they do not significantly align in the presence of an external magnetic field; however, molecules with much larger magnetic anisotropies, such as filamentous phage, liquid-crystals, bicelles, do partially align with this field, creating an anisotropic environment which transfers some of its alignment to the protein under study. Under these conditions magnetic dipole–dipole interactions no longer average to zero, and RDCs are created. Details are beyond our scope but the net result is that the directions can be determined of many N–H vectors within the molecule, such as those in the polypeptide backbone. RDCs are not restricted to static 3D structure determination and can also provide significant dynamical information across a wide range of time constants. All these constraints on distances, directions, and dynamics are used to calculate the overall 3D structure and its dynamics.

The dipole–dipole interactions revealed by NOESY spectra are a specific example of magnetization transfer (MT), perhaps the key phenomenon that has enabled modern NMR approaches. MT refers to conditions under which nuclei with nonzero spin can interact with other such NMR-active nuclei. These interactions result in the transfer of nuclear spin coherence or polarization from one population of nuclei to another. An important example is denoted *J*-coupling, in which the transfer of magnetization occurs through chemical bonds rather than through space as in NOESY spectra. *J*-coupling arises between two nuclei in a molecule that are connected by one, two, or three bonds, and is manifested by the splitting of peaks in the NMR signal. This splitting represents the periodic exchange of magnetization energy between the nuclei. As in the case of coupled harmonic oscillators, the frequency at which energy passes back and forth between oscillators (here, nuclear spins) can be dramatically lower than the Larmor frequencies of the individual nuclei. The coupling constant *J* describes the separation of the components of the split signal and is normally measured in Hz. The origin of this coupling involves both nuclear spin and the spin of the valence electrons. In the case of two nuclei, this involves four different types of interactions, leading to the splitting of the peaks associated with both nuclei. The value of *J* is sensitive to the dihedral angles of the bonds connecting the two nuclei but is independent of the magnetic field. The size of the coupling depends on many structural properties. A common spectrum in protein NMR denoted ^3J involves coupling through the three covalent bonds that connect the backbone amide proton and the Cα-proton of the same amino acid. Such spectra are characteristic of the φ dihedral angle in the Ramachandran plot of backbone angles, and therefore specify the secondary structure of the amino acid. The famous Karplus equation relates the measured value of *J* to the value of φ:

$$^3J_{HNH\alpha}(\phi) = A\cos^2(\phi) + B\cos(\phi) + C.$$

Here ^3J is the coupling constant in Hz where the superscript (here, 3) gives the number of bonds through which coupling is taking place, and the subscript (here, HNHα) identifies the hydrogen atoms involved. A, B, and C are empirical constants that depend upon the atoms involved in the three-bond linkage.

Knowledge of backbone angles greatly increases the power of static structure determination based on NOESY spectra. It does not take a great stretch of the imagination to see that dynamic information can also be obtained. If a protein is undergoing conformational fluctuations that alter φ, this could be observable by NMR and the time constants of the fluctuations estimated.

An important example of *J*-coupling enables the so-called Heteronuclear Single-Quantum Coherence (HSQC) correlation spectrum, another workhorse of modern protein NMR. It makes use of the incorporation of ^{15}N into proteins to generate a 2D spectrum that displays the chemical shift of ^{15}N and ^1H nuclei bonded to each other. The shape and strength of these peaks contain critical information about the dynamics of the protein on a residue-by-residue basis. The x-axis displays the chemical shift of each proton attached to a nitrogen atom, while the y-axis displays the chemical shift of that nitrogen. Procedures exist for assigning peaks to individual residues in the sequence and analyzing the HSQC spectrum residue by residue. The general principle for generating HSQC spectra is related to amplitude modulation (as in AM radio). A given nitrogen nucleus is illuminated for varying times by RF corresponding to its Larmor frequency, which is about 10X lower than that of a proton. At a later time it transfers magnetization via *J* coupling to its bonded hydrogen. The hydrogen NMR signal will be modulated by the lower frequency signal originating in the nitrogen. A Fourier transform of the variation in proton intensities yields the nitrogen frequency.

Although there has been great progress in recent years, limitations on molecular mass still affect many forms of static NMR spectroscopies and structure determination. The determination of full 3D structures of single-chain proteins becomes very challenging for molecules heavier than 50 kDa. However, selective labeling can allow the study of selected regions such as binding interfaces up to one MDa (Gans et al. 2010).

8.5.2 Dynamic Information

We give only a flavor of the ways in which the nature of changes in structure, their time-scales, and atomic motions are estimated by NMR techniques. This is a rapidly evolving field.

In contrast to crystallography, NMR is not well suited to probe structural changes driven by energetic pumps such as binding or light activation. For example, there is no ready NMR analog of the DED map in X-ray crystallography that might reveal only the pump-induced differences in chemical shifts exceeding a specified magnitude. Since NMR signals are intrinsically weak, data acquisition times are often long by crystallographic standards. Only relatively long-lived, quasi-static 3D structures can be determined and the time scales accessible by kinetic studies may be limited. However, fast dynamic aspects can be readily detected by specific NMR approaches. A comprehensive review (Sekhar and Kay 2019) emphasizes the dynamical measurements made possible by specific NMR approaches over a very wide time range from 1 ps to 1 s.

A common dynamic feature in proteins known as exchange appears when an atom or group flips back and forth between two locations or states that display different chemical shifts, where the time spent executing the flip is very small relative to the time spent in each location. Exchange operates at slower timescales, exemplified by a 180° ring flip or by the presence or absence of a bound ligand. These transitions are manifested in different ways in the NMR spectrum depending on the frequency of the flip, compared to the difference in chemical shift of the two states in Hz. A flip in the kHz frequency range (on the ms time scale) is typically much slower than the chemical shift difference of the two states. A proton involved in such a flip will spend significant time in each of the two distinct states and environments and will give rise to two peaks in a spectrum, one for each value of the chemical shift determined by its environment. The weights of these peaks are proportional to the occupancy of the two states and provide a hint at the equilibrium constant of the flipping reaction. In the opposite extreme, when the time spent in each state is small compared with chemical shift difference, a single peak appears at the average of the endpoint chemical shifts. In the intermediate state when the flip frequency is within 10X of the difference in chemical shift in Hz, two broadened peaks appear, connected by a valley.

A particularly powerful tool introduced by Kay and colleagues, termed chemical exchange magnetization transfer (CEST), probes exchanging systems in the time window $2 \text{ ms} < t_e < 25 \text{ ms}$. CEST is capable of detecting otherwise NMR-silent (invisible) states present in concentrations as low as 0.5% (Vallurupalli et al. 2012). In the CEST protocol, the depletion of an invisible state resonance I is measured in response to the application of a weak radiofrequency field pulse B_1 that scans through the resonance frequency of a spin probe such as ^{15}N. A typical pulse will lie in the range of 20–50 Hz. When the frequency of B_1 matches a resonance in an invisible state, there is transfer to the ground state, since these two states are in exchange. Thus, resonance I decreases. A parallel effect is seen when the ground state is in resonance with B_1. The net result is a CEST profile with minima at the frequencies of both the ground state and the invisible state. This information can be used to extract parameters such as the chemical shift for the invisible state.

In the μs–ms time range, Carr-Purcell-Meiboom-Gill (CPMG) experiments (Farber and Mittermaier 2015) play a major role. We mentioned above that peaks are broadened when exchange processes in certain frequency ranges are in play. A CPMG experiment quantifies such broadening of NMR resonance lines. The excited states not directly measurable in the NMR experiment can be indirectly detected by measuring how much the line width of the ground state changes due to their presence. CPMG experiments are sensitive enough to detect excited states present only at the 1% level.

The "Model Free" analysis (Lipari and Szabo 1982a, 1982b; Fenwick and Dyson 2016) of ^{15}N or ^{13}C relaxation allows the determination of local order parameters denoted S, which measures how dynamic a portion of a protein is with respect to its overall tumbling. Lipari and Szabo proposed a mathematical model of the spectral density function upon which T_1, T_2, and heteronuclear NOEs depend. The spectral density function is the Fourier transform of the set of fluctuating magnetic fields caused by the rotational diffusion of the molecule under study; that is, the autocorrelation function of the motion. In the original Model Free analysis, the autocorrelation function is assumed to have two components. The first arises from the overall tumbling of the biological molecule, assumed to much slower than the second, which arises from fast "internal motions." Fast motions are modeled with an internal correlation time much shorter than that due to the overall tumbling, in which the order parameter S takes on values between 0 and 1. When S is 1, the portion of the molecule under study is rigid with respect to the overall tumbling. When S is 0, the portion moves quickly and isotropically with respect to the overall tumbling. These parameters are typically studied in the backbone ^{15}N of proteins by measuring ^{15}N T_1, T_2, and ^{1}H–^{15}N heteronuclear NOEs as a function of magnetic field. They provide a detailed view of local dynamics in proteins. For example in ^{15}N relaxation, proteins have much lower values of S in dynamic loops than in rigid α-helices or β-sheets. Figure 8.13 illustrates these and related scenarios.

NMR analyses operate principally on a near-equilibrium energy landscape where peaks and valleys are modest multiples of k_BT. Here NMR can estimate the populations of the two states from peak amplitudes, and their rates of interconversion. For conformations differing in free energy by ΔG_{AB} and separated by a barrier $\Delta G^{\neq}$, one can determine relative populations from the Boltzmann distribution and estimate the lifetimes of the states from transition state theory. This situation is exemplified by slower transitions such as partial unfolding of a protein, which often have biological significance and have observably different conformations. The application of NMR to enzyme dynamics has been the subject of an excellent review (Palmer III 2015).

One area where NMR stands out strongly from other methods is in the study of intrinsically disordered proteins, or of molecules having disordered regions or segments (Gibbs et al. 2017). Such flexibility makes crystallization very chancy. If crystals are formed there

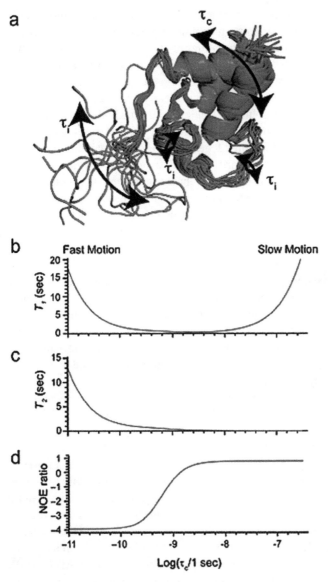

Figure 8.13 NMR relaxation. NMR relaxation is driven by molecular motions at the atomic position under study. (a) A small protein with both rigid (blue) and flexible (red) regions. The correlation time for the rotational diffusion of the rigid section of the protein, τ_c, is primarily determined by its size and correlates with the molecule's molecular mass. Internal motions on top of the overall tumbling with a correlation time τ_i cause deviations of the measured relaxation rates of nuclei in those regions. (b) T_1 as a function of $\log(\tau_c)$. T_1 has a single minimum. T_1 is long for both small and large molecules c) T_2 as a function of $\log(\tau_c)$. T_2 is a monotonically decreasing function of correlation time, so that NMR line width increases with the size of the molecule. d) HN heteronuclear NOE ratio as a function of $\log(\tau_c)$. Slowly tumbling molecules have a heteronuclear NOE ratio of about 0.8. Smaller values indicate internal motion in addition to the overall tumbling of the protein. A negative heteronuclear NOE is a clear sign of sub nanosecond time scale motion. Figure by Joseph Sachleben.

are immediate concerns that conformational selection has excluded more biologically relevant, less ordered structures. An early example of this limitation was encountered in calmodulin. Calmodulin consists of two calcium-binding domains, each containing two Ca^{++} binding sites within the so-called EF hand motifs, where the domains are connected by a linker region. In the first crystal structure of calmodulin the linker was a well-ordered α-helix, clearly separating the EF-hand motifs. However, early evidence from SAXS suggested that this region is not helical in solution (Heidorn and Trewhella 1988). A review (Kursula 2014) notes that NMR and other studies show that the linker region adopts a variety of disordered or ordered conformations, depending on solution conditions and the presence or absence of calmodulin binding partners. The range of calmodulin structures illustrates the need for understanding the dynamic behavior of disordered regions. More generally, the constraints imposed on the structure and its dynamics under the specific experimental conditions must be evaluated, and the biological relevance of the structure critically determined.

8.5.2.1 Example: Zinc Finger Dynamics

Zinc finger domains, both free and bound to their target DNA, represent a revealing and technically accessible story of the application of NMR to a system that depends upon flexible regions for its function. Zinger fingers (ZFs) are small, modular protein domains in which the coordination of Zn^{++} greatly stabilizes the fold. They often appear as repeating elements in multidomain proteins and are widely distributed in life, displaying many functions. One of the most important is as transcription factors where the ZFs bind in the major groove of the DNA and typically recognize three adjacent base pairs. Amino acid sequence differences among ZFs create significant preferences for particular base triplets, and thus many ZFs have strong DNA binding specificity. ZFs have received much attention in protein design.

A classic study (Hyre and Klevit 1998) on the yeast ZF transcription factor ADR1 exemplifies the use of NMR in dealing with disorder. ADR1 regulates transcription of the gene encoding the alcohol dehydrogenase ADH2 and of other downstream enzymes. Genetic analysis identified two tandem ZFs spanning residues 103–155 plus a 20 amino acid leader as the minimal sequence capable of binding to the relevant promoter sequence UAS1, the DNA binding domain. This unit is denoted ADR1–DBD. The study measured parameters from ^1H–^{15}N heteronuclear NOEs, ^1H T_1 relaxation times, and other experiments to probe the dynamics and solvent exchange of both the apo and DNA-bound states of ADR1–DBD. In the apo state the leader appears to be largely unstructured and shows almost no protection from solvent exchange (discussed in Chapter 8.6). The two ZFs tumble as folded domains whose motions are partially correlated. DNA binding produces dramatic condensation of the system. Changes in the pattern of NMR exchange broadening indicate that the region putatively involved in DNA binding becomes more protected from solvent exchange. This and other measurements further indicate that this region undergoes a disorder-to-order transition on DNA binding, and that the DNA–ADR1–DBD complex forms a single molecular complex.

Important advances that couple modeling to NMR observations have resulted in more detailed analysis of flexible systems. A study of the two-domain enzyme prolyl isomerase Pin1 (Bouchard et al. 2018) uses Langevin dynamics to produce conformational ensembles that largely preserve the internal, tertiary structures of the individual domains while generating collective inter-domain motions. These motions are subject to constraints revealed by NMR measurement, particularly those from RDCs. The analysis reveals clear differences between wild-type Pin1 and a dysfunctional I28A mutant. The ability to create ensembles consistent with experiment provides a new tool for protein design and function analysis.

8.6 Hydrogen–Deuterium Exchange

Hydrogen–deuterium exchange (HDX) is an NMR-based method that uses experimental approaches far different from those discussed above and offers different domains of application.

Dynamical behavior in proteins and other macromolecules occurs on all time scales and on all spatial scales. Many studies focus on local processes that take place in a spatially defined region at a specific locus, such as the active site of an enzyme or the bond undergoing trans-to-cis isomerization in a prosthetic group. Protein folding is an example of the other extreme, a global process. It involves all residues in a protein or domain, and the reaction coordinate may range from a simple, rapid, two-state process to serial steps that operate on an extended time scale up to seconds. These two examples fall on the extremes of a continuum of behaviors that are of biological importance. For example, some enzymes have a tighter, less flexible structure when binding a competitive inhibitor than in the inhibitor-free, looser, more flexible form (Williams et al. 2003). This observation may be weakly supported by lower temperature factors in the crystallographic analysis of the inhibitor-bound form, but it is challenging to get a precise molecular picture of these tighter structures, and how they differ from looser ones.

Another example is the molecular nature of intermediates in protein folding or in processes involving intrinsically disordered proteins. Descriptive phrases such as random coil are often invoked to describe intermediates, but a stretch of chain that occupies only three or four conformations will seem random to most methods of observation. Having a clear picture of the range of actual structures assumed by the chain makes for a deeper understanding of the process under study. Protein–protein interactions are defined by regions on the surface of two molecules that interact and can be regarded as buried. The interacting surfaces exclude solvent and during their interaction, effectively become part of the protein interior – an echo of protein folding. This burial may be transitory or long-lived. Finally, normal dynamical fluctuations in a folded protein are manifested by many types of structural perturbations over a wide range of timescales. The technique of HDX provides a powerful tool for investigating these classes of dynamical and structural behaviors.

For each amino acid in a peptide chain (except for proline), the amide hydrogen is free to exchange with protons in the solvent. If H_2O in the solvent is replaced by D_2O, then the amide hydrogen in each position will over time be replaced by deuterium. This replacement is effectively a label that can be detected both by NMR because a deuteron has spin 1 whereas a proton has spin 1/2, and by mass spectrometry (MS) because a deuteron is one atomic mass unit heavier than a proton. NMR and MS are often used in tandem with HDX. In recent years MS has become a powerful stand-alone tool for monitoring and interpreting HDX in protein complexes that are too big for NMR studies.

The rate at which hydrogen is exchanged for deuterium in a particular amino acid depends upon the solvent accessibility at that site. An important reference point is provided by the exchange rate when the amino acid is placed in the center of a small peptide where there is no possibility of hydrogen bonding within the peptide chain. The amide can form an H-bond only to a solvent water. The exchange rate of the amide hydrogen thus defines the most rapid possible rate in an HDX experiment. This value is divided by the rates observed in a protein sample to give the *protection factor*, which can be as large as 10^9 for a residue deeply buried in the protein whose exchange is severely limited. Site-specific maps of the rate at which amide hydrogen is exchanged thus provide powerful dynamic information about the solvent accessibility throughout the molecule, on a residue-by-residue basis.

The experimental protection factors also provide detailed targets to test the accuracy of computer simulations of the dynamics of that protein. We discussed in Chapter 8.3.2 that protein molecules at equilibrium undergo large energy fluctuations, 150 kJ mol^{-1} or more (Cooper 1976, 1984), through their interactions with the solvent thermal bath. Events that violate intuition gathered from examination of a static structure are thus quite common. For example, ^{19}F NMR shows that a bulky, planar side chain such as that of a specific fluorinated tryptophan, which appears in a crystal structure to be fully buried in the protein interior, can undergo a 180° flip thousands of times per second (Li and Frieden 2007). The COREX algorithm developed by Hilser and many colleagues (Hilser and Whitten 2014) has been quite successful in predicting protection factors. COREX uses an ensemble approach in which the free energies of many partially unfolded states are evaluated using the calculated free energies of transfer of chain segments from the interior to solvent. Using the Boltzmann factor, COREX then estimates the fraction of time that a chain segment is solvent-exposed and uses that fraction to estimate the protection factor. The program represents individual microstates by localized folding–unfolding reactions embedded in a molecule that is otherwise in its native state. These microstates are randomly combined producing a comprehensive library of partially folded states. This enumeration of states effectively introduces entropy into the calculation and leads to an estimate of the free energy of unfolding.

HDX is well described by a two-step model in which each amide hydrogen is almost constantly participating in a hydrogen bond, either with an H-bond acceptor within the protein or with a solvent water. The random opening of this bond in an equilibrium process offers the opportunity for HD solvent exchange, with the subsequent re-formation of the hydrogen bond. The chemical equilibrium scheme in Equation 8.4 summarizes this model.

$$
\left(R_{cl}\right)_H \underset{k_{op}}{\overset{k_{cl}}{\rightleftharpoons}} \left(R_{op}\right)_H \overset{exchange}{\underset{k_{ch}}{\rightarrow}} \left(R_{op}\right)_D \underset{k_{cl}}{\overset{k_{op}}{\rightleftharpoons}} \left(R_{cl}\right)_D .
$$

$$(8.4)$$

Here k_{cl}, k_{op}, and k_{ch} are the rate coefficients for the five reactions, and (R) represents a particular amide hydrogen in protein R. This model contains the simplifying assumption that only the open (op) state of the protein can exchange; in the closed (cl) state of the protein, the amide proton does not exchange. The model has proved durable in analyzing large-scale opening and closing events. Eigen first modeled the exchange process as one that is acid- or base-catalyzed, depending on the pH. His critical result is that $k_{ch} = k_A[H^+] + k_B[OH^-]$, where the subscripts A and B denote acid and base. This formalism gives rise to a "V"-shaped plot for the dependence of $\log k_{ch}$ on pH, with a minimum near pH 3 (Connelly et al. 1993). Thus at pH 7, exchange is dominated by base catalysis. This plot reveals the basis for important classes of experiments in which exchange processes can be effectively frozen by a rapid transition to low pH.

Equation 8.4 is often applied in one of two extreme – yet common – cases. In the regime termed EX1, $k_{ch} \ll k_{cl}$. In this regime virtually every time an amide opens, it exchanges. Thus $k_{op} \approx k_{HD}$, where k_{HD} is the overall rate of HDX exchange. The EX1 regime occurs when the folding equilibrium of the native protein has been perturbed, but EX1 may also occur at higher pH or for slow transitions. In the regime termed EX2, the opposite extreme holds: $k_{ch} \gg k_{cl}$, meaning that the probability of exchange during a single opening event is small. The EX2 regime reflects behavior under near-physiological conditions of pH, which leads to the simplification $k_{HD} = \dfrac{k_{op}}{kcl} k_{ch} = K_{op} k_{ch}$. Here K_{op} is the equilibrium constant for the opening–closing reaction. The value of k_{ch} is known from exchange experiments on fully unfolded short peptides, and k_{HD} is measured from the HDX experiment on the protein itself. Thus, one can calculate K_{op}, which yields the free energy of the opening reaction. If these experiments are carried out at varying concentrations of denaturants or chaotropic reagents, the change in free energy on opening can be determined for many individual residues as a function of this concentration. This generates pictures of the absence or presence of hydrogen-bonded structures throughout the protein.

In HDX there are two main approaches for labeling the protein: continuous labeling and pulsed labeling. Continuous labeling is well suited to study conformational fluctuations near equilibrium. A near-native protein is exposed to D_2O and the incorporation of deuterium is measured as a function of time. In current practice, the amide nitrogen atoms in proteins are often pre-labeled with ^{15}N to allow 2D NMR methods to monitor amide resonances on a residue-by-residue basis. Resonances for the native protein must be assigned beforehand. The experiment begins with the transfer of the protein into (say) 90% D_2O. The zero-time point includes this transfer time plus the time needed to record the NMR spectrum, typically a few minutes. The effective time resolution can be improved by rapidly dropping the pH, thus quenching the HDX reaction. Some amides will exchange too rapidly to be observed, while others may exchange very slowly. However, HDX titrations can often be obtained for more than 50% of the residues in the protein.

A transformative use of HDX arose in the dissection of protein folding intermediates in cytochrome c (Bai et al. 1995). The authors carried out experiments in the presence of increasing amount of the denaturant guanidinium hydrochloride, which revealed the presence of a series of unfolding intermediates. The occupancy of these states could be further manipulated by temperature so that a series of HDX experiments revealed the hydrogens exposed in these higher energy intermediates. Combining these data with 3D structural information allowed the free energy, structure, and solvent-exposed surface of each intermediate to be modeled.

The use of mass spectrometry to detect the exchange of D for H has been an important complement to the use of NMR, especially for larger proteins and complexes where NMR assignments become difficult. The classical method using liquid chromatography–mass spectrometry analysis of proteolytic fragments has provided spatially resolved and even residue-specific information. In the study of folding, MS is particularly valuable in revealing the process of HDX in individual molecules, instead of averaging over all molecules in the sample as in NMR. In cooperative transitions for example, the deuteration states of spatially nearby residues may be highly correlated. The extent of correlation is revealed in the differing masses measured for a given peptide. In a conceptual example, if the deuteration of two amides in this peptide is perfectly correlated, then at half occupancy there will be two mass states seen for the peptide, an all-H mass and an all-D mass, two atomic mass units higher.

Pulse-labeling HDX experiments allow the dynamic, temporal sequence of events during folding to be probed. Commonly, an initially denatured protein is quickly transferred to a new environment that initiates folding. For example, dilution reduces the concentration of denaturant. Using standard rapid mixing technology, the protein is exposed to a short, ms HDX pulse at various times after the induction of folding. Such experiments are conducted at a high pD (8–10) where the large rate coefficient permits significant labeling during the short pulse. Rapid transfer to low pH after the pulse is sometimes used to quench the reaction, allowing more time for measurements.

HDX experiments which utilize both NMR and MS for deuterium monitoring have demonstrated the existence of many classes of intermediates in protein folding, and have led to conceptual syntheses such as the foldon hypothesis developed in the Englander group (Englander and Mayne 2014). However, such studies are limited to proteins of modest molecular mass. The *top–down* method of HDX/MS (Masson et al. 2019) explores entirely different realms. This method bypasses the fractionation of the protein into peptides and instead cleaves each protein only once, creating a complementary N- and C-terminal pair to be analyzed by MS. Because the cleavage point is largely random, peptide ladders differing by a single residue are obtained, thus providing sequence resolution in an entirely different way.

References

Alonso-Mori, R., Asa, K., Bergmann, U. et al. (2016). Towards characterization of photo-excited electron transfer and catalysis in natural and artificial systems using XFELs. *Faraday Discussions* 194: 621–638.

Alonso-Mori, R. and Yano, J. (2018). X-ray spectroscopy with XFELs. In: *X-ray Free Electron Lasers* (ed. S. Boutet, P. Fromme and M.S. Hunter), 377–399. Springer.

Bai, Y., Sosnick, T.R., Mayne, L. et al. (1995). Protein folding intermediates: native-state hydrogen exchange. *Science* 269 (5221): 192–197.

Benkovic, S.J., Hammes, G.G., and Hammes-Schiffer, S. (2008). Free-energy landscape of enzyme catalysis. *Biochemistry* 47 (11): 3317–3321.

Berg, H.C. (2018). *Random Walks in Biology*. Princeton University Press.

Bock, L.V. and Grubmüller, H. (2022). Effects of cryo-EM cooling on structural ensembles. *Nature Communications* 13 (1): 1–13.

Bouchard, J.J., Xia, J., Case, D.A. et al. (2018). Enhanced sampling of interdomain motion using map-restrained Langevin Dynamics and NMR: application to Pin1. *Journal of Molecular Biology* 430 (14): 2164–2180.

Cavanagh, J., Skelton, N.J., Fairbrother, W.J. et al. (2010). *Protein NMR Spectroscopy: Principles and Practice*. Elsevier.

Chance, B., Gibson, Q.H., and Eisenhardt, R.H. (2013). *Rapid Mixing and Sampling Techniques in Biochemistry*. Elsevier.

Chen, K. and Tjandra, N. (2011). The use of residual dipolar coupling in studying proteins by NMR. *NMR of Proteins and Small Biomolecules* 47–67.

Cheng, Y. (2018). Single-particle cryo-EM—How did it get here and where will it go. *Science* 361 (6405): 876–880.

Chergui, M. and Collet, E. (2017). Photoinduced structural dynamics of molecular systems mapped by time-resolved X-ray methods. *Chemical Reviews* 117 (16): 11025–11065.

Connelly, G.P., Bai, Y., Jeng, M.F. et al. (1993). Isotope effects in peptide group hydrogen exchange. *Proteins: Structure, Function, and Bioinformatics* 17 (1): 87–92.

Cooper, A. (1976). Thermodynamic fluctuations in protein molecules. *Proceedings of the National Academy of Sciences* 73 (8): 2740–2741.

Cooper, A. (1984). Protein fluctuations and the thermodynamic uncertainty principle. *Progress in Biophysics and Molecular Biology* 44 (3): 181–214.

Copperman, J., Dashti, A., Mashayekhi, G. et al. (2018). Energy landscapes from single-particle imaging of biological processes in and out of equilibrium. *Bulletin of the American Physical Society* 63.

Copperman, J., Hosseinizadeh, A., Mashayekhi, G. et al. (2017). Manifold-embedding methods for extracting continuous conformational ensembles of biological molecules from single-particle measurements using X-ray free electron lasers. *APS* 2017: M1. 274.

Cox, N., Pantazis, D.A., and Lubitz, W. (2020). Current understanding of the mechanism of water oxidation in photosystem II and its relation to XFEL data. *Annual Review of Biochemistry* 89.

Dandey, V.P., Budell, W.C., Wei, H. et al. (2020). Time-resolved cryo-EM using spotiton. *Nature Methods* 17 (9): 897–900.

Dashti, A., Mashayekhi, G., Shekhar, M. et al. (2020). Retrieving functional pathways of biomolecules from single-particle snapshots. *Nature Communications* 11 (1): 1–14.

Dashti, A., Schwander, P., Langlois, R. et al. (2014). Trajectories of the ribosome as a Brownian nanomachine. *Proceedings of the National Academy of Sciences of the United States of America* 111 (49): 17492–17497.

Dashti, A., Schwander, P., Langlois, R. et al. (2018). Trajectories of the ribosome as a Brownian nanomachine. In: *Single-Particle Cryo-Electron Microscopy: The Path toward Atomic Resolution: Selected Papers of Joachim Frank with Commentaries* (ed. J. Frank), 463–475. World Scientific.

Davis, K.M., Sullivan, B.T., Palenik, M.C. et al. (2018). Rapid evolution of the photosystem II electronic structure during water splitting. *Physical Review X* 8 (4).

Dill, K.A. and MacCallum, J.L. (2012). The protein-folding problem, 50 years on. *Science* 338 (6110): 1042–1046.

Doucleff, M., Hatcher-Skeers, M., and Crane, N.J. (2011). *Pocket Guide to Biomolecular NMR*. Springer Science & Business Media.

Englander, S.W. and Mayne, L. (2014). The nature of protein folding pathways. *Proceedings of the National Academy of Sciences* 111 (45): 15873–15880.

Fan, X., Wang, J., Zhang, X. et al. (2019). Single particle cryo-EM reconstruction of 52 kDa streptavidin at 3.2 Angstrom resolution. *Nature Communications* 10 (1): 1–11.

Farber, P.J. and Mittermaier, A. (2015). Relaxation dispersion NMR spectroscopy for the study of protein allostery. *Biophysical Reviews* 7 (2): 191–200.

Fenwick, R.B. and Dyson, H.J. (2016). Classic analysis of biopolymer dynamics is model free. *Biophysical Journal* 110 (1): 3–6.

Frank, J. (2017). Time-resolved cryo-electron microscopy: recent progress. *Journal of Structural Biology* 200 (3): 303–306.

Frank, J. (2018). New opportunities created by single-particle cryo-EM: the mapping of conformational space. *Biochemistry* 57 (6): 888-888.

Frank, J. and Ourmazd, A. (2016). Continuous changes in structure mapped by manifold embedding of single-particle data in cryo-EM. *Methods* 100: 61–67.

Fransson, T., Chatterjee, R., Fuller, F.D. et al. (2018). X-ray emission spectroscopy as an in situ diagnostic tool for X-ray crystallography of metalloproteins using an X-ray free-electron laser. *Biochemistry* 57 (31): 4629–4637.

Frazier, M.N., Pillon, M.C., Kocaman, S. et al. (2021). Structural overview of macromolecular machines involved in ribosome biogenesis. *Current Opinion in Structural Biology* 67: 51–60.

Fuller, F.D., Gul, S., Chatterjee, R. et al. (2017). Drop-on-demand sample delivery for studying biocatalysts in action at X-ray free-electron lasers. *Nature Methods* 14 (4): 443–449.

Gans, P., Hamelin, O., Sounier, R. et al. (2010). Stereospecific isotopic labeling of methyl groups for NMR spectroscopic studies of high-molecular-weight proteins. *Angewandte Chemie* 122 (11): 2002–2006.

Geloni, G. (2020). Self-seeded free-electron lasers. In: *Synchrotron Light Sources and Free-Electron Lasers: Accelerator Physics, Instrumentation and Science Applications* (ed. E.J. Jaeschke, S. Khan, J.R. Schneider and J.B. Hastings), 191–223. Springer.

Gibbs, E.B., Cook, E.C., and Showalter, S.A. (2017). Application of NMR to studies of intrinsically disordered proteins. *Archives of Biochemistry and Biophysics* 628: 57–70.

Glaeser, R.M. (2018). Proteins, interfaces, and cryo-EM grids. *Current Opinion in Colloid & Interface Science* 34: 1–8.

Grime, G.W., Zeldin, O.B., Snell, M.E. et al. (2019). High-throughput PIXE as an essential quantitative assay for accurate metalloprotein structural analysis: development and application. *Journal of the American Chemical Society* 142 (1): 185–197.

Heidorn, D.B. and Trewhella, J. (1988). Comparison of the crystal and solution structures of calmodulin and troponin C. *Biochemistry* 27 (3): 909–915.

Henderson, R. (1995). The potential and limitations of neutrons, electrons and x-rays for atomic-resolution microscopy of unstained biological molecules. *Quarterly Reviews of Biophysics* 28 (2): 171–193.

Herzik, M.A., Wu, M., and Lander, G.C. (2019). High-resolution structure determination of sub-100 kDa complexes using conventional cryo-EM. *Nature Communications* 10 (1): 1–9.

Hill, T.L. (1960). An introduction to statistical thermodynamics. Reading, Massachusetts, Addison-Wesley.

Hilser, V.J. and Whitten, S.T. (2014). Using the COREX/BEST server to model the native-state ensemble. In: *Protein Dynamics* (ed. Livesay, D.), 255–269. Springer.

Hyre, D.E. and Klevit, R.E. (1998). A disorder-to-order transition coupled to DNA binding in the essential zinc-finger DNA-binding domain of yeast ADR1. *Journal of Molecular Biology* 279 (4): 929–943.

Ibrahim, M., Fransson, T., Chatterjee, R. et al. (2020). Untangling the sequence of events during the S2 → S3 transition in photosystem II and implications for the water oxidation mechanism. *Proceedings of the National Academy of Sciences* 117 (23): 12624–12635.

Jensen, S.C., Davis, K.M., Sullivan, B. et al. (2017). X-ray emission spectroscopy of biomimetic Mn coordination complexes. *The Journal of Physical Chemistry Letters* 8 (12): 2584–2589.

Jobe, A., Liu, Z., Gutierrez-Vargas, C. et al. (2019). New insights into ribosome structure and function. *Cold Spring Harbor Perspectives in Biology* 11 (1): a032615.

Kern, J., Hattne, J., Tran, R. et al. (2014). Methods development for diffraction and spectroscopy studies of metalloenzymes at X-ray free-electron lasers. *Philosophical Transactions of the Royal Society B: Biological Sciences* 369 (1647): 20130590.

Kinschel, D., Bacellar, C., Cannelli, O. et al. (2020). Femtosecond X-ray emission study of the spin cross-over dynamics in heme proteins. *Nature Communications* 11 (1): 4145.

Kok, B., Forbush, B., and McGloin, M. (1970). Cooperation of charges in photosynthetic O2 evolution–I. A linear four step mechanism. *Photochemistry and Photobiology* 11 (6): 457–475.

Kursula, P. (2014). The many structural faces of calmodulin: a multitasking molecular jackknife. *Amino Acids* 46 (10): 2295–2304.

Levantino, M., Lemke, H., Schirò, G. et al. (2015). Observing heme doming in myoglobin with femtosecond X-ray absorption spectroscopy. *Structural Dynamics* 2 (4): 041713.

Li, H. and Frieden, C. (2007). Comparison of C40/82A and P27A C40/82A barstar mutants using 19F NMR. *Biochemistry* 46 (14): 4337–4347.

Li, Y., Cash, J.N., Tesmer, J. (2020). High-throughput cryo-EM enabled by user-free preprocessing routines. *Structure* 28 (7): 858–869. e853.

Lipari, G. and Szabo, A. (1982a). Model-free approach to the interpretation of nuclear magnetic resonance relaxation in macromolecules. 1. theory and range of validity. *Journal of the American Chemical Society* 104 (17): 4546–4559.

Lipari, G. and Szabo, A. (1982b). Model free approach in the interpretation of nuclear magnetic resonance in macromolecules. II. Analysis of experimental results. *Journal of the American Chemical Society* 104: 4558–4570.

Lyumkis, D. (2019). Challenges and opportunities in cryo-EM single-particle analysis. *Journal of Biological Chemistry* 294 (13): 5181–5197.

Maji, S., Liao, H., Dashti, A. et al. (2020). Propagation of conformational coordinates across angular space in mapping the continuum of states from cryo-EM data by manifold embedding. *Journal of Chemical Information and Modeling* 60 (5): 2484–2491.

Marion, D. (2013). An introduction to biological NMR spectroscopy. *Molecular & Cellular Proteomics* 12 (11): 3006–3025.

Masson, G.R., Burke, J.E., Ahn, N.G. et al. (2019). Recommendations for performing, interpreting and reporting hydrogen deuterium exchange mass spectrometry (HDX-MS) experiments. *Nature Methods* 16 (7): 595–602.

Noble, A.J., Dandey, V.P., Wei, H. et al. (2018). Routine single particle cryoEM sample and grid characterization by tomography. *Elife* 7: e34257.

Ourmazd, A. (2019). Cryo-EM, XFELs and the structure conundrum in structural biology. *Nature Methods* 16 (10): 941–944.

Pais, A. (1982). *Subtle Is the Lord: The Science and the Life of Albert Einstein: The Science and the Life of Albert Einstein*. USA: Oxford University Press.

Palmer, A.G., III (2015). Enzyme dynamics from NMR spectroscopy. *Accounts of Chemical Research* 48 (2): 457–465.

Post, C.B. (2003). Exchange-transferred NOE spectroscopy and bound ligand structure determination. *Current Opinion in Structural Biology* 13 (5): 581–588.

Prabhu, N.V. and Sharp, K.A. (2005). Heat capacity in proteins. *Annual Review of Physical Chemistry* 56 (1): 521–548.

Prestegard, J.H., Bougault, C.M., and Kishore, A.I. (2004). Residual dipolar couplings in structure determination of biomolecules. *Chemical Reviews* 104 (8): 3519–3540.

Samoylova, L., Shu, D., Dong, X. et al. (2019). Design of hard x-ray self-seeding monochromator for European XFEL. In: *AIP Conference Proceedings* (2054). AIP Publishing LLC.

Sauter, N.K., Kern, J., Yano, J. et al. (2020). Towards the spatial resolution of metalloprotein charge states by detailed modeling of XFEL crystallographic diffraction. *Acta Crystallographica Section D-Structural Biology* 76: 176–192.

Sayre, D. (1952). Some implications of a theorem due to Shannon. *Acta Crystallographica* 5 (6): 843-843.

Scheres, S.H. (2010). Classification of structural heterogeneity by maximum-likelihood methods. *Methods Enzymol, Elsevier* 482: 295–320.

Scheres, S.H. (2012). RELION: implementation of a Bayesian approach to cryo-EM structure determination. *Journal of Structural Biology* 180 (3): 519–530.

Sekhar, A. and Kay, L.E. (2019). An NMR view of protein dynamics in health and disease. *Annual Review of Biophysics* 48: 297–319.

Sierra, R.G., Weierstall, U., Oberthuer, D. et al. (2018). Sample delivery techniques for serial crystallography. In: *X-ray Free Electron Lasers* (ed. P.F.S. Boutet and M.S. Hunter), 109–184. Springer.

Stabrin, M., Schoenfeld, F., Wagner, T. et al. (2020). TranSPHIRE: automated and feedback-optimized on-the-fly processing for cryo-EM. *Nature Communications* 11 (1): 1–14.

Suga, M., Akita, F., Hirata, K. et al. (2015). Native structure of photosystem II at 1.95 Å resolution viewed by femtosecond X-ray pulses. *Nature* 517 (7532): 99–103.

Vallurupalli, P., Bouvignies, G., and Kay, L.E. (2012). Studying "invisible" excited protein states in slow exchange with a major state conformation. *Journal of the American Chemical Society* 134 (19): 8148–8161.

Van Gunsteren, W.F., Daura, X., and Mark, A.E. (2002). Computation of free energy. *Helvetica Chimica Acta* 85 (10): 3113–3129.

Vinyard, D.J. and Brudvig, G.W. (2017). Progress toward a molecular mechanism of water oxidation in photosystem II. *Annual Review of Physical Chemistry* 68: 101–116.

Voss, J.M., Harder, O.F., Olshin, P.K. et al. (2021). Rapid melting and revitrification as an approach to microsecond time-resolved cryo-electron microscopy. *Chemical Physics Letters* 778: 138812.

Williams, D.H., Stephens, E., and Zhou, M. (2003). Ligand binding energy and catalytic efficiency from improved packing within receptors and enzymes. *Journal of Molecular Biology* 329 (2): 389–399.

Yip, K.M., Fischer, N., Paknia, E. et al. (2020). Atomic-resolution protein structure determination by cryo-EM. *Nature* 587 (7832): 157–161.

Zhu, D., Cammarata, M., Feldkamp, J. et al. (2013). Design and operation of a hard x-ray transmissive single-shot spectrometer at LCLS. *Journal of Physics: Conference Series* 425 (5): 052033.

Zivanov, J., Nakane, T., Forsberg, B.O. et al. (2018). New tools for automated high-resolution cryo-EM structure determination in RELION-3. *Elife* 7: e42166.

9

Looking Forward

9.1 Overview: Unraveling Function and Mechanism

As currently used, the phrase *Structural Biology* in our title means that it is concerned with spatially compact systems such as single proteins, stable assemblies, and molecular machines, at the molecular level. The phrase has often implicitly excluded dynamics. In the realm of such compact systems, function and mechanism are straightforward concepts. Function is what the system does, and mechanism is how it does it (Chapter 1). A phrase such as "XYZ is a serine protease" tells us that the function of XYZ is to break peptide bonds, and that its mechanism involves a buried serine with an anomalous pK_a. Yet such a shorthand description of function and mechanism is an example of the reductionism that has taken hold in many areas of biology. A major goal of many fields is to undo this reductionism intelligently, to understand function and mechanism at many levels of integration while fully incorporating the understanding and predictive capacity embodied in molecular structures.

We have discussed how functional trajectories, ELA analysis, and other approaches can help to undo one aspect of this reductionism, the projection of the ensemble of protein motions onto a static PDB file. But this still represents a tightly molecular view. Drugs illustrate well the ramifications of the idea of function. Drugs must interact with their desired targets, but also avoid interactions with other undesired targets. Function includes specificity. It seems to us a distinct possibility that AI/ML approaches (imagine generalizations of AlphaFold2) will allow the prediction of protein–protein or protein–drug interactions, even weak ones, on a genome-wide basis. Such a capability would, for example, act as a hypothesis-generating engine, translating genome-wide association studies (GWAS) into functional networks, or making molecular sense out of observed pleotropic effects.

Meanwhile, the field of synthetic biology rapidly advances towards the understanding of relation between global phenotype and genome sequence, especially in the case of partially or fully synthetic organisms (Riolo and Steckl 2022). One may fantasize a "killer app" in which the response of a cell to a chosen environmental change is monitored at levels ranging from the organismal to the molecular, in dynamic detail.

This chapter briefly identifies some future directions that we see as promising. Without a doubt these are speculation; readers will choose their own directions. This presentation itself lies on the trajectory of structural dynamics as a critical component of *Structural Biology*.

Dynamics and Kinetics in Structural Biology: Unravelling Function Through Time-Resolved Structural Analysis, First Edition. Keith Moffat and Eaton E. Lattman.

9.2   Single Particle Imaging, Energy Landscape Analysis, and Functional Trajectories

In Chapter 8.3 we discussed the use of ELA based on single-particle images as a potentially powerful tool for visualizing and analyzing dynamics. ELA for protein dynamics is in a rapidly developing phase, and questions remain about its applicability and generality. Almost all ELA analyses are based on cryo-EM images, and the highest effective global resolution of which we are aware is 4.5 Å (Dashti et al. 2020), limited by the number of snapshots. Individual cryo-EM images now achieve a resolution just better than 2.0 Å. The fundamental limitation of cryo-EM for dynamics lies in the "cryo" prefix. Frozen particles are inactive and incapable of exhibiting structural dynamics. Approaches based on freeze-trapping are limited at present to the ms time range by the time taken to freeze the entire sample including its mount and surrounding liquid. In contrast, single-particle X-ray images obtained from scattering of fs XFEL pulses can examine active samples at near room temperature. We see the retention of activity as a potential advantage over the scattering of electrons in cryo-EM. A roadmap for the development of X-ray-based single particle methods, their promises and difficulties has been critically discussed by several groups (Ayyer et al. 2019; Chapman 2019; Bielecki et al. 2020). In a prototypical experiment, unfrozen nanodroplets containing an active single particle are injected into the XFEL chamber and (sometimes) are intersected by an XFEL pulse. Particles are thus freed from the constraints of lattice contacts and of freezing. The "diffraction before destruction" principle operates to reduce radiation damage. The resultant scattering pattern is a 2D Ewald section through the continuous X-ray diffraction pattern of the particle, heavily contaminated by scattering from the solvent in the droplet and from other sources in the beamline. However, the high pulse rate of the EuXFEL (Chapter 6.3) enables as many as 10^7 individual X-ray scattering patterns to be collected in a typical user shift. These patterns can be combined to create a complete 3D scattering pattern from the particle, using methods based on those used in assembling cryo-EM images. The phase problem for continuous intensity distributions is more straightforward than for the Bragg patterns from crystals. Phasing algorithms are reviewed by (Poudyal et al. 2020), who also discuss how many snapshots are needed to achieve a chosen resolution. Issues of sample delivery, high repetition-rate data collection, and others remain. Yet we are optimistic about the ultimate power of the method, and of the desirability of studying particle dynamics in active samples.

9.3   Artificial Intelligence and Machine Learning

9.3.1   AI/ML and Physical Chemistry Approaches to Folding and Structure

The highly successful AI/ML approach to the accurate prediction of protein folds differs radically in the input information from the long explored – but apparently less successful – approaches based on physical chemistry. The AlphaFold2 system (Jumper et al. 2021; Varadi et al. 2022), closely followed by the RosettaFold program of the Baker group at the University of Washington (Baek et al. 2021), assesses information from AI/ML algorithms and strategies, and inputs data from large-scale, multiple sequence alignments (MSAs).

A limited number of experimental structures, largely from the PDB, train the system. Foundational chemical principles such as the covalent, linear connectivity of a polypeptide chain and its self-avoiding fold are studiously avoided and only introduced in the final stages of the prediction. In marked contrast, physical chemistry approaches assess information from numerous experimental structures related at some level in sequence to the target sequence, the folding environment, and the laws of physical chemistry contained in potential functions that govern interatomic interactions. These give rise to constraints such as the absence of inter-penetrating atoms and acceptable backbone torsion angles in the polypeptide backbone evident in Ramachandran plots.

Proteins in nature fold, briskly and successfully. Folding begins as the mRNA is translated and protein is synthesized on the ribosome. Folding depends only on the ribosomal environment, the local, immediately preceding amino acid sequence and the local, succeeding mRNA sequence. The overall time to fold is usually not much longer than the time to complete translation. Certain native sequences require protein chaperones to fold within a specified time and avoid degradation. Mutant sequences that do not fold under any experimental circumstances are promptly degraded. To state the obvious, the principles of physical chemistry are *explicit* throughout the natural folding process.

It is hard to conceive of two more radically different information sources than the inputs of AI/ML to AlphaFold2 and of physical chemistry to natural folding. Nevertheless, both information sources largely ensure folding. How does this come about?

The MSA-based approach embodies the belief that homologous sequences differing by mutation have folded into homologous structures. Since all members of the MSA have folded in accord with physical chemistry, the principles of physical chemistry are *implicit*. The proteins represented in the MSAs have survived to the present day over evolutionary time precisely because they are fit: they exhibit fitness. Evolutionary biology indicates that fitness does not depend explicitly on structure but rather, on mechanism and function. Three-dimensional structure may be thought of as a means of expressing mechanism and function, which is then judged to be fit or unfit.

Fitness requires first, the ability of a protein to fold to a static structure that resists degradation; and second, the ability to execute a biological function through the coordinated series of dynamic changes in structure that constitute a mechanism. Mutant sequences that cannot fold, or if after folding have seriously compromised function, will be judged unfit over evolutionary time. They will not be present in today's MSAs. The interesting class of intrinsically unfolded proteins have functions tailored to this characteristic (Nishikawa 2009), and represent a notable exception to this point. From our viewpoint of structural dynamics, the second aspect of fitness is of much greater importance: all proteins in the MSAs are active. They retain their biological function and at least qualitatively, the chemical mechanism that gives rise to that function.

9.3.2 AI/ML and Structural Dynamics

At present, the output of both AlphaFold2 and Rosetta AI/ML prediction programs is static: a single set of time-independent atomic coordinates presents the best prediction (or the mean of a small set of best predictions). Restriction to a single set has been presented as a major limitation (Lane 2023). We emphatically agree and predict that dynamics will come to the fore. AI/ML approaches will play dramatic new roles in extending predictions to structural dynamics and hence directly to mechanism. We set as a goal the ability to predict and understand the range of structures accessible to a single protein molecule; to separate purposeful motions along a functional trajectory between states from purely random, non-functional motions; and to identify and predict the structures of short-lived intermediates along a reaction coordinate.

Do continuous, predicted structural distributions represent a mapping of the energy landscape? Can Alphafold2 be further extended to predict not just structural distributions, but the rates of interconversion among the structures? We predict that the answer to the first question is "yes," and suggest that the extension proposed in the second question would be highly desirable.

As we have just seen, the major information input to Alphafold2 in MSAs harnesses the results of nature's experiment: evolution based on selection for fitness. This viewpoint is critical. Since the MSAs incorporate evolutionary fitness, they contain *implicit* information on folding, mechanism, and function. As we have emphasized throughout, mechanism and function in turn depend on both static structure and on the dynamic changes in structure essential to mechanism. Sequences naturally evolve by random mutation; some variants turn out to be fit and survive because they retain folding, function, and mechanism. That is, MSAs contain information on both static structure and on those dynamic structures associated with mechanism.

This leads us to ask: are today's predictions of structure by AlphaFold2 in fact purely static? Do they implicitly contain dynamic information? Preliminary answers to these two questions appear to be "no" and "yes." Modification of the input MSAs and training structures to AlphaFold 2 can emphasize and direct multistate modeling. For example, modification generates predictions that distinguish between active and inactive states of G-protein coupled receptors and transporters (Del Alamo et al. 2022; Heo and Feig 2022) and different structures of the activation loop in the Abl tyrosine kinase (Lane 2023).

This promising beginning suggests a third question: How might dynamic information appear in AlphaFold predictions? Consider how the accuracy of structure predictions is estimated. In the challenging 14th Critical Assessment of Protein Structure Prediction (CASP14), accuracy was judged by comparison of the predicted structure against a blind set of known – but carefully undisclosed – experimental atomic structures, largely obtained by cryocrystallography. Differences between the predicted and experimental structures proved markedly smaller for Alphafold2 than for all other prediction programs competing in CASP14 (Jumper et al. 2021). That is, AlphaFold2 provided more accurate predictions.

Naturally, most emphasis has been placed on the accurate Alphafold2 predictions, evident in these small differences. Larger differences between predicted and experimental structures were flagged in CASP14 as "inaccuracies." However, larger differences could arise from one, two, or all of three causes. First, there are limitations of the Alphafold2 prediction approach – a genuine inaccuracy. Second, the experimental structures could be perturbed by their determination at cryo temperatures ~100 K, where molecules are no longer dynamic and completely lack activity – a confounding feature. Third, the predicted structure could inherently contain two or more component structures – a hitherto unsuspected dynamic contribution. If the third cause contributes significantly to the larger differences, the "inaccuracies" are only apparent. They would in fact be representations of reality and most interesting for our argument: Alphafold2 predictions contain both static and dynamic components.

We suggest that "inaccurate" structures should be examined more closely for dynamic features. These may be evident, for example, around the active site, in surface loops as in the Abl kinase prediction (Lane 2023) or near chain termini, or in motion of a domain as a rigid body. Some of the less accurate predictions for otherwise accurate structures – prediction outliers – should also be examined. For example, where AlphaFold2 predictions are made of a protein that undergoes a dynamic, monomer–dimer equilibrium, individual predictions may reveal monomer or dimer components, with slightly different tertiary structures for the monomer in each. Alternatively, the dynamic equilibrium may be intramolecular, among a number of components differing slightly in tertiary structure – a structural distribution. Predictions made in the absence or presence of a cofactor or inhibitor may distinctly shift this structural distribution.

More practically, it may be feasible in future CASP competitions to use experimental structures determined at near-room temperature where molecules are demonstrably active, or where crystal structures are available in space groups with different intermolecular contacts and solvent conditions, or where structures of inhibitor or intermediate complexes are available.

If despite such efforts few authentic, dynamic components can be identified in today's "static" Alphafold2 predictions, then a large base of dynamic, short-lived, intermediate structures will have to be developed to be used as an AI/ML training set. The base could be experimental, from X-ray or EM experiments on active samples at near-room temperature. This is a tall order. However, the challenge may be met with EM and crystallography of *active* samples, and with SPI by X-ray approaches (Chapter 9.2). The base could also be computational, derived from long-time MD simulations founded on X-ray structures.

9.3.3 AI/ML and De Novo Protein Design

De novo protein design requires a qualitatively different energy landscape for "folders," "binders," and "actives," differing in complexity. A successful folder design will have a single energy well in the energy landscape; its folding funnel may be deep. A successful binder design (for example, a kinase enzyme intended to specifically bind a candidate drug) will have two energy wells, associated with the free and bound states. The wells will be adjacent in structural space but distinct. Importantly, they are unlikely to be very deep and must be separated by only a modest energy barrier, to ensure rapid dynamic transitions between them. A successful active design will have a more complicated energy landscape: a connected series of wells, each arising from an intermediate on the reaction coordinate. Each is separated from its flanking neighbors by modest energy barriers which will again ensure rapid transitions between them. AI/ML approaches to de novo protein design will have to take the principles governing the energy landscape into account. A recent example of a binder design is Praetorius et al. 2023.

The energy landscape for design of photoreceptors is quite different. The design of a binder for a chromophore can be similar to that above, though the energy well for the apo form need not be deep and that for the chromophore-bound form may be deep. The differences arise from the large energy available on absorbing a visible photon (compared with the net energy from binding a small molecule non-covalently), which drives the energy to a mountaintop. The reactions between intermediates in a photocycle are all substantially downhill from the mountaintop.

9.4 Experimental Approaches

9.4.1 Time-resolved Electron Microscopy of Active Samples

Structural biology of proteins, nucleic acids, and their complexes has been transformed by atomic-level structure determination by cryo-EM, largely conducted on single, isolated particles. However, there are two limitations: the particles, surrounding solvent, and the EM grid are frozen to ~100 K by plunging into a cryogen before introducing them to the microscope; and as a consequence of freezing, the particles are completely inactive. However, an advantage is that radiation damage of single particles by an electron beam is substantially reduced from that in a hard X-ray beam (Henderson 1995).

Prior to the introduction of cryo-EM, electron microscopy of biological samples was conducted on samples at near-room temperature. Protein samples were encased in negative stain containing heavy metals. This greatly increased the imaging contrast and minimized the effects of radiation damage, but rendered the proteins inactive and limited the resolution to a few nm.

We ask: can room temperature EM of active, unstained, single particles at near-atomic resolution be developed? If so, can extension from initial, static measurements to dynamic measurements on light-activated systems could be envisaged? Development would encounter problems similar to those now successfully attacked at XFELs in X-ray-based SPI: introducing single particles to the electron beam; addressing radiation damage; degradation of the high vacuum, the electron beam and the scattering background by the sample and its aqueous environment; and the accurate imaging of particles at different locations and depths as they traverse the electron beam.

9.4.2 The Compact X-ray Free Electron Laser Source

We've emphasized the transformative role of XFEL sources in the study of dynamics at many points. Evolution of such sources continues to open new and exciting experimental options. For example, Chapter 6.3 presents the MHz pulse rate available at the European XFEL. For the most part, new and proposed XFEL sources are gigantic facilities with construction and commissioning costs not far from $1 billion. This limits the number of XFELs that can be built worldwide, the number of users per year that can conduct their experiments, and the classes of experiments that can be carried out.

As a radical alternative to this very large-scale model, a team at Arizona State University is constructing a compact X-ray laser source (CXFEL) that will bring exciting new capabilities to bear on dynamical experiments across a wide range of disciplines (Graves et al. 2020). The physical extent of the entire source, including the accelerator, is only about 10 meters – basement-sized. X-rays are generated by inverse Compton scattering, where a patterned beam of electrons is scattered from the coherent electromagnetic field of a powerful optical laser, which plays the role of the permanent magnet array in an undulator in conventional facilities. The short period of this optical lattice requires an operating voltage in the MeV range and the electron accelerator is correspondingly compact. The CXFEL achieves high coherence and lasing using the technique of emittance exchange in which the electron beam is diffracted into a series of spatial orders by a silicon grating, and these orders are swapped into a series of microbunches in the time domain by sophisticated electron optics. Pulses as short as 100 as are anticipated. The CXFEL will operate at 1 kHz pulse frequency and unlike SASE sources based on the amplification of noise, will provide the exacting pulse-to-pulse reproducibility particularly well suited to dynamic experiments. Of course, the total energy per pulse will be only a small fraction of that at convectional XFELs. As pointed out by John Spence (personal communication), the uncertainty principle ensures that as pulses naturally possess a wider X-ray bandwidth, better suited to obtaining integrated intensities by Laue diffraction.

A preliminary, non-lasing version of the source has been completed. The lasing (CXFEL) version enters its second phase of construction in 2023. Its full capabilities will not be apparent until it is commissioned and used for experiments. We cannot resist mentioning one tremendously exciting application: two-color core-hole spectroscopy. CXFEL's as pulse duration outruns the primary X-ray photoionization damage processes, thereby allowing the visualization of oxidation state changes of important metal centers in biology. In this core-hole application, a core electron is ejected by a sub-fs X-ray pulse and the molecule is probed by a second X-ray pulse of different energy, which can monitor metal–metal charge transfer or other electronic processes (Serkez et al. 2020). CXFEL potentially allows the interval between pump and probe pulses to be less than 1 fs. Some argue that such ultrashort events are the province of chemistry rather than of biology. Yet highly evolved metal cluster sites enable sophisticated biological processes. What could be more biological than photosynthesis or vision?

9.4.3 Weak fs Pulses

XFEL experiments to date use an X-ray beam that is only modestly attenuated. Although the radiation damage from a single pulse is sufficient to ultimately destroy the sample, diffraction before destruction (Chapter 4.3.3) nevertheless allows a single diffraction or scattering pattern of good quality to be obtained prior to destruction, in "pump once, probe once" experiments. Acquiring a complete set of crystal diffraction data is both extremely demanding of sample amount and subject to errors arising from variation between crystals.

We ask: can XFEL sources be sufficiently attenuated to minimize radiation damage and permit a "pump once, probe several" experiment? The attenuation would have to be substantial – a factor of 100x or more by comparison with today's modestly attenuated beams – to sufficiently lower radiation damage and allow time points up to (say) 1 ns. The consequent weakness in the scattered intensities would be compensated by a larger sample of 100x higher scattering power, although each sample would be more effectively used.

Two large technological challenges would have to be overcome. The probe pulses would have to be efficiently distributed in time e.g. uniformly in log t; and the 2D detector would have to discriminate in time between photons arriving from each probe pulse.

The distribution of probe pulses in time might be obtained by modulating the temporal patterning of the new CXFEL source (Chapter 9.4.2). This source has the further advantages of being intrinsically weak, likely to require no (or at most modest) attenuation, and offering pulses in the as to few fs time range. Finally, the data would also contain time-dependent radiation damage, which analysis would have to distinguish from the desired time course of structural changes associated with the mechanism.

9.4.4 Temporal Noise and Chirped X-ray Pulses

The time resolution of ultrafast crystallographic experiments on isomerization of the chromophore of PYP has been greatly enhanced by exploiting the jitter in the pump–probe time delay and the spikiness of the intensity distribution inherent in the XFEL pulses generated by the SASE process (Chapter 6). The time resolution achieved is much less than the duration of the probe X-ray pulse.

This raises a provocative question: can the time resolution of SR-based pump–probe experiments be reduced far below the duration of the X-ray SR pulse of ~100 ps, for example by introducing time jitter or spikiness? Discussions with Abbas Ourmazd suggest that the addition of time jitter to SR pulses would not be effective. Exploitation of the time jitter in XFEL pulses is derived from the quantum coherence of the structural changes associated with isomerization. All the statistically large number of molecules in a crystal behave synchronously, as though they were in effect a single gigantic molecule. The coherent, synchronous behavior persists through the fs range. However, the coherence has decayed by about 1 ps, a time much shorter than the 100 ps duration of SR pulses.

However, the time resolution achievable could be improved if energy-chirped or spatially-chirped X-ray pulses could be obtained (Moffat 2003) . The time resolution is then established by the nature of the chirp, rather than by the total duration of the pump or probe pulses and the jitter or drift in the time delay between them. For energy-chirped pulses, the radius and center of the Ewald sphere varies in time with energy/wavelength, across the duration of the pulse. The scattering angle of all Laue spots also varies with time; each spot moves radially to a larger scattering angle as the energy decreases.

It remains to be established whether energy-chirped pulses can be obtained from SR sources or the CXFEL. If so, what rates of energy variation (in units of, say, eV/ps) can be achieved? Are the maximum rates sufficient to generate measurable energy variation across a spot over the duration of the pulse? Is the loss of time resolution arising from the radial dimensions of the spot a serious limitation? Only an integrating detector is adequate; no sensitivity to X-ray energy is required.

Chirped pulse amplification is being studied as a means of further enhancing the total pulse energy at XFEL sources (Li et al. 2022), and the technology may be adaptable to the generation of X-ray pulses chirped in energy.

9.4.5 New Light-dependent Systems

Light is an experimentally convenient means of reaction initiation but the number of naturally light-dependent systems is small (Chapter 3.3.2.3). More natural systems may be discovered, and sensitivity to light may be conferred on normally light-independent systems by optogenetic and photopharmacological approaches.

It is striking that all natural signaling photosensors known to date are proteins (Moffat et al. 2013). There is no obvious chemical reason why this is so. We ask: do RNA-based signaling photoreceptors exist? If so, in what systems might they be located? A signaling photoreceptor must satisfy two prime requirements: it can absorb a photon in the visible; and undergo a structural change that alters the intramolecular affinity of the photoreceptor domain for another domain, or the intermolecular affinity for a downstream binding partner. Although RNA bases absorb in the near-UV, no posttranscriptionally modified bases have been discovered (yet) whose absorption has been significantly red-shifted to the blue region of the visible. The shift would require adding a substituent such as a vinyl group to increase the extent of electron delocalization; chemically plausible. Modification will increase its bulk and alter hydrogen-bonding capability. A more straightforward route is to seek a natural RNA that has acquired the ability to bind a chromophore non-covalently. FMN is an obvious chromophore candidate: readily bioavailable, absorbing in the blue, sensitive both to light and to redox state through alteration of its hydrogen bonding capabilities, and with many protein-based analogs that are authentic signaling photoreceptors such as LOV proteins. Protein-based flavoenzymes are numerous and more promisingly, FMN-dependent RNA riboswitches in bacteria regulate the biosynthesis and transport of riboflavin and related compounds (Serganov et al. 2009; Crielaard et al. 2022). Curiously, the light- or redox-dependence of the affinity of the riboswitch for FMN has not been emphasized.

Regulation of RNA modification or metabolism offers a natural system in which an RNA-based signaling photoreceptor might exist. It would be interesting to conduct a systematic search.

In the photopharmacology area, design, synthesis, and characterization studies are needed to extend the number of compounds which can be caged or uncaged by light-dependent isomerization or electron transfer. Both reactions are very rapid and lack side products, two major advantages. If widely available cofactors or reactants such as NAD, NADP, ATP, or cAMP could be uncaged, the range of reactions to which they could be applied would be greatly increased. The caged compounds are likely to be used in biophysical studies at mM concentrations on samples of tens of μm path length. This suggests that the absorption coefficient of their reactant, ground state be low. A further advantage is present if the absorption coefficient of the first excited state is even lower, to minimize two photon absorption (Chapter 3.2.3.1). In contrast, the dark reactions after initial covalent bond rupture of a caged compound such as the classical example of caged ATP are relatively slow and inherently irreversible. The dark reactions also involve intermediates such as carbenes and nitrenes whose high reactivity makes them prone to side reactions.

9.4.6 Advantage in Dynamic Crystallography of Multiple Copies in the Asymmetric Unit

As noted in Chapter 4.1, crystals often contain more than one molecule per asymmetric unit: the crystal possesses non-crystallographic symmetry. The molecules in the asymmetric unit differ in intermolecular contacts and in the constraints imposed by these contacts.

A consequence for dynamic experiments is that certain molecules in the asymmetric unit may be inactive. Other molecules may be active and proceed down the reaction coordinate, but with differing rate coefficients. That is, non-crystallographic symmetry acquires a temporal component. A diffraction pattern at a single physical time point then contains data from different reaction time points, one per molecule. The more molecules in the asymmetric unit, the more time points are simultaneously sampled. When an entire data set is acquired on such crystals, systematic errors in the time domain will be minimized and the dynamic structural changes will be more accurately determined. The increase in accuracy will be most evident in those time ranges when individual molecules in the asymmetric unit have progressed to different points along the reaction coordinate, and populate different intermediates. The advantages do not appear to have been explored.

9.4.7 Exploiting Correlations in DED Maps

A time-dependent set of DED maps covering much of the course of a reaction contains a series of structures, from reactant through intermediates to product (and back to reactant if a cycle is involved). Assume that the chemical kinetic mechanism is unknown at this stage. The DED map at a single time point may contain one or more structures. If adjacent maps are close in time, most (or all) of the same structures are likely to be present in these maps, though with slightly different populations and fractional occupancies. That is, there is likely to be a high correlation in DED features between temporally adjacent DED maps, but a progressively diminishing correlation as the temporal distance between the maps increases. If the correlation is constant across (say) four contiguous maps, then the existence of a stationary state spanning these four time points is strongly indicated. Conversely, if the correlation varies sharply with time, this suggests the breakdown of one intermediate and the formation of the next intermediate during these four time points. In other words, the time dependence of the correlation coefficients across the entire time series gives information about the general nature of the mechanism, in the absence of any details. Temporal correlations between entire DED maps remain to be fully exploited (Wickstrand et al. 2020).

Spatial correlation between DED features is more straightforward. For example, suppose the displacement of an α-helix as a rigid body spans several time points, which gives rise to an array of positive and negative DED features along the backbone of the helix and some of its side chains. Any one feature may be statistically insignificant i.e. its magnitude is less than 3σ where σ is the rms value of the DED across the asymmetric unit. However, when spatial correlations among the DED features are considered, the displacement of the helix between intermediates is strongly indicated. By considering the DED features collectively, the magnitude and direction of the displacement can be accurately estimated . The existence and nature of spatial correlations ultimately assist in refinement of the time-independent structures of the underlying intermediates.

A related form of combined temporal and spatial correlations exists. For example, suppose that some of the residues in a domain (a sub-domain) move as a rigid body. The DED features associated with each residue in this rigid body will display the same time course. Although the magnitude of the features varies spatially from residue to residue in the sub-domain, their time courses will be spatially correlated. The time course of DED features associated with the other residues in the domain will not be correlated. The classification into correlated and non-correlated regions defines the boundary between moving and non-moving regions of the domain. Initial studies of model DED maps confirmed the principle (Kostov and Moffat 2011) but the study has not been extended to experimental DED maps.

9.5 Evolutionary Relevance of Trajectories

To end, we step back and take a broad evolutionary perspective on kinetics and dynamics. The question arises: at the molecular level, is evolutionary selection applicable to both dynamics and kinetics, or only to kinetics?

As we have seen (Chapter 2), chemical kinetics considers states and the transitions between them. The properties of states do not depend on the functional trajectory (more casually, the pathway) by which one state changes to another. That is, the "history" of the trajectory is not evident in the final state. In marked contrast, dynamics considers both the states themselves and the functional trajectory between them. From the viewpoint of energy landscape analysis (Chapter 9.2), kinetics deals entirely with a very small number of selected points within the energy landscape. Points situated at local minima of free energy – the valley bottoms in the landscape – correspond to the stable or metastable structures of the reactant, each intermediate, and product. Other points situated at the saddle points or mountain passes in energy – at the lowest maxima of the large number of possible pathways between adjacent states – correspond to transition states. We cautioned earlier that unlike the states associated with the local minima, the so-called "transition states" are not authentic states in the strict thermodynamic sense. Each transition state is assumed to be in equilibrium with the two states that flank it, one preceding and the other following it in the overall mechanism, a concept that has proved very useful.

However, dynamics deals with all points in the energy landscape. It is not restricted to the very small subset of points associated with the valley bottoms and the saddle points, but also considers the points describing all possible functional trajectories between these points without consideration of their energy. For example, certain trajectories between valley bottoms may traverse mountain peaks of very high energy, or be detoured into side valleys – box canyons – from which exit is challenging. These trajectories too are relevant. A point on the landscape on the side of a mountain, far from normal functional trajectories, may nevertheless correspond structurally to the binding site for a novel drug. This point offers a "druggable target" of clear biomedical relevance (Chapter 8.3.3).

Each state is associated with a 3D structure or more strictly, a family of structures within a few kT of each other. In kinetics, two rate coefficients (forward and reverse) are associated with every transition between adjacent states. Rate coefficients describe the increase or decrease of the populations of statistically very large numbers of molecules. Individual molecules within the population adopt different functional trajectories that require different times to pass from one state to the next. That is, they have different transit times. States and their properties such as 3D structure and the magnitude of rate coefficients associated with their interconversion clearly determine intracellular processes and how cells respond to their environment. States and rates appear to be integral to mechanism and function.

As we noted above (Chapter 9.3.1), selection during evolution evaluates fitness of the organism as a whole. Selection influences states, their structures, and the rate coefficients of individual reactions, but only if they contribute to fitness of the organism. It seems clear that selection can be – and has been – applied to kinetics, and that the structures of intermediates and the rate coefficients with which they interconvert are highly relevant to mechanism and function. What is not so clear is whether dynamics – the intimate details of functional trajectories – are also influenced by selection. In short, does the entire energy landscape contribute to, or even control, fitness? Or, do only specific points on that landscape contribute?

For example, suppose a reaction clearly contributes to the fitness of the organism, and a mutation introduces a modest energetic bump into the functional trajectory of that reaction. The conventional view is that this mutation will have no effect on the overall rate of the reaction unless the difference in energy between the reactant and the transition state is altered. In the absence of such an alteration, fitness is unchanged. If however the energetic bump is near the transition state, that state may itself be affected (or even replaced by a new transition state) and fitness may indeed be changed.

We are inclined to accept the conventional view: a kinetic view of mechanism is highly informative and may be sufficient. Experimental efforts to determine dynamics and functional trajectories in active, single molecules may be most productive when they attack the limitations of kinetics. For example, they could be applied to fast reactions that follow slow reactions, in which populations of the intermediate do not accumulate and the structure of the intermediate is invisible to kinetics. However, all molecules pass through this intermediate and with sufficient time resolution, its structure and properties can be determined in a dynamic experiment.

References

Ayyer, K., Morgan, A.J., Aquila, A. et al. (2019). Low-signal limit of X-ray single particle diffractive imaging. *Optics Express* 27 (26): 37816–37833.

Baek, M., DiMaio, F., Anishchenko, I. et al. (2021). Accurate prediction of protein structures and interactions using a three-track neural network. *Science* 373 (6557): 871–876.

Bielecki, J., Maia, F.R., and Mancuso, A.P. (2020). Perspectives on single particle imaging with X rays at the advent of high repetition rate X-ray free electron laser sources. *Structural Dynamics* 7 (4): 040901.

Chapman, H.N. (2019). X-ray free-electron lasers for the structure and dynamics of macromolecules. *Annual Review Biochemistry* 88 (1): 35–58.

Crielaard, S., Maassen, R., Vosman, T. et al. (2022). Affinity-based profiling of the flavin mononucleotide riboswitch. *Journal of the American Chemical Society* 144 (23): 10462–10470.

Dashti, A., Mashayekhi, G., Shekhar, M. et al. (2020). Retrieving functional pathways of biomolecules from single-particle snapshots. *Nature Communications* 11 (1): 1–14.

Del Alamo, D., Sala, D., Mchaourab, H.S. et al. (2022). Sampling alternative conformational states of transporters and receptors with alphafold2. *Elife* 11: e75751.

Graves, W., Fromme, P., Holl, M. et al. (2020). The ASU compact XFEL project. *Bulletin of the American Physical Society* 65: Abstract.

Henderson, R. (1995). The potential and limitations of neutrons, electrons and X-Rays for atomic-resolution microscopy of unstained biological molecules. *Quarterly Reviews of Biophysics* 28 (2): 171–193.

Heo, L. and Feig, M. (2022). Multi-state modeling of G-protein coupled receptors at experimental accuracy. *Proteins: Structure, Function, and Bioinformatics* 90 (11): 1873–1885.

Jumper, J., Evans, R., Pritzel, A. et al. (2021). Highly accurate protein structure prediction with AlphaFold. *Nature* 596 (7873): 583–589.

Kostov, K. and Moffat, K. (2011). Cluster analysis of time-dependent crystallographic data: direct identification of time-independent structural intermediates. *Biophysical Journal* 100: 440–449.

Lane, T.J. (2023). Protein structure prediction has reached the single-structure frontier. *Nature Methods* 20: 170–173.

Li, H., MacArthur, J., Littleton, S. et al. (2022). Femtosecond-terawatt hard X-ray pulse generation with chirped pulse amplification on a free electron laser. *Physical Review Letters* 129 (21): 213901.

Moffat, K. (2003). The frontiers of time-resolved macromolecular crystallography: movies and chirped X-ray pulses. *Faraday Discussions* 122: 65–77.

Moffat, K., Zhang, F., Hahn, K. et al. (2013). The biophysics and engineering of signaling photoreceptors. In: *Optogenetics* (ed. P. Hegemann and S. Sigrist), 7–22. Berlin: De Gruyter.

Nishikawa, K. (2009). Natively unfolded proteins: an overview. *Biophysics* 5: 53–58.

Poudyal, I., Schmidt, M., and Schwander, P. (2020). Single-particle imaging by X-ray free-electron lasers—how many snapshots are needed? *Structural Dynamics* 7 (2): 024102.

Praetorius, F., Leung, P.J.Y., Tessmer, M.H. et al. (2023). Design of stimulus-responsive two-state hinge proteins. *Science* 381 (6659): 754–760.

Riolo, J. and Steckl, A.J. (2022). Comparative analysis of genome code complexity and manufacturability with engineering benchmarks. *Scientific Reports* 12 (1): 1–9.

Serganov, A., Huang, L., and Patel, D.J. (2009). Coenzyme recognition and gene regulation by a flavin mononucleotide riboswitch. *Nature* 458 (7235): 233–237.

Serkez, S., Decking, W., Froehlich, L. et al. (2020). Opportunities for two-color experiments in the soft X-ray regime at the European XFEL. *Applied Sciences* 10 (8): 2728.

Varadi, M., Anyango, S., Deshpande, M. et al. (2022). AlphaFold protein structure database: massively expanding the structural coverage of protein-sequence space with high-accuracy models. *Nucleic Acids Research* 50 (D1): D439–D444.

Wickstrand, C., Katona, G., Nakane, T. et al. (2020). A tool for visualizing protein motions in time-resolved crystallography. *Structural Dynamics* 7 (2): 024701.

Appendix A

Review of Crystallography

A TR crystallography experiment is underlain by a set of conventional crystallography experiments carried out at a series of time points. This appendix provides a review of the elements of crystallography that presumes some previous exposure to the material. There are many excellent introductions to crystallography, ranging from short books (Lattman and Loll 2008; Rhodes 2010) to monographs (Drenth 2007) and tomes (Blundell and Johnson 1976; Rupp 2010). A wonderfully clear, very brief introduction appears in (Eisenberg and Crothers 1979), Chapter 17.

A.1 The Electron Density Function

In 1822 Fourier published *The Analytical Theory of Heat* in which he elaborates his transformative idea that any well-behaved periodic function can be expanded in a convergent series of sines and/or cosines (Baron Fourier 2003). Structural crystallography is based on the use of Fourier series to represent the atomic structure of a crystal in terms of the 3D distribution of electrons within it. A standard form of this equation in crystallography is

$$\rho(\boldsymbol{r}) = \frac{1}{V} \sum_{\boldsymbol{h}} F(\boldsymbol{h}) cos(2\pi \boldsymbol{hr} - \alpha(\boldsymbol{h})). \tag{A.1}$$

Here $\rho(\boldsymbol{r})$ is the density of electrons per Å^3 and represents the image. V is the volume of the crystal unit cell, the box that is repeated in all three directions to create the periodic structure. Each term in this series is termed a *Fourier component*. The vector $\boldsymbol{h}$ represents a triple of integers $h\,k\,l$ that indexes the cosine terms and ensures that the periodicity of each cosine agrees with that of the crystal lattice. The components of the vector $\boldsymbol{r}$ are the spatial coordinates expressed in fractions of the edges of the unit cell. Thus, $\boldsymbol{hr} = h\frac{r}{a} + k\frac{s}{b} + l\frac{t}{c}$, where a, b, c are the lengths of the edges of the unit cell, and $r\,s\,t$ are the linear distances along these axes. See Figure A.3a. $F(\boldsymbol{h})$ is the weight for each cosine term in the series, and we shall see below that the intensity of each spot in an X-ray diffraction pattern is proportional to $F(\boldsymbol{h})^2$. The quantity $\alpha(\boldsymbol{h})$ defines the position of the origin of the cosine wave and is called the phase. Finding the phase is a critical step in the determination of the structure,

Dynamics and Kinetics in Structural Biology: Unravelling Function Through Time-Resolved Structural Analysis, First Edition. Keith Moffat and Eaton E. Lattman.
© 2024 John Wiley & Sons Ltd. Published 2024 by John Wiley & Sons Ltd.

but once both the $F(h)$ and the $\alpha(h)$ are known, one can calculate the electron density function.

The electron density function is often said to exist in *real space* – a representation of the physical structure of the crystal. Correspondingly, the diffraction spots are often said to exist in *diffraction space*, which is also termed *reciprocal space* for reasons discussed below.

Equation A.1 has been written using cosines, but in practice almost all sources rewrite this equation using the Euler relationship that connects complex exponentials with trigonometric functions,

$$\exp(ix) = \cos(x) + i\sin(x); \cos(x) = \frac{1}{2}\{\exp(ix) + \exp(-ix)\}. \tag{A.2}$$

Using this formalism, we can rewrite Equation A.1 in a way that separates the phase α from the spatial factor,

$$\rho(r) = \frac{1}{V}\sum_h F(h)\exp(i\alpha(h))\exp(-i2\pi h \cdot r). \tag{A.3}$$

We combine the amplitude F and the phase factor $\exp(i\alpha(h))$ into single complex number $F(h) = F(h)\exp(i\alpha(h))$ and further condense Equation A.3 to

$$\rho(r) = \frac{1}{V}\sum_{all\,h} F(h)\exp(-i2\pi h \cdot r). \tag{A.4}$$

Fourier also provided a method of inverting Equation A.4. He showed that

$$F(h) = \int_{\substack{unit\\cell}} \rho(r)\exp(i2\pi h \cdot r)dV. \tag{A.5}$$

We will invoke Equation A.5 to understand how to calculate the diffraction pattern from a model of a crystal.

Equations A.4 and A.5 have counterparts for nonperiodic samples such as those used in single particle imaging.

$$\rho(r) = \int_{-\infty}^{\infty} F(q)\exp(-2\pi iqr)dV \tag{A.6a}$$

$$F(q) = \int_{-\infty}^{\infty} \rho(r)\exp(2\pi iqr)dV. \tag{A.6b}$$

This formulation is called a Fourier transform, and displays an elegant symmetry between the expressions for ρ and F. Because ρ is not periodic, all possible frequencies of the Fourier components are required, so that the integer triple h is replaced by the continuous variable q. Correspondingly, the real space variable r used just above is no longer in fractional coordinates, but rather carries units such as Å. The Fourier transform is a powerful tool in many areas of crystallography.

A.2 Bragg's Law

The X-ray diffraction pattern from a crystal consists of many point-like maxima often called spots or Bragg reflections. As we discuss below each spot represents diffraction from a single fluctuation $exp(-2\pi i hr)$, and the number of photons in the spot is proportional to $F^2(h)$. The square root of this quantity provides the coefficient $F(h)$ in Equation A.3, but the critical phase $\alpha(h)$ has been lost. Thus, a crystal serves as an analog computer or Fourier analyzer, providing by direct measurement the modulus of the mathematically defined Fourier coefficients $F(h)$.

Bragg's Law provides a convenient way of visualizing the connection between the reflections and the density fluctuations. Bragg approximates the geometry of a crystal by using families of equally spaced, parallel planes running through the crystal in different directions. A Bragg plane family is a simplified representation of the cosine density fluctuation in Equation A.1. Like the cosine, a family of planes divides the x-axis into h equal lengths, the y-axis into k equal lengths, and the z-axis into l equal lengths: $h\ k\ l$ are often called Miller indices. They also represent the three components of the normal vector to the planes. Note that the families of planes have periodicity that agrees with that of the crystal lattice. A sample family is shown in Figure A.1 left. Figure A.1 right shows a cosine density fluctuation with the same values of $h\ k\ l$, and the same period d as the Bragg planes.

Figure A.2 shows the Bragg construction. A.2a shows two Bragg planes edge on, which are imagined to be slightly reflective, redirecting a tiny fraction of the two incoming X-ray beams into emerging scattered beams on the upper right. The lower reflected ray travels path that is AB+BC longer than the path of the upper beam. It is easy to show that this additional path is equal to $2dsin\theta$, where d is the spacing between planes and θ, the angle shown in the diagram, is known as the Bragg angle. Bragg suggested that there would be constructive interference between the X-rays scattered by adjacent planes only when the path length difference was equal to an integral multiple n of X-ray wavelength λ. This condition is shown in Figure A.2b, and is summarized in Bragg's Law

$$n\lambda = 2dsin(\theta). \tag{A.7}$$

Experimentally, as one scans through all values of θ, there will be no observable diffraction except at discrete values of θ that satisfy this equation. These diffracted beams will thus form

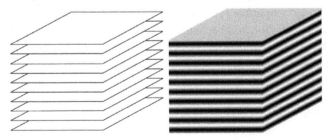

Figure A.1 Left, family of Bragg planes. Right, cosine density fluctuation with same *hkl* as the planes on the left. Reproduced from (Lattman and Loll 2008) with permission.

(a)

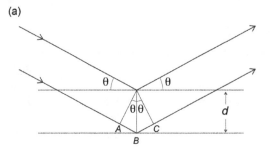

Figure A.2 (a) Bragg construction shown parallel incident X-ray beams impinging from the left onto two Bragg planes, and the reflected rays emerging from the right; (b) shows that reflection can take place when incident beam comes from $-\theta$ giving reflection $-h-k-l$.

(b)

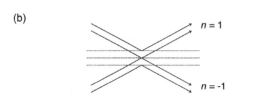

a row of spots on the detector surface. In this context we can say that the diffraction pattern of a family of Bragg planes is a row of points in diffraction space, each spot indexed by the corresponding value of n. The name *reciprocal space* arises from Equation A.7, because large values of the interplanar spacing d correspond to small values of the diffraction angle θ.

A.3   Bragg's Law for Cosine Fluctuations

A cosine density fluctuation can be regarded as a large set of interleaved Bragg plane families. See Figure A.1. Each set has its own weight determined by the cosine variation, and each set gives Bragg scattering independently. However, it turns out that scattering from all the sets together – which form the cosine fluctuation – cancels except for the special case of $n = \pm 1$ in Bragg's Law. Thus, the scattering from a cosine fluctuation comprises only two spots, unlike the row of spots from the Bragg planes. Further, as shown in Equation A.2, a cosine is the sum of two complex exponentials, so that *one can regard each spot as being diffracted by the periodic fluctuation exp($2\pi i\mathbf{hr}$)*. The amplitude of the scattered wave is proportional to the weight of the cosine density fluctuation, so that a density fluctuation $F\exp(2\pi i\mathbf{hr})$ will scatter a wave with amplitude F. If we now imagine a density fluctuation to be translated an amount $\Delta\mathbf{r}$ its equation will become $F\exp(2\pi i\mathbf{hr}-2\pi i\mathbf{h}\Delta\mathbf{r})$, and we can see that a scattered wave will be advanced or retarded compared to the one emerging under the initial conditions. This change is called a phase shift, and is given by $2\pi\mathbf{h}\Delta\mathbf{r}$.

A.4   Diffraction Patterns

Equation A.5 allows us to calculate the diffraction pattern with the phase if we know the electron density. If we recognize that matter is composed of atoms, we can represent the electron density within a sample by the sum of the electron densities of all the constituent atoms

$$\rho(\mathbf{r})= \sum_{all\ atoms\ j} a_j(\mathbf{r}-\mathbf{r}_j).$$

(A.8)

Here a_j is the electron density profile of the j-th atom in the particle, and r_j is the position of its center. All atoms of a given type, say carbon, have the same diffraction patterns, except for the phase shift introduced by translation. We can precalculate these patterns by applying Equation A.5 to a single atom. This gives

$$f_a(h) = \int a(r)exp(2\pi ihr)dV. \tag{A.9}$$

Here f_a is the amplitude of scattering for an atom of type a. Curves for all atom types have been tabulated. They are called atomic scattering factors. With this definition we rewrite Equation A.5 as

$$F(h) = \sum_{all\,atoms\,j} f_j(h)exp(ihr_j). \tag{A.10}$$

This equation introduces atomic coordinates into the representation of the structure and allows for models to be built, and to be refined by adjusting the atomic coordinates r_j.

Equation A.11 reminds us that we measure not $F(h)$ but the square of its magnitude I, thus losing the phase.

$$I(h) = |F(h)|^2 = F(h)F(h)^*. \tag{A.11}$$

X-ray detectors have no sense of when wave crests arrive (phase) but only record the number of arriving photons, intensity. $F(h)^*$ is the complex conjugate of $F(h)$ and the product $F(h)F(h)^*$ is a convenient form for the calculation of the derivatives that arise in Section A.7 on refinement.

A.5 The Crystal Lattice and Its Diffraction Space Counterpart

Crystal diffraction patterns often show the Bragg reflections in a lattice-like arrangement. This pattern is known as the reciprocal lattice. It has an intimate connection with the lattice upon which the crystal itself is built. Understanding it greatly simplifies dealing with the geometry of diffraction. We have just seen that each reflection arises from a family of Bragg planes, but these families are difficult to visualize. Much easier is to represent the (hkl) family of planes by a vector normal to them. This vector has components along the a, b, c unit cell axes that are proportional to $h\,k\,l$. It will be convenient to define the length of this vector to be $1/d$, often denoted d^*, where d is the Bragg plane spacing. Now, instead of families of Bragg planes running in many directions we can visualize an array of vectors running in defined directions out from the crystal. As we shall present next, the tips of these vectors map out a lattice, called the reciprocal lattice, in which each point represents one family of Bragg planes and therefore one Bragg reflection.

Figure A.3a shows a schematic diagram of a crystal lattice. The unit cell of the lattice is defined by the vectors a, b, c. The corners of the unit cells are called lattice points, and the coordinates of every lattice point are of the form $ma+nb+pc$, where m,n,p are integers. The unit cell vectors shown as orthogonal in the diagram, but this is not generally the case. Their lengths are typically presented in Angstrom units. The molecular contents of one unit cell are also displayed. We make no attempt to discuss the symmetry of the arrangement of molecules in the unit cell, described by a space group.

Figure A.3b outlines in 2D how the reciprocal lattice is developed. The four blue squares represent four unit cells of a 2D crystal. The dotted lines represent Bragg planes whose names are given in parentheses, as (1,1). The red vectors represent normal to these families of Bragg planes (or equivalently to cosine fluctuations), and the red dots at termini of these

(a)

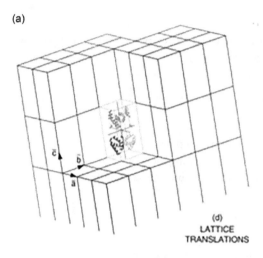

(d)
LATTICE
TRANSLATIONS

(b)

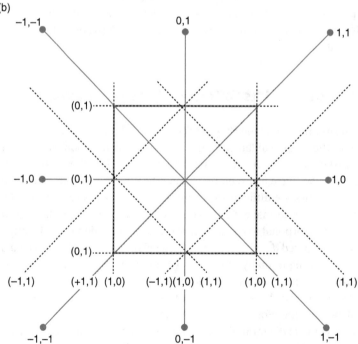

Figure A.3 (a) Lattice of a protein crystal showing unit cell translations and one molecule illustrative of the those that compose the crystal. (b) Development of the reciprocal lattice in 2D. Shows one unit cell of a crystal in black solid, plus a few Bragg lines dotted. The normal vectors to the Bragg lines are shown in red, with length $1/d$. The termini of these vectors are marked with red dots that map out the reciprocal lattice points. 3D lattice would have planes and reciprocal lattice vectors coming out of the plane of the page.

vectors represent the reciprocal lattice points, which are labeled with h and k. Note that these red dots lie on a regular net, and that additional families of Bragg planes would contribute more points to this net. In 3D Bragg planes would point out of the plane of the page, as would the resultant additional reciprocal lattice vectors, giving rise to a third index l.

We have presented a geometrical rationale for introducing the reciprocal lattice, but it arises in more fundamental contexts. When analyzing X-ray scattering from a crystal using quantum mechanics, the equivalent of the Bragg condition emerges as a constraint on the changes of momentum that a scattered X-ray photon can undergo after interacting with the periodic potential of the lattice. These allowed momentum changes are quantized, and are described exactly by the reciprocal lattice vectors.

The patterns of spots we see on our detectors form a snapshot of a portion of this reciprocal lattice. This identification is made clear in a construction devised by Ewald, which uses the reciprocal lattice to determine when the (hkl) family of Bragg planes is oriented to satisfy Bragg's Law. In Figure A.4a we see in cross section a sphere of radius $1/\lambda$. The X-ray beam is incident from the left, and passes through the origin of the sphere O, and through the point O*, which is the origin of the reciprocal lattice. A family of Bragg planes (hkl) satisfying Bragg's Law is sketched near O, showing the Bragg angle θ and the reflected ray OP that continues onto the point P' at the detector. In addition, the reciprocal lattice vector for these planes, of length ($1/d$), is drawn from point O*. It is easy to show that this vector also terminates at the point P, which lies on the surface of the sphere. The right triangle APO* gives $(1/d) = (2/\lambda)\sin\theta$, which can be rearranged to give Bragg's Law $\lambda = 2d\sin\theta$. Thus, whenever a family of Bragg planes satisfies Bragg's Law, the reciprocal lattice vector corresponding to it lies on the surface of Ewald's sphere, and direction of the outgoing or reflected ray is along OP.

For clarity A.4a shows only a single reciprocal lattice point. Figure A.4b elaborates (4a) by showing many reciprocal lattice points lying in the plane of the page. Again, there are reciprocal lattice points above and below the page as well. Imagine that the crystal is rotated through a small angle. Such a rotation takes place around an axis coming out of the page at O*, and will carry the reciprocal lattice with it, bringing other reciprocal lattice points into the reflecting conditions as they pass through the sphere. In the case of protein crystals, which have large unit cells, the reciprocal lattice points are much more closely spaced than they appear in (4b). The program XRayView developed by George Phillips and colleagues provides a wonderful visual introduction to the reciprocal lattice and related topics. As this is written the program is available for download from http://www.phillipslab.org/downloads.

The Ewald construction is easy to elaborate to account for departures for the idealized assumptions made above. To allow for a spread of wavelengths in the incoming beam, the sphere is thickened into an envelope delimited by spheres of radius $1/\lambda_{min}$ and $1/\lambda_{max}$. Allowance must be made for the flux of the beam at the various wavelengths between these. These considerations are particularly critical in the Laue method, which uses a relatively broad range of wavelengths. Lack of parallelism in the incoming beam means that the line AO* representing it is broadened into a cone, and the Ewald sphere is convolved with this cone, further thickening it. Real crystals always have mosaic spread, meaning that they diffract over a range of Bragg angles even with an idealized beam. This effect is represented by an angular broadening of the reciprocal lattice points corresponding to the mosaic spread. Other refinements are possible.

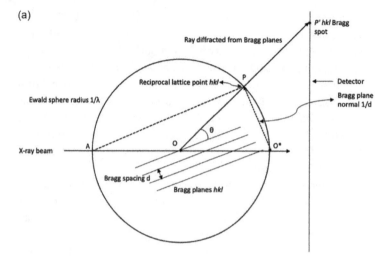

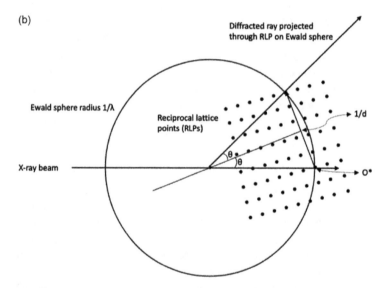

Figure A.4 (a) As described in the text, the Ewald construction illustrated for a single Bragg plane family *hkl*. (b) As described in the text, elaborates on figure 4a to show a set of reciprocal lattice points. Appropriate rotation of the crystal about *O* brings each of these points onto the surface of the sphere, where they satisfy the reflecting condition.

A.6 Determination of the Phase

The structure factor amplitudes $F(\boldsymbol{h})$ needed to calculate the electron density function are obtained directly from the observed diffraction patterns: what of the phases $\alpha(\boldsymbol{h})$? Historically phase determination has been a continuing obstacle that has been overcome in various ways in different branches of crystallography, leading to several Nobel Prizes. In small molecule crystallography so-called direct methods introduced by Karle and Hauptmann and others (Ladd and Palmer 1980) have led to software packages that determine phases directly from structure factor amplitudes when the molecule weight of the molecules in the crystal is below a thousand or so. In protein crystallography the attachment of electron-rich atoms such as mercury to the protein molecule led to the phasing of the first protein crystal structures by the method of isomorphous replacement (Blundell and Johnson 1976). Later this method was expanded to harness atoms such as selenium that display strong variation of scattering across absorption edges – so-called anomalous scattering – to provide phase information. Isomorphous replacement is based entirely on the additivity property of Fourier transforms. This property also allows heterogeneous mixtures of structures, provided they occur randomly in real space, to be handled in time-resolved experiments. Currently by far the most common method of phase determination is called molecular replacement, which relies on the likelihood that the structure under study – the target – is similar to one or more of the 160,000 known structures in the Protein Data Bank – the probe. Molecular replacement uses search methods to create a model of the target crystal structure in which the probe molecule is posed in the target crystal lattice in place of the target molecule. Such a hybrid model can then (usually) be refined to give an unbiased model of the target. The program Phaser has been particularly powerful here (McCoy et al. 2007).

A.7 Building and Refining a Model

The electron density function, Equation A.1, is typically plotted as a contour map, and does not come with the names and coordinates of the constituent atoms already in place. The clarity and interpretability of electron density maps varies widely, depending on the resolution and on the accuracy of the phases. Methods such as solvent flattening can be invoked to improve experimental maps. If, as is usually the case, the amino acid sequence is known, current software can usually fit a preliminary atomic model into a good quality electron density map without human intervention (Morris et al. 2003). Typically, there are regions of the map that are less well defined, and these are reinterpreted in during refinement, as described below. Of course, there are still cases in which intense human effort is needed to build a model into a marginal electron density function, but it seems likely that AI approaches here will gradually supersede human intervention. Especially in TR experiments a number of intermediate structures may be simultaneously present in the electron density map. The interpretation of such maps is discussed in Chapter 7.

An initial atomic model for a crystal structure can be elaborated and improved by a process called refinement, which aims to make appropriate changes in the model structure that systematically decrease the discrepancy between the observed and calculated structure factor amplitudes. This discrepancy is generally presented using a residual

$$R = \frac{\sum_{h} |F_{obs}(\boldsymbol{h}) - F_{calc}(\boldsymbol{h})|}{\sum_{h} F_{obs}(\boldsymbol{h})}. \tag{A.12}$$

Here F_{obs} and F_{calc} are the amplitudes of the observed structure factor and of the one calculated from the current model. In the most straightforward approach refinement seeks to reduce R by finding appropriate small changes Δr, Δs, Δt in the coordinates of each atom. For each $\boldsymbol{h}$ one can write an equation designed to render F_{obs} and F_{calc} equal by adding a differential term. Thus,

$$|F_{obs}(\boldsymbol{h})| = |F_{calc}(\boldsymbol{h})| + \Delta|F_{calc}(\boldsymbol{h})| \tag{A.13a}$$

$$|F_{obs}(\boldsymbol{h})| - |F_{calc}(\boldsymbol{h})| = \Delta F = \sum \frac{\partial|F_{calc}(\boldsymbol{h})|}{\partial p_j} \Delta p_j. \tag{A.13b}$$

Here the Δp_j are shifts in the atomic coordinates or other parameters. The derivatives on the right-hand side of Equation A.11b can be calculated by using Equations A.10 and A.11 to re-express $|F_{calc}(\boldsymbol{h})|$. This formalism gives rise to a set of N_h equations, one for each reflection $\boldsymbol{h}$, in N_j variables, one for each Δp_j. If the number of equations is larger than the number of parameters, these equations can in principle be solved by standard methods to provide the Δp_j. Then one can update the parameter value by adding the shifts to the current values: $p_{j,new} = p_j + \Delta p_j$. It will come as no surprise that the new parameter values do not actually reduce ΔF to zero. There are experimental errors in F_{obs}; the linear power series expansion is a weak approximation. Also, in practice, solutions to such a large set of equation are unstable unless the number of observations greatly exceeds the number of parameters. This is very rarely the case in protein crystallography, where resolution is usually limited. Even in the best cases in small molecule crystallography, dozens of cycles of refinement, each updating the p_j, are required before the process converges.

Konnert and Hendrickson introduced known amino acid chemistry into the refinement equations to provide additional observations (Hendrickson 1985). The length of a carbon–carbon single bond is known to lie within certain rather tight limits, for example; other bond lengths and angles are also constrained. By adding such restraints into the set of refinement equations, the system become much better behaved, and structures even at 3Å resolution can be reliably refined. As a caution, the energy of binding a ligand may be used to distort it far from its native geometry, so that it may be cleaved for example. The more heavily one relies on restraints, the more readily one may refine away such a meaningful distortion. Such departures from canonical geometry are particularly common in TR studies in which a protein is passing over an energy barrier.

The process of refinement just described requires that the R value decrease monotonically. If the model is in a false minimum, the refinement cannot fix it. A simple example is an interior aromatic side chain that is flipped 180 degrees from its correct position. The model will look a lot worse when the side chain is misoriented by 90 degrees, so refinement will not go there. The process of simulated annealing, introduced by Brunger, heats the protein model to high temperatures in the computer, encouraging large motions, and then gradually cools it while refining (Brünger and Rice 1997). Simulated annealing has greatly reduced, but not eliminated, the issues of false minima. In the end, difference electron density maps, discussed in Chapter 7, are usually invoked. Such a map, given by

$$\Delta\rho(\boldsymbol{r}) = \frac{1}{V}\sum_{\boldsymbol{h}}(F_{obs}(\boldsymbol{h}) - F_{calc}(\boldsymbol{h}))exp(i\alpha(\boldsymbol{h}))exp(-2\pi i\boldsymbol{h}\boldsymbol{r}), \tag{A.14}$$

should be featureless for a well-refined structure. As discussed elsewhere, peaks and valleys in these maps suggest needed alterations. Adding an atom or group where a valley lies will fill it in.

References

Baron Fourier, J.B.J. (2003). *The Analytical Theory of Heat*. Courier Corporation.

Blundell, T.L. and Johnson, L.N. (1976). *Protein Crystallography*. Academic Press.

Brünger, A.T. and Rice, L.M. (1997). Crystallographic refinement by simulated annealing: methods and applications. *Methods Enzymol* 277: 243–269.

Drenth, J. (2007). *Principles of Protein X-ray Crystallography*. Springer Science & Business Media.

Eisenberg, D. and Crothers, D. (1979). *Physical Chemistry with Applications to the Life Sciences*. Meno Park, CA: Benjamin/Cummings.

Hendrickson, W.A. (1985). Stereochemically restrained refinement of macromolecular structures. *Methods Enzymol* 115: 252–270.

Ladd, M.F.C. and Palmer, R.A. (1980). *Theory and Practice of Direct Methods in Crystallography*. Springer.

Lattman, E.E. and Loll, P.J. (2008). *Protein Crystallography: A Concise Guide*. JHU Press.

McCoy, A.J., Grosse-Kunstleve, R.W., Adams, P.D. et al. (2007). Phaser crystallographic software. *Journal of Applied Crystallography* 40 (4): 658–674.

Morris, R.J., Perrakis, A. and Lamzin, V.S. (2003). ARP/ wARP and automatic interpretation of protein electron density maps. *Methods Enzymol* 374: 229–244.

Rhodes, G. (2010). *Crystallography Made Crystal Clear: A Guide for Users of Macromolecular Models*. Elsevier.

Rupp, B. (2010). *Biomolecular Crystallography: Principles, Practice, and Application to Structural Biology*. New York, Abingdon, Garland Science.

Index

Dynamics and Kinetics in Structural Biology: Unravelling Function Through Time-Resolved Structural Analysis,
First Edition. Keith Moffat and Eaton E. Lattman.
© 2024 John Wiley & Sons Ltd. Published 2024 by John Wiley & Sons Ltd.

Printed in the USA
CPSIA information can be obtained
at www.ICGtesting.com
LVHW080811151123
763324LV00002B/2